Riparian Management in Forests of the Continental Eastern United States

Riparian Management in Forests of the Continental Eastern United States

Edited by
Elon S. Verry
James W. Hornbeck
C. Andrew Dolloff

LEWIS PUBLISHERS
Boca Raton London New York Washington, D.C.

Library of Congress Cataloging-in-Publication Data

Riparian management in forests of the continental Eastern United States / edited by Elon S. Verry, James W. Hornbeck, C. Andrew Dolloff.
       p.   cm.
     Includes bibliographical references (p.) and index.
     ISBN 1-56670-501-0 (alk. paper)
     1. Riparian forests--East (U.S.)--Management. 2. Riparian areas--East (U.S.)--Management. 3. Forested wetlands--East (U.S.)--Management. I. Verry, Elon S. II. Hornbeck, James. W. III. Dolloff, Charles Andrew, 1952-
SD144.A112R56 1999
333.91'62'0974—dc21

99-048938

Cover photo by Peter L. Card III: The Mississippi River, Mile 10, Clearwater County, Minnesota

Editors

Elon S. Verry, Ph.D., is a Research Forest Hydrologist in the Riparian Ecology and Management unit of the North Central Research Station, USDA Forest Service, Grand Rapids, Minnesota. Dr. Verry received a B.S. in Forest Management and an M.S. in Water Resources from the University of Illinois and his Ph.D. in Earth Resources is from Colorado State University. He is past Chair of the National Atmospheric Deposition Program, and an Adjunct Professor at the University of Minnesota. He received the USDA National Honor Award for personal and professional excellence, and the Medallion of the University of Helsinki for life-time achievement in peatland hydrology and ecology.

He has been stationed at the Forestry Sciences Laboratory in northern Minnesota his entire career where his interests are: peatland hydrology and ecology, impact of forest management on water resources, selection and management of waterfowl impoundments, nutrient cycling, atmospheric deposition, global warming, riparian management and ecology, and stream geomorphology. He is an associate editor of *Silva Fennica*.

James W. Hornbeck, Ph.D., is a Research Forester in the Ecological Processes unit of the Northeastern Research Station, USDA Forest Service, Durham, New Hampshire. Dr. Hornbeck received a B.S. in forestry and M.S. in watershed management from Pennsylvania State University and a Ph.D. in forestry from State University of New York, College of Environmental Science and Forestry.

His research interests include the effects of forests and forest disturbances, especially harvesting and atmospheric deposition, on hydrologic and nutrient cycles and water quality. In the course of his 38-year career with the US Forest Service, he has conducted research at the Leading Ridge Watershed Research Unit in Pennsylvania, the Fernow Experimental Forest in West Virginia, and the Hubbard Brook Experimental Forest in New Hampshire. He has participated in the Hubbard Brook Ecosystem Studies since 1967. He currently serves as Associate Editor for the *Northern Journal of Applied Forestry*.

C. Andrew Dolloff, Ph.D., is Project Leader and Research Fisheries Scientist in the Coldwater Streams and Trout Habitat Research Unit, USDA Forest Service Southern Research Station in Blacksburg, Virginia. The mission of the Unit is to acquire new knowledge about factors that influence the distribution, abundance, and production of trout and other coldwater fish and to provide the technical basis for protecting, enhancing and restoring coldwater streams and their fauna. Dr. Dolloff received his B.S. in Wildlife Science from the University of Maine, his M.S. in Zoology from North Carolina State University, and his Ph.D. in Fisheries and Wildlife Management from Montana State University.

Before he assumed his present duties, Dr. Dolloff was a research biologist for the Pacific Northwest Research Station in Juneau, Alaska where he conducted research on the ecology of coarse woody debris and habitat relations of juvenile salmonids. He actively participates in the teaching and research programs at several universities and currently holds adjunct faculty appointments at Virginia Tech and Clemson. Dr. Dolloff's research interests include the influence of natural and anthropogenic disturbance on stream communities, stream fish population ecology, and ecology and management of coarse woody debris. In addition to these research pursuits, he is involved in the development of management strategies and practices for the protection or recovery of aquatic and riparian ecosystems.

A step/pool intermittent bedrock stream in West Virginia.

Contributors

Warren E. Archey
Massachusetts DEM
Division of State Forests and Parks
PO Box 1433
Pittsfield, MA 01240

Matthew E. Baker
University of Michigan
School of Natural Resources
and Environment
430 East University
Ann Arbor, MI 48109

Burton V. Barnes
University of Michigan
School of Natural Resources
and Environment
430 East University
Ann Arbor, MI 48109

Susan C. Barro
USDA Forest Service
North Central Research Station
845 Chicago Avenue, Suite 225
Evanston, IL 60202-2357

John E. Baumgras
USDA Forest Service
Northeastern Research Station
180 Canfield St.
Morgantown, WV 26505-3101

Jerry Bernard
USDA Natural Resources
Conservation Service
PO Box 2890
Washington, DC 20013

Charles R. Blinn
University of Minnesota
1530 Cleveland Ave. North
St. Paul, MN 55108

David W. Bolgrien
USEPA Mid-Continent Ecology Division
6201 Congdon Blvd.
Duluth, MN 55804

Donald Brady
USEPA Headquarters
401 M Street, S. W.
Washington, DC 20460

Richard R. Buech
USDA Forest Service
North Central Research Station
1831 Highway 169 East
Grand Rapids, MN 55744

Thomas R. Crow
University of Michigan
School of Natural Resources
and Environment
430 East University
Ann Arbor, MI 48109

Contributors

Richard M. DeGraaf
USDA Forest Service
Northeastern Research Station
201 Holdsworth Hall
University of Massachusetts
Amherst, MA 01003

C. Andrew Dolloff
USDA Forest Service
Southern Research Station
Virginia Polytechnic Institute
and State University
140 Cheatham Hall
Blacksburg, VA 24061

John F. Dwyer
USDA Forest Service
North Central Research Station
845 Chicago Avenue, Suite 225
Evanston, IL 60202-2357

Clayton Edwards
USDA Forest Service
North Central Research Station
5985 Highway K
Rhinelander, WI 54501-9128

Pamela J. Edwards
USDA Forest Service,
Northeastern Research Station
PO Box 404, Nursery Bottom
Parsons, WV 26287-0404

Samuel Emmons
USDA Forest Service, Region 9
300 W. Wisconsin Ave.
Milwaukee, WI 53203

Steven T. Eubanks
USDA Forest Service
Tahoe National Forest
631 Coyote Street
Nevada City, CA 95959

John G. Greis
USDA Forest Service, Region 8
1720 Peachtree Road NW
Atlanta, GA 30367

Craig W. Hedman
International Paper Company
Southlands Experimental Forest
Bainbridge, GA 31717

Dale Higgins
USDA Forest Service
Chequamegon National Forest
1170 4th Avenue South
Park Falls, WI 54552

Bob Hollingsworth
USDA Forest Service, Region 9
310 W. Wisconsin Ave.
Milwaukee, WI 53203

James W. Hornbeck
USDA Forest Service,
Northeastern Research Station
PO Box 640
Durham, NH 03824

Bonnie L. Ilhardt
USDA Forest Service, Region 9
310 W. Wisconsin Ave.
Milwaukee, WI 53203

Contributors

Pamela J. Jakes
USDA Forest Service
North Central Research Station
1992 Folwell Avenue
St. Paul, MN 55108

Phillip Janik
USDA Forest Service
Office of the Chief
201 14th St. S. W. at Independence Ave.
Washington, DC 20250

James N. Kochenderfer
USDA Forest Service
Northeastern Research Station
PO Box 404, Nursery Bottom
Parsons, WV 26287-0404

Timothy K. Kratz
Center for Limnology
University of Wisconsin-Madison
10810 County Highway N
Boulder Junction, WI 54512-9733

Russell A. LaFayette
USDA Forest Service
Riparian Management and
Watershed Improvement
201 14th St, S. W. at Independence Ave.
Washington, DC 20250

James A. Mattson
USDA Forest Service
North Central Research Station
410 MacInnes Drive
Houghton, MI 49931

Keith R. McLaughlin
USDA Forest Service
Water Quality & Hydrology Program
201 14th St. S. W. at Independence Ave.
Washington, D.C. 20250

Joseph C. Mitchell
University of Richmond
Department of Biology
Richmond, VA 23173

John J. Moriarty
Minnesota Frog Watch and
Ramsey Co. Park and Recreation Dept.
2015 N. Van Dyke Street
Maplewood, MN 55109-3796

Brian J. Palik
USDA Forest Service
North Central Research Station
1831 Highway 169 East
Grand Rapids, MN 55744

Harry Parrott
USDA Forest Service, Region 9
310 W. Wisconsin Ave.
Milwaukee, WI 53203

Thomas K. Pauley
Marshall University
400 Greer Blvd.
Huntington, WV 25755

Heather A. Pert
2101 West 42nd Avenue
Pine Bluff, AR 71603

Contributors

Michael J. Phillips
Minnesota Dept. of Natural Resources
500 Lafayette Road
St Paul, MN 55155-4044

Carl Richards
Natural Resources Research Institute
University of Minnesota
Duluth, MN 55811

Michael J. Solomon
USDA Forest Service
Huron-Manistee National Forest
1755 S. Mitchell
Cadillac, MI 49601

Gordon W. Stuart
USDA Forest Service (Retired)
97 Seavey Street
Westbrook, ME 04092

Lloyd W. Swift, Jr.
USDA Forest Service
Southern Research Station
3160 Coweeta Lab Road
Otto, NC 28763

Michael A. Thompson
USDA Forest Service
North Central Research Station
410 MacInnes Drive
Houghton, MI 49931

Elon S. (Sandy) Verry
USDA Forest Service
North Central Research Station
1831 Highway 169 East
Grand Rapids, MN 55744

Jackson R. Webster
Department of Biology
Virginia Polytechnic Institute
and State University
Blacksburg, VA 24061

David J. Welsch
USDA Forest Service, Northeastern
Area State and Private Forestry
PO Box 640
Durham, NH 03824

Mariko Yamasaki
USDA Forest Service
Northeastern Research Station
PO Box 640
Durham, NH 03824

John C. Zasada
USDA Forest Service
North Central Research Station
1831 Highway 169 East
Grand Rapids, Minnesota 55744

Preface

Riparian areas in forests of the Eastern United States provide a variety of products and amenities, among the most important being protection of water quality and aquatic habitats. Management of riparian areas has moved to the forefront because of conflicts between their protection capabilities and the ever-increasing use they receive. In 1995 the three of us naively agreed to a request from the Eastern Region of the USDA Forest Service to summarize the state-of-the art in managing forested riparian areas. It quickly became clear that this task was beyond the capabilities of three individuals and one organization. The issues are too complex and the information base too large and diverse.

Thus, we turned to a common solution to such problems and organized a conference titled "Riparian Management in Forests of the Continental Eastern U.S."

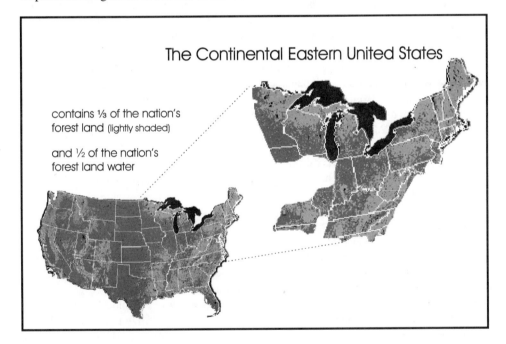

We invited 49 authors from universities; county, state, and federal agencies; forest industries; and natural resource endeavors to help us get a handle on the unique needs of riparian forest management in the Continental Eastern United States. Why the Continental

Eastern United States and not all of the Eastern United States? The ocean coast, estuary, coastal plain and piedmont regions are nearly as large in area as the Continental Eastern United States, and their inclusion more than doubles the animal and plant species considered; each region needs a separate treatment. The Continental Eastern United States is separated by its humid, temperate, continental climate. It contains ⅓ of the nation's forest land, and over ½ of the water within all of the nation's forest land. The tools for management presented here will help managers to find their own solutions for local conditions within this region.

Each day, thousands of natural resource managers face the challenge of managing eastern riparian areas characterized by fragmented ownership, fragmented ecosystems, and diverse interest groups. The reader of this book will be challenged by the diversity of conditions and the diversity of values that must be addressed in Eastern United States forests. That diversity is brought to the book by the authors whose own diversity nearly spans that of the eastern population. Be aware that the fragmented ownership of the East presents a different management problem than the large-block government ownership patterns of the West. The mixtures of people and land uses found in the east require the manager to adapt their knowledge to each location. Management in the east will always need to deal with professional managers from many disciplines, with multi-disciplinary teams, and with stake holder groups that differ from watershed to watershed. You will not find the single answer to apply everywhere. You will find both a set of tools and a set of values proffered by diverse authors. The resource manager and those who advise them can adapt this knowledge and their experiences to our diverse conditions and complex social dimensions. We realize that land management in the east is bound to these realities, and we have built this book to help us manage in a responsible way.

Symposium topics were chosen to represent all major issues and aspects of managing riparian areas. We wanted to reach natural resource administrators, educators, and on-the-ground managers from industry, consulting firms, and municipal, state, and federal agencies who routinely face the complex problems of caring for, protecting and teaching about riparian areas. It was obvious from the outset that the subject was timely and popular when the allotted meeting room space for 250 attendees was quickly filled.

Our fondest hope is that this book, prepared from the experiences and ideas of chapter authors, reviewers, and symposium attendees, does indeed summarize the state-of-the art regarding management of forested riparian areas in the Continental Eastern United States, and that it will serve as a desktop reference for those charged with caring for this very important resource.

Elon S. Verry   James W. Hornbeck   C. Andrew Dolloff

Acknowledgments

Planning for this book included the preparation of 19 chapters by invited authors who presented their work at a conference in Columbus, Ohio, March 23-25, 1998. The conference was organized to allow the authors of each chapter to lead a separate one hour discussion session after a block of four chapters had been presented. Audience members could pick and choose which of the concurrent discussion sessions to participate in. The discussion sessions were divided into two half-hour parts so that every attendee had ample opportunity to question and pass comments to each author. In addition, two or more subject-area specialists reviewed each chapter between receipt of the first drafts in January and the conference in March. Thus, the chapters comprise not only the authors' ideas, but also many insights gained from an audience dealing with riparian management on a daily basis.

The credit for this approach and the planning of the book chapter goes to the Book and Conference Planning Committee:

Andy Dolloff	Southern Research Station, Blacksburg, VA
John Greis	USFS, Region 8, Atlanta, GA
Bob Hollingsworth	USFS, Region 9, Milwaukee, WI
Jim Hornbeck	Northeastern Research Station, Durham, NH
Bonnie Ilhardt	USFS, Region 9, Milwaukee, WI
Harry Parrott	USFS, Region 9, Milwaukee, WI
Sandy Verry	North Central Research Station, Grand Rapids, MN
Nancy Walters	Southern Research Station, Asheville, NC
Dave Welsch	Northeastern Area State & Private Forestry, Durham, NH

We especially thank Nancy Walters, who brought organizational and administrative skills to bear, and Tracey Lupien and Nancy Dudrey, who performed numerous tasks during the planning and conference endeavors. We also especially thank the conference participants and 35 peer reviewers who freely gave of their thoughts and time, and four copy editors who shaped the final product: Lucy Burde, Barb Winters, Bob Wray, and Aimée Alger.

Acknowledgments

We sincerely appreciate the insights given by peer reviewers at the conference, and two peer reviewers for each chapter that gave of their time:

Tim Adams, South Carolina Forestry Commission, Columbia, SC
Stan Adams, North Carolina Division of Forest Resources, Raleigh, NC

Susan Alexander, Pacific Northwest Research Station, Corvallis, OR
Ted Angradi, Northeastern Research Station, Parsons, WV

Mike Aust, Virginia Tech. Blacksburg, VA
Peter Bisson, Pacific Northwest Research Station, Olympia, WA

Bernard Bormann, Pacific Northwest Research Station, Corvallis, OR
Richard Buech, North Central Research Station, Grand Rapids, MN

Marilyn Buford, U.S. Forest Service, Washington, DC
Richard Burns, National Forests in North Carolina, Asheville, NC

Nancy Burt, R9, Green Mountain National Forest, Rutland, VT
Glenn Chen, U.S. Forest Service, Logan, UT

Charles Cogbill, Consultant and Historian, NH
Peter Dillon, Ontario Ministry of Environment, Dorsett, ON

Gary Donovan, Champion International, Bucksport, ME
Pam Edwards, Northeastern Research Station, Parsons, WV

Andy Egan, University of Maine, Orono, ME
Dave Funk, Northeastern Research Station (ret.), Durham, NH

Frank Golet, University of Rhode Island, Kingston, RI
Carola Haas, Virginia Tech, Blacksburg, VA

Dale Higgins, Chequamegon National Forest, Park Falls, WI
Jack Holcomb, U. S. Forest Service, Region 8, Atlanta, GA

Jim Hornbeck, Northeastern Research Station, Durham, NH
George Ice, National Counsel for Air and Stream Improvement, Corvallis, OR

Acknowledgments

Scott Jackson, University of Massachusetts, Amherst, MA
Jim Keyes, U. S. Forest Service, Region 8, Atlanta, GA

Dennis Lemly, Southern Research Station, Blacksburg, VA
Steve McMullin, Virginia Tech, Blacksburg, VA

Dale Nichols, North Central Research Station, Grand Rapids, MN
Jim Petranka, University of North Carolina, Asheville, NC

John Potyondy, U. S. Forest Service, Stream Team, Ft. Collins, CO
Steve Roy, Green Mountain National Forest, Rutland, VT

Mary Sheremeta, Northeastern Research Station, Amherst, MA
John Stanturf, Southern Research Station, Stoneville, MS

Fred Swanson, Pacific Northwest Research Station, Corvallis, OR
Rick Swanson, U. S. Forest Service WO, Lands, Missoula, MT

Linda Thomasma, Northeastern Research Station, Parsons, WV
Albert Todd, U.S. Forest Service, S&PF Chesapeake Bay Program, Washington, DC

Carl Trettin, Southern Research Station, Charleston, SC
Jim Vose, Southern Research Station, Otto, NC

Donald Weller, Smithsonian Environmental Research Center Edgewater, MD
Jack Williams, U.S. Forest Service, Boise National Forest, Boise, ID

Sailing on Leech Lake reservoir, Chippewa National Forest, Minnesota

An intermittent stream in West Virginia.

Sandy Verry

Foreword: A State Perspective

Warren E. Archey

Chief, Bureau of Forestry, Massachusetts, and
President, Northeastern Area Association of State Foresters

Not long ago interest in riparian forests was seen as an esoteric pursuit; an academic and arcane subject that only theoretical ecologists could fully appreciate. Only recently has an understanding of riparian forests been seen as integral to a full understanding of vital forest processes — clearly including the hydrologic/geomorphologic regime — but also extending to nutrient cycling, chemical interaction, biological relationships and energy transfer. The ultimate utility of this understanding though is to assist in treating riparian forests with a thoughtful hand by establishing enhanced management and restorative protocols.

In state government, the greatest focus of forest cutting practices legislation and development of best management practices (BMPs) has been in the area of riparian protection. That certainly is a theme seen by the Water Resources Committee of the Northeastern Area Association of State and Private Foresters in assisting states in the development of BMPs. The wide scope of BMP development has been documented by the National Association of State Foresters (NASF). Also held in common among all states is the notion that there is some mystery as to the effectiveness of the implementation of BMPs. The Southern Group of State Foresters established a "Regional BMP Compliance Monitoring Task Force" to examine this issue. The Northeastern Area State and Private Forestry, Water Resources Committee, has done similarly and is examining BMP compliance protocols, region-wide, with more efforts to follow.

Whether our focus is forest preservation/protection or utilitarian, it is clear that healthy riparian forests are vital as filters and nutrient attenuators to protect water quality for drinking, fisheries, and recreation. It is essential to develop an analytical framework for decisionmaking by modeling riparian processes. That framework needs the capacity to quantify space and time impacts of management alternatives that assist in selecting the desired future condition in riparian forests. Understanding hydrologic/geomorphic processes that give rise to riparian habitats, physiographically classifying habitats, identifying active biochemical processes, and finally, conducting analyses, will allow us to chart the course to attain that desired future condition. Most often it is not the absolute numbers that are desirable, but the effect of relative change, especially as we consider management protocols.

The outcome of such analyses may result in the need for modified or restorative actions (e.g., reducing sediment, controlling stream temperatures, reducing nutrient loading).

In this book, riparian forests are examined from the standpoint of definitions, linkages, classification schemes, functions, ecology, management, restoration, monitoring, and the societal context. The organizers have done a remarkable job of covering the depth and breadth of the topic, the authors have treated their chapters with particular skill, and I commend it to all who would seek greater understanding of riparian forest management. I am hopeful that by spanning the broad range between policy making and management, this book will hold and focus your vision on managing at the water's edge.

A bedrock falls in northwest Wisconsin's Pattison State Park.

Foreword - A National Perspective

Philip Janik

Deputy Chief, State and Private Forestry
(now Chief Operating Officer) US Forest Service

The timing could not be better for addressing riparian area management, and resulting impacts on surface waters. The National Clean Water Action Plan has raised the profile of water issues, and the Forest Service Natural Resource Agenda features water as one of four major points of emphasis. Also, the Forest Service leadership team has identified water and watershed management as the issue of the upcoming decade — comparable to the prominence of old-growth forests as the issue of the past decade.

The emphasis on water issues points up how strongly we have come to value water for municipal supplies, hydropower, fish and wildlife relationships, irrigation, recreation, and sustaining water associations and cultures. There are several current examples of how the Forest Service, together with federal, state, tribal and private partners, is addressing broad watershed issues:

- Chesapeake Bay Program to "Save the Bay"
- Inland West Water Strategy
- New York City Watershed Project
- The Pacific Habitat Strategy
- The Anadromous Fish Habitat Assessment

In addition to the above, high-profile examples, the Forest Service continues to deal with other regional and local challenges such as: how to deal with mixed ownerships of watersheds; what incentives are appropriate and effective in helping change voluntary management on private watersheds; how much area of a watershed can be intensively managed and still provide for long-term sustainability of water resources; which risk assessment and predictive models should be used; what is needed regarding management of forest roads and protecting water quality; how can we best sell and have the public accept the prevention approach as compared to generating problems that later require watershed restoration?

Riparian area management is an intimate concern within each of the above programs and challenges. Fortunately our nation has come to understand the values of riparian areas, and the importance of providing proper management. But we must continue to improve our knowledge and understanding of riparian areas. This book, with its objective of summarizing state-of-the-art science and management for riparian areas, will help to insure the health of our nation's waters and the lands that border them.

Autumn seeps in to a Virginia mountain stream.

Contents

Contents xviii

Contents   xix

Contents

Roger Bay

Memories in the Boundary Waters Canoe Area Wilderness,
Superior National Forest, Minnesota.

Sandy Verry

Learning to read the river. Allegheny National Forest, Pennsylvania.

Chapter 1

The Challenge of Managing for Healthy Riparian Areas

Elon S. Verry and C. Andrew Dolloff

"Learn to read the land [the river], and when you do I have no fear of what you will do with it: indeed, I am excited about what you will do for it."
Aldo Leopold, 1966 — A Sand County Almanac

"There is a need to place such common resources as water, land, and air on a higher plane of value and to assign them a kind of respect that Aldo Leopold called the land ethic, a recognition of the interdependence of all creatures and resources."
Luna Leopold, 1997 — Water, Rivers and Creeks

Clearly, Aldo Leopold, a forest ranger, a wildlife biologist, a director of wood-use research, and a small woodland owner, called those in natural resource management to do two things: read the land and manage it — ". . . I have no fear of what you will do for it." Thirty-one years later, Luna Leopold, Chief Hydrologist with the U.S. Geological Survey, Dean of Geology at The University of California-Berkley, and river restoration advocate, emulated his father's "Round River" when he called us to recognize the interdependence of land, water, air, and all creatures. Neither rejected their past, and neither rejected the many uses nor many users of our forest lands. Aldo Leopold worked to restore his cut-over and worked-out Sand County farm and woodland to take a productive place in their Wisconsin community. At the heart of managing for healthy riparian areas is seeing them with this same sense of community. A central challenge to many of us is managing with shared decisions. This may be the hardest task we have. To paraphrase Gifford Pinchot, first Chief of the U.S. Forest Service and Dean of the Yale School of Forestry, be absolutely honest and sincere, learn to recognize the point of view of the other person, and meet each other with arguments you will each understand.

The challenge of managing for healthy riparian areas means coming to grips with our heritage, understanding how the land and streams change, dealing with diverse and divisive

1

issues, learning to read the land and rivers, expanding our set of management tools and most important, seeing with the vision of community. It is our purpose to bring an understanding of riparian values and riparian functions to a community vision of place, a landscape that holds ponds and lakes, grows forests, and gives of itself in a river that runs through it. This is a place where living includes both work and play, where working the land means improving the watershed's landscape.

Understanding Our Heritage

Streams, rivers, and lakes have influenced our lives in North America for more than 20,000 years. Since Ice Age nomads traversed the Beringia Plain, travel to, along, and on streams and rivers has formed the basis of family ties, commerce, and the flow of new ideas. Indeed, the history of North America and many fortunes and fates have been defined by how we used water-travel routes and the riparian lands beside them.

Native Americans and early settlers had a special relationship with streams, depending on them for bathing, fishing, hunting, travel, and trade. As populations expanded in the 1700s, this dependency slowly began to erode. Stream-side forests were cleared both to make way for towns and fields and later to fuel the fleets of steamboats that ferried cargos of people and goods. Even the rivers themselves were affected as snags and deadheads were removed from harbors and channels, and stream bottoms were dredged to allow passage of ever larger vessels. Slowly forests gave way to agriculture, and the accompanying erosion on the land (and in the channel) produced new sediment that reduced our capacity to navigate small rivers. On large rivers the sediment required constant dredging to keep them open to commerce. By the mid-1800s, railroads tied regions of the continent together. Rivers, although still of enormous everyday consequence, were frequently viewed as obstacles to be bridged, straightened, or used to lay track in. By the beginning of the 20th century, the special relationship among humans, rivers, and streamside forests was forever changed. Today, even though highways now connect us to virtually any place, our social collective has awakened a desire to regain the cultural, family, and personal ties lost when the steam-driven locomotive put railroads and highways on an equal footing with rivers. River walks, lake walks, and trail systems everywhere accentuate our riparian lands in both urban and rural settings. When we stop to rest, we stop at the water's edge.

Our heritage is one of natural resource exploitation, of wresting from the forest a family living, one of population movement, one of balancing the need for food, fuel, and transportation. Our heritage is also one of constant change: from hunter-gatherers to farmers, from pioneers to settlers, and from loose bands to complex tribes and villages. Our heritage now demands that we come to grips with sustainable resource use. Today we highly value streams and lakes and the land at their edge. Nevertheless, we either take for granted, or cannot see, the ties that bind a watershed to its streams and lakes.

Each of us sees riparian land and water differently; some of us see with eyes focused on opportunities for commerce, water supply, harvest of trees, fish, or waterfowl. Others see song birds, willows, beavers and the subtle harmony of a natural community. Some see the power of water and sediment to shape channels into predictable patterns of stream and valley geometry (stream habitat and geomorphology). Some of us — a growing number — see people and burgeoning demands for goods, services, and amenities. Our challenge is to see with a community vision rather than the vision of single use. We must see beyond the simple juxtaposition of trees and water. Our challenge is to understand how current forest and stream conditions have come to be, how the land and water function together, how their functions can be optimized, and how we can manage for a community vision of future condition.

But how are we to see? "There is no one alive today that saw what rivers, streams, and brooks looked like prior to being cleaned for barge, log and steamboat transportation (Rector 1951)." How do we judge current riparian conditions if we cannot see the past? Few, if any, of the world's river systems can be considered truly undisturbed, and they are too small to be representative. Even if we knew the pristine condition of streams and their forests, replicating all of their structure and composition may not be practical or desirable in today's society.

Fortunately, natural processes have restored many examples of healthy riparian areas in today's forests. From these we can derive the information we need to manage degraded sites. Our challenge is to understand the difference between degraded sites (the current condition at many sites) and those sites with a full suite of healthy functions. We need to learn all we can about these areas, for they are the basis of riparian restoration, and we need to use the eyes of many disciplines to build the foundation. We begin by examining how we developed rivers, streams, lakes, and riparian forests. The issues of today are old ones, formed and then debated throughout the last 200 years. How the debates have changed and how we have changed is a history of the riparian conditions we now have to work with.

The examples below are chosen not to second-guess how we developed resources, but to illustrate how our social institutions became established and to show what physical conditions we have to work with. They include resource depletion, resource regulation, habitat alteration, natural resource development, marshaling of a transportation workforce, a changing landscape with changing uses, squatters' rights and private owner rights, watershed analysis, watershed restoration, natural versus catastrophic fire, forest type and structure changes, forest harvesting and floods, and state subsidies for private land enhancement.

The Primary Forests of the Continental Eastern United States and How They Changed

Environmental change since Colonial times from the East Coast to the tall-grass prairies has been thoroughly summarized by Whitney (1994). We have borrowed liberally from his accounts and extended some of his time series to show how primitive forests changed. The trees the first settlers saw were big and covered with epiphytes, and mosses grew on the forest floor. They were generally 200 with some even over 300 years old. Coarse woody debris on the forest floor is estimated at 6 to 12 tons/acre for the central hardwoods, and 16 to 21 tons/acre in the cooler hemlock/northern hardwood stands. The English colonists found the dense forests a stark contrast to the sparsely wooded landscapes of their homeland where, in 1696, less than 10% of the land was wooded. Pictures of old-growth remnants taken 200 years later in the Eastern U.S. confirm the large tree diameters ranging from 3 to 15 feet (Table 1.1). Except for jack pine on droughty sand, black spruce on peatland, and red spruce at high elevations, virtually all primary forest stands were a mixed type. Stand volumes ranged from 3,000-25,000 board feet/acre with a maximum development in multiple canopied white pine / hemlock forests of 100,000 board feet/acre (Table 1.2). Tree heights ranged from 70 to 130 feet tall, and basal areas from 120 to 392 ft^2/acre (Table 1.3).

Table 1.1 Diameters of old-growth trees in the Eastern U.S. (Based on photographs taken from 1875 to 1915)

Tree Species	Diameter Ft.	Location
Red spruce	3	White Mtns., NH
Sugar maple	3	Petoskey, MI
Hemlock	4	Ashuelat, NH
White oak	4	Allegheny Plateau, WV
Yellow birch	5	Colbroke, CT
Chestnut	6	Graham Co., NC
White pine	8	Cornwall, CT
Yellow poplar	8	Dickenson Co., VA
Yellow poplar	10	Vincennes, IN
Sycamore	15	Mt. Carmel, IL

Frequent natural disturbances ensured that most trees did not exceed 300 years of age. Fires on droughty sand and shallow soil over bedrock kept early succession birch, aspen, and jack pine stands in the 50 to 100 year range. They were most prominent in west-central and northeastern Minnesota. Red pine (often mixed with jack pine) and super canopy white pine stands ranged from 100 to 250 years old with the older stands protected from fire by water bodies. In southern New England, hurricanes kept tree ages between 100 and 150 years, while in central and northern New England, trees lived from 150 to 300 years. Along the Northeast Coast, the birch and spruce trees were 100 to 200 years old. Many of the hemlock, yellow birch, and sugar maple stands in Michigan and Wisconsin were 140 years old, and hardwoods in the rich Ohio and Wabasha River valleys were 100 to 200 years old.

Table 1.2 Average forest volumes of Eastern United States, primary forests circa 1800 (Whitney 1994). With permission.

State	B.F./Acre
VT (Williams 1794)	3000-11,800
ME	7700
IL	8000
OH	13,000
PA	17,500
IN- mixed hdwds	25,000
IN-W.Pine/Hemlock	100,000

Forests blanketed 454 million of the 552 million total acres in the Continental Eastern United States. With the exception of 98 million acres in western Minnesota, western Missouri, Iowa, and the prairie peninsula of central and northern Illinois, at least 95% of every state was forested. The conversion of forest land to agriculture in the eastern United States was perhaps the largest that has ever occurred in the span of nearly 200 years (Figure 1.1a-c Greeley's 1620, 1850, 1920 maps, Greeley 1925).

The loss of forests and subsequent reforestation in three states is shown in Table 1.4. Of the 454 million acres of original forest land, 99.999% was cutover; only 0.001% remains in primary (ca. 1650) forest stands (Whitney 1994). Forty-nine percent (224 million acres)

Table 1.3 Average values of basal area and height, measured in remnant old growth, mixed forests of the Eastern U.S. 1904 to 1988 (after Whitney 1994). With permission.

Mixed Type	Location	Basal Area ft.^{2} / ac.	Height feet
Beech/Sugar maple	OH, IN, S. MI	130	114
Central Applach. hdwds	E. KY, S.E. OH, S. IN, S. IL	152	66
Northern hardwoods	Cent. NH, NH, MI-U.P.	157	82
Sugar maple/Basswood	MO, WI, S. MN	178	98
Oak/Hickory	KY, IN, NJ, S. IL	122	115
White/Red pine	MI, N. NY	209	118
Hemlock/N. hardwoods	VT, PA, MI-U.P., WI	235	-
Hemlock/White pine	PA, MI-L.P. S. NH,	283	-
Red spruce	NH	152	-
Red spruce/Balsam fir	ME, NH	174	-
W. Pine/Hemlock (gone)	Pisgah, NH (Foster 1988)	392	115
White pine (still there)	Hartwick Pines, MI (Rose 1984)	318	118

of the cutover forests regenerated to second growth forests, and 51% (231 million acres) were converted to and remained in agricultural or urban land as of 1990 (FIA 1998). The western Great Lake States, the Appalachian Mountain States, and the Great River States each account for about 31% of the converted forest land. The New England States and the Chesapeake/Delaware Bay States each accounted for about 4% of the converted forest land.

Overall, forest land in the Eastern United States has been reduced to 40% of its total area (Figure 1.2) How much of this forest cutting and conversion affected the riparian forests of How much of this cutting and conversion affected the riparian forests of the Continental Eastern United States? Virtually all of it! All riparian forests were cut and half of them

Table 1.4 Changes in forest land (%) for three states (adapted from Whitney (1994), Frelich (1995)and USDA Forest Service FIA Eastern Forest Database (1998))

Rhode Island			
1767	1875	1972	1990
77	32	67	74

Ohio			
1853	1883	1979	1990
54[1]	18	27	29

Michigan			
1820	1870	1980	1990
85	77	51	36

[1]In the 1600s, Ohio was at least 85% forested (Greeley 1925).

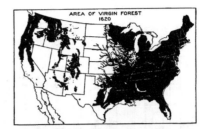

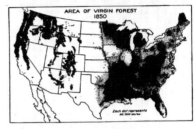

Figure 1.1a-c Virgin forest area in 1820, 1920, and 1950 (Greeley 1925). With permission.

were completely converted to agriculture. Many areas such as Cadiz township in Wisconsin, were nearly all converted to agriculture (Figure 1.3).

Throughout the Eastern United States, the old growth forests were replaced by early succession species. The aspens and birches, with light seed and root suckers or stump sprouts, and the cherries, with long-lived seeds invaded many of the cut-over acres. In shady areas sugar maple and balsam fir, both with wind-dispersed seeds, took advantage of canopy openings. White pine provided the super nurse canopy for hemlock in primary forests, but both declined from overlogging, the spread of blister rust, and harsh exposure. Beech, with limited reproduction in the first 50 years, declined while vigorous black cherry stump sprouts occupied the site. Similarly, short-interval harvest for pulpwood changed the ratio of red spruce to balsam fir in Maine. In 1902, the volume ratio of red spruce to fir was 7:1, but

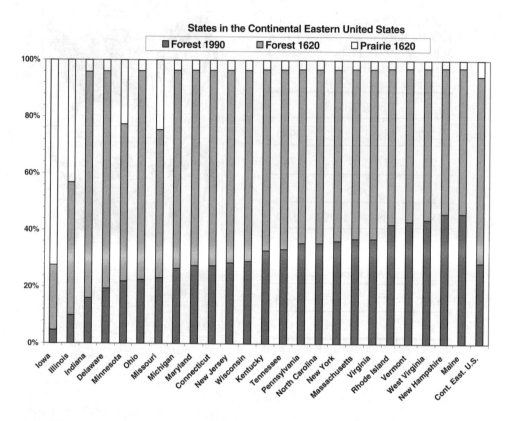

Figure 1.2 Land use in the Continental Eastern United States, 1620 and 1990 (derived from Greeley 1925; Whitney 1994; and USDA Forest Service 1998).

by 1972 it was only 1:1. This change in ratio resulted from repeatedly cutting second-growth stands and was exacerbated by slash-fed fires in the early 1900s. Wide-spread fires following original logging were often severe because fuel on the ground over large areas greatly exceeded that in old-growth forests and more-restricted, blowdown areas. Fire managers today call them catastrophic because of their size and the heat generated on the ground that could consume the entire forest floor. Although difficult to document, forest floors once 4 to 14 inches thick are now only 1 to 3 inches thick. Finally, Whitney (1994) considered grazing to be a problem for riparian forests in the eastern United States. It was ubiquitous; pig and cattle grazing were widespread in the Eastern United States and was the preferred method to complete the conversion from forest to pasture.

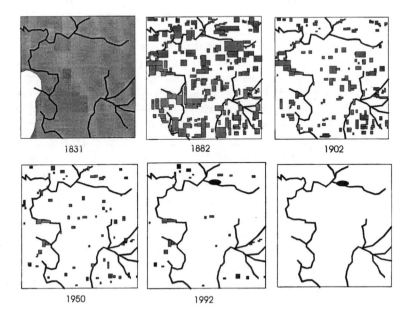

Figure 1.3 Forest area (shaded) in Cadiz Township, south central Wisconsin (derived from Curtis 1956 and DeLorme 1992). With permission.

The Primary Rivers of the Continental Eastern United States and How They Changed

Early conditions on the rivers of North America are not well documented, but one example from the western wilderness comes from the journals of Daniel Greysolon, Sieur du Luth, who was dispatched from Quebec City in 1678 by the Governor of Canada to take possession of the land west of Lake Superior and find "The Great River." He describes his portage from the western end of Lake Superior to the Mississippi River: "In June 1680 . . . I took two canoes with an Indian, who was my interpreter, and four Frenchmen. I entered a river (northwestern Wisconsin's blue ribbon trout stream the Bois Brule). . . When, after having cut some trees and broken about 100 beaver dams, I reached the upper waters . . . and made a portage of half a league to reach a lake, the outlet of which fell into a very fine river (the St. Croix) which took me down into the Mississippi." (Winchell and Upham 1884).   Two hundred years later in 1880, beaver dams still occurred every third mile, and lake trout, brook trout, and the coaster brook trout of the Bois Brule River were carried in bushel baskets to residents of the new logging town named in the French explorer's honor (Duluth, Minnesota). The standing stock of mature trees, a mature fishery, and a mature beaver community were severely depleted over the next 50 years until hunting and fishing seasons, limits, fire protection, and forest management were begun in the 1930s. Sieur du Luth's route to the Great River bisects the St. Croix drainage that divides Wisconsin and Minnesota and produced a peak white pine yield of 8.9 billion board feet in a single year (1892); perhaps the largest annual harvest of wood ever (Rector 1951).

Exploitation of the St. Croix (named after Maine's St. Croix) repeated the pattern of land-use change witnessed in New England, New York, and Pennsylvania in the 70 years preceding North America's peak annual wood harvest — river and stream cleaning, bank-side harvest of big trees, harvest of most large trees within 5 miles of the stream, homesteading, conversion of half the cleared forest land to agriculture, and development of permanent dams for navigation, trade and the generation of electricity. Harnessing water to generate electricity mimics the spring surge release and the fall and winter head-building of splash dams, but changed the frequency from annual to daily. The first hydroelectric plant in the world was built on the Fox River at Appleton, Wisconsin, in 1882. It has taken a century to acknowledge what dropping the water levels each day (and building them each night) to provide for peak periods of power generation has done to stream habitat. Even in the humid East, water withdrawal for irrigation, snow-making, and water supply play a role in changing stream and lake habitats.

Logging has long been a feature of the Eastern North American landscape. The first sawmill was built in 1623 on the Salmon Falls (present day Piscataqua) River in what became the state of Maine. Splash dams, first employed centuries earlier in Europe and Asia, were built all over Eastern North America to expedite the transportation of logs to mills and markets downstream. Dam construction techniques ranged from crude piles of logs and debris fashioned for one-time "splashes" of logs during spring floods in smaller streams and brooks to precisely engineered structures of concrete, steel, and wood designed to allow year-round floods on larger streams and rivers. Log splashes (Figure 1.4) wreaked havoc not only on fish and their habitat, but also on riparian farm land. Farmers in many states vigorously opposed these violent drives as logs overran their fields, fences, and outbuildings. Many such dams operated into the 1920s in eastern Kentucky's Red River basin before road construction and dwindling timber supplies signaled their end (Coy, et al. 1992).

Logging in the first 200 years was largely restricted to forests near settlements on the fringe of the Atlantic shore. Many trees were cut simply to clear land for homes and fields and to provide fuel for cooking and heating. Logging provided timber both for local use and for trade with Europe. Rivers were the only means of transporting logs long distances, and logging along rivers was restricted to those areas within a day's pull for oxen and horses.

Figure 1.4   Splashing logs through the Breaks of the Sandy. Boulder fragments after dynamiting rock in the Russel Fork of the Big Sandy River near the   present community of Haysi in Dickenson County, VA (just upstream of the Kentucky state line). Note the person near the center of the photo (American Lumberman, March 19, 1910).

In the early 19th century, logging was one of the major industries of the new country. In the 1810s the Saco River in New Hampshire (and Maine) was the first center of commercial logging. Bangor, Maine on the Penobscot River led the eastern states (colonies) in timber harvest from 1820 to 1850. White pine was selectively cut from Maine forests up to 1840 after which red spruce dominated lumbering in Maine into the 1900s. Steam-powered   sawmills,   centered around the Hudson River near Glenn Falls, allowed New York to surpass Maine in lumber production in 1850. Williamsport,   Pennsylvania   on   the Susquehanna River was the top producer in 1860 (hemlock and pine). Timber production in the Lake States began in the 1840s and followed suit with an initial assault by river improvement companies to clean the rivers (nearly all of them). Production surged after the Civil War and peaked in 1892. Lake States white pine built mills and towns from Albany, NY, to Denver, CO, during the second great surge of immigration across the Midwest. Although never perceived as a lumbering center, the Ohio and central Mississippi Valleys (oaks, hickories, and sycamores) comprised 1/3 of the original forest land converted to agriculture before and after the Civil War. Thereafter, transportation changed the rivers forever.

Logging in the Appalachians (beech at first) served local populations, but in the late 18th and early 19th centuries, logging for the manufacture of charcoal used in iron smelting, and for iron and coal mine shaft supports, changed the impact of logging on the landscape. Large areas were cut over and then abandoned when the cost of transporting wood rose too high. But it was during the post-Civil War years of the 1880s when, prompted by the looming shortages resulting from over exploitation of Lake

States and New England forests, loggers moved both west and south into the Appalachian forests in search of red spruce, yellow poplar, and American chestnut.

Before the 1850 census in the United States, at least 100 million acres of forests had been converted to agriculture east of the Mississippi River (Williams 1989). George Perkins Marsh was the first to speak for land and water conservation in 1864 when he published *Man and Nature*. This early call for resource conservation was taken seriously by James D. Porter, Assistant Secretary of State in 1886, when he asked his ambassadors in Europe to trace what was happening to forests and forestry there. It had long been the practice in Europe to clean big trees from rivers, both large and small. As a result, rivers and streams have changed from structurally diverse systems to aquatic highways. "The intensity and ubiquity of human influence on the amounts, dynamics, and functional importance of [coarse woody debris] have been tremendous." (Harmon et al. 1986). The removal of logs and limbs from streams has occurred for hundreds, even thousands, of years. In the Medieval Ages (500 to 1500 A.D.), the Volga, Elbe, Tiber, Po, and Rhine Rivers in Europe were extensively used for log transportation.

As recently as a century ago in Europe, concepts of land and water stewardship were widely expanded, not unlike that occurring in North America today. The following are paraphrased from various parts of Europe in 1886 and capture the origin of many issues, values, and processes debated today (underlined below for emphasis) (Porter 1887).

> From the Consul of Thuringia (Germany): "Forestry is pursued (1886) in so careful and scientific a manner that not even ponds or marshes are allowed to be drained. It was here among the Teutonic Tribes (who terrorized Rome about 300 A.D.) that the practice of "squatters rights" and thus private rights began. Cleared land on the river and the adjacent forests became a right of ownership for the heads of families and tribes. By the early 1700's, in Europe, more wood had been consumed than could be grown in several centuries. Tree planting was well established in the mid 1700's, and the regulation of tree harvest by age and acreage (as opposed to wood volume) was well established at the end of the 19th Century in parts of Germany.

> From the Consul of Italy: "Forests have been destroyed to gain lumber, pasture, and arable land. . . . The ususal results have followed: a decrease in the depth of navigable streams, an increase in rainy season floods, avalanches, landslides and denudation of mountains." Flooding and sedimentation.

> The Consul to Austria-Hungary: Sad experience has taught the necessity of the greatest stringency in the forest laws in the mountain districts: ... even on the steep rivers, stonework can not withstand the torrents. Whenever a communal forest borders on these rivers, its maintenance is held to be of especial importance. In all these forests, therefore, not a tree can be felled without the consent of the state foresters.

No animals are allowed to pasture, and the greatest precautions are taken to guard against fire. State regulation of land use with mandatory "BMPs."

From the Consul at Nice, France: The law of 1882 provides for both stream and slope work to prevent sedimentation. The work is to be done directly by the state and by landowners with or without state aid. "Excessive grazing . . . (that) causes denudation has caused incalculable damage in the great mountain regions of France. When hillsides are covered with trees, the winter snowpack melts slowly in their shade; but when the trees are gone, the full force of the sun produces flood peaks similar to heavy rain storms." State and private landowner mitigation (with or without state aid – subsidies, tax breaks, or grants).

The Consul of Palermo (Switzerland): Stream restoration, and the concept of government-private cooperation in achieving a common public welfare, was debated in Switzerland as reported by civil engineer, Robert Lauterburg, in 1885: "It is well known . . . what great inroads upon river banks and what heavy landslides have been caused by running water." While our engineers and foresters have tried to check the devastation of floods and erosion, it is not within their province to compel owners to look after their own interests (private rights), even in cases where the public welfare is endangered.The stream Nolla near Thusis in the canton Graubünden: "has lapsed into such a condition that after *every* thunderstorm it pours down immense masses of earth, mud, and rock into the valley. . . . the rainfall, instead of being absorbed by the trees as formerly, pours in simultaneously from all sides in great masses into the bed of the stream, washes it away, undermines the banks, and finally sweeps away the entire mass of rubbish down into the valley." Regional sedimentation.

Herein lies a central argument that is being debated in North America today -- the impact of forest harvesting on floods. Chapter 5 will directly address what we now know about cause and effect in both flat and mountainous sections of the Eastern U.S. and the relative importance of catastrophic floods versus the approximately annual bankfull discharge that is incipient to flooding.

Although the preceding accounts are by our standards anecdotal, they clearly demonstrate that our 19th century counterparts were keen observers and highly perceptive. Many of the foresters, farmers, engineers, and politicians of the day understood the basic relationships between land use and water. Of course other, more contemporary arguments, such as the debates over recreation use by broad segments of society, were not part of the last century's debate on forest and water in Europe, but they are part of today's debate in both America and Europe. Now we will consider how European and Asian heritage was applied to the forests of the Eastern United States.

All across Eastern North America, log transportation played a major role in river widening, loss of sinuosity (the river's length divided by the straight line valley length), and a loss of habitat. The first major use of log rafts occurred on the Delaware River during the 1750s. The advantages, particularly to part-time loggers and farmers, were quickly perceived and log rafts soon appeared on the Susquehanna, Allegheny, and other rivers in nearby New York and New Jersey (Williams 1989). Perhaps surprisingly, log drives, the transport of unbound logs en masse down a watercourse, were not common until at least 60 years later; the first recorded drive occurring in a tributary of the Hudson in New York in 1810.

Unlike log rafts, which required water deep enough to float the largest logs in the raft (and hence were confined to large, main-stem rivers), log drives occurred throughout entire watersheds. Trees were felled and rolled to the banks of tributaries during the fall and winter, ready for dumping into the high flows of spring to take their chilling ride to downstream sawmills. After the timber supply in a watershed was exhausted, the logging camps typically moved on. However, depending on the size and extent of the watershed, some rivers supported log drives for many decades. The Kennebec Log Driving company in Maine holds the record for longevity and intensity of log drives. Starting in 1835 when it served 63 sawmills, and for the next 141 years until 1976, the year of the last river drive in Maine, the company moved its logs down the Kennebec River. Between those years, nearly every major river in the Continental Eastern U.S., from the St. Croix in Maine to the French Broad in North Carolina and from the St. Croix in Minnesota and Wisconsin down the mighty Mississippi, streams and rivers were used by river-drivers to transport logs or log rafts to mills.

The ecological consequences of these drives and the stream improvements they made necessary were doubtlessly devastating. Rector's remark on river stream and brook cleaning illustrates the prevailing mission of those whose job it was to prepare rivers for log drives: "Most of the rivers were not suitable for driving logs in their natural condition . . . Fallen trees must be removed from the channel: snags, rocks, and shoals needed to be cleared with ax, pick, and in the later decades, dynamite (Rector 1951) (Figure 1.4)." This was mean work. "The accumulations of centuries in the form of driftwood and fallen trees frequently covered the streambed for miles, and all had to be cleared away. If the ground was swampy, horses could not work in it, and everything had to be done by hand. Tree by tree, stick by stick, the obstructions had to be lifted out and put far enough away so high water would not float them back in again. Islands and shoals had to be dug away; stumps, bedded in the mud, had to be grubbed out; embankments had to be made at sharp bends. The sunken tree at the bottom of the stream had to be cut and worked loose in its bed (Pike 1967)." Although we have no ecologically based accounts of the impact that these stream improvements had on aquatic biota, the preceding quotations describe the complexity that once characterized, and is now largely absent, from most aquatic habitats. Figure 1.5 may be the oldest preserved photograph of "Carding the Ledges" for a Brook Drive in Maine (Smith 1972). (We have it on good faith from one astute reviewer that carding the ledges may have derived from the task of carding wool with a brush having short steel bristles (card) to straighten the fiber.

Although we focus on forest land, in fact many of our watersheds are mixtures of forest, farm and urban land uses. Observations in Pennsylvania at the turn of the 20th century

Figure 1.5 "Brook Drive - Carding the Ledges" in Maine. Note that only axes, picks, and one-man saws were available for clearing the riparian forest. From the Larson Collection, University of New Brunswick. Smith 1972. With permission.

illustrate the interaction of widespread logging and riparian agriculture development. J. T. Rothrock (1902), Pennsylvania's first Director in the Department of Forestry, writes after literally watching how land use changed stream habitat: "In all our alluvial valleys the frequent freshets work greater or less damage to the farm land. In fact, it can hardly be said that the beds of any of our rivers, which flow through wide valleys, are constant. They not only have entirely deserted the ancient water courses, leaving them off as back channels to one side or the other, but they are changing them from year to year before our eyes. . . Whilst it is true that a large quantity of valuable soil is sometimes deposited by these freshets on the surface of the land, it is also equally true that this same soil has come from the margin or river bank of somebody else's holding." The concept that river flow would change with land-use change and that the impact on stream habitat was the result of in-channel erosion and deposition was readily accepted by Rothrock.

Land clearing and farming in the Chesapeake Basin caused a fourfold increase in deposition, in the Gunpowder Estuary, shallowing or filling parts for over two miles (Brush 1986). Sediment export into southern Lake Michigan increased tenfold following conversion to agriculture (Davis 1976 as reported in Whitney 1994). A quote from Whitney (1994) details the impact of forest to agriculture conversion in southwestern Wisconsin:

"James Knox (1977) was able to demonstrate an increase in the width of many of the headwater and tributary stream channels of the Platte River system . . . and a decrease in the width of the main channel downstream. The wider and shallower upstream channels were associated with bank erosion and an increase in the bedload of the streams following settlement. The changes reflect the geomorphic response of the channel to the occurrence of large and more frequent floods since settlement (Knox 1977). The reduced width of the main channel downstream was due to excessive overbank sedimentation and the deposition of the finer, suspended particles downstream. At their peak in the 1920s and the 1930s, historic rates of overbank floodplain sedimentation exceeded their presettlement rates by two orders of magnitude (Knox 1987)."

Curtis (1951) detailed the shortening of a stream with a conversion from forested to dairy farming over a span of 104 years in Jordan County in southern Wisconsin. In 1831 the county was 98% forested with 40 miles of stream, in 1902 it was 9% forested with 30 miles of stream, and in 1935 it was 4% forested with 25 miles of stream: a 36% loss of stream habitat.

The impacts of river cleaning, original logging, catastrophic fire, and conversion to agriculture can accumulate over the span of a century. We must deal with these conditions, even in forest regions today. Recent measurements of stream sinuosity on the southern Lake Superior Clay Plain illustrate how a sequence of land-use change can reduce the stream habitat. This is a landscape checkered by agriculture, mature forests, and recently cut forests, strung with roads large and small. Tree ages were used to date when the channels changed. They were aged in old channels, on existing flood plains, and abandoned flood plains on the first terrace to date channel down-cutting and a progressive loss of stream sinuosity. The tree dates group into periods associated with landscape changes: original stream cleaning (late 1840s), original logging (1870-1905), the catastrophic Hinckley Fire (1918) (it consumed the forest floor — once 4 inches thick, and now recovered to 1 inch thick), and agriculture development from two periods: the late 1920s to the early 1930s and the mid 1950s following the Korean War. While channel sinuosity does not indicate land-use changes in highly entrenched streams, all the channels in this study are E channel types with broad floodplains and a modal sinuosity of two (Rosgen 1994). Figure 1.6 shows the change in sinuosity (and a representative sketch) associated with cumulative land-use change. Percentage-wise, stream length (as reflected in their sinuosities) is shortened by 6, 30, 40, and 45% in the last time periods over 110 years (Figure 1.6). The streams still run in the same valley, so an eye not perceptive to habitat change may see no change.

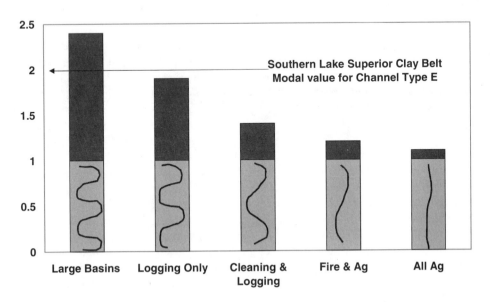

Figure 1.6 Cumulative changes in channel sinuosity caused by land use changes in basins of the Lake Superior Clay Plain. Sketches simulate channel length in each land use type to visually show how channel habitat is lost as streams shorten even though they flow through the same length of valley.

In a similar way, forest clearing for mining and for mine supports in the coal mining Appalachian Region, along with the development of roads that crowded streams into narrower meander belts, caused both longitudinal and cross sectional changes in stream habitat. Finally, mining which results in acid drainage to streams, can eliminate entire fish populations. For many resources, we simply cannot begin management with a clean slate because we have damaged these resources through ignorance or indifference. We cannot practice forestry beginning with the best stand conditions. We cannot practice fisheries management beginning with the best habitat conditions. We cannot practice farming beginning with the best fields. We cannot maintain ecosystem diversity beginning with fully diverse communities. The conditions developed in the Continental Eastern United States over the last 200 years, and the divisive practices pursued over those centuries frame the challenge of managing for healthy riparian areas.

The Issues Facing Riparian Area Managers Today

In a world characterized by uncertainty, a few facts are clear: more people are demanding healthier environments, greater recreational opportunities, and more products derived from wood. The demand for wood and wood products has increased dramatically since 1980. The rate of harvest (projected through 2005) is reminiscent of the wave of logging that surged through the Eastern States from 1865 through 1905. Today, as in 1905, the public debate revolves around questions of land use vs. protection. At the turn of the last century, many people were aware of the benefits of well-managed forests: "If you live at a distance [from the proposed Adirondack Forest Preserve], your benefits consist of not only wood in the form of houses, barns, furniture, paper or the cheerful fire in the grate, for there is no substance so widely used as wood, but the air you breathe and, in this instance, the stream that flows by or carries you or your product or turns wheels to give you light, transportation and large variety of other things." (O'Neil 1910). The difference today is that more of us are closer, both physically and philosophically, to the water's edge in our eastern forests. From the viewpoint of society as a whole, the issues are deceptively simple. We want it all: intact, functioning ecosystems; continuous supplies of high-quality water; and lumber, paper, and other forest products. But as individuals, we see things differently. As individuals we fear losses.

If we protect riparian areas, we fear the loss of:

- Wood and wood products
- Access to minerals and mining
- Opportunities for hydropower
- Grazing and cropland
- Water withdrawals
- Freedom to manage private land

If we do not protect riparian areas, we fear the loss of:

- Water quality and quantity
- Habitats for plants and animals
- Native plant and animal species
- Recreation and aesthetic qualities
- Natural filtering of sediment
- Connectivity with other landscapes

Quite naturally, we want the opposite of these losses. As we struggle to achieve our goals for riparian area management, we need to know what the rules are. We need to know:

- How to define riparian areas
- How to classify waters and valleys
- How to assess the impacts that may have accumulated within a watershed
- How have riparian functions been impaired (what's the problem?)
- What silviculture is appropriate for riparian forests
- How does forest and water management influence animal populations
- How to balance & sustain agriculture, forestry, recreation, and urban land uses
- How to recognize and evaluate a healthy functioning condition of riparian areas
- How to plan for desired future conditions
- How to work together across landscapes with many ownerships
- What techniques can we use to restore riparian ecosystems
- How we can enhance natural processes to manage the routing of water and sediment

The chapters that follow will give you some of the answers to this "need to know" list. In summary, how do we establish the ethic and practice of land and water stewardship that protects riparian functions in the Continental Eastern United States? The fears and wants are similar to those mentioned previously by George Marshall in 1864 and those recorded by James Porter for Europe in 1897. We cannot answer all these questions to everyone's satisfaction. However, in the chapters that follow, we present a set of tools and concepts that will help in the search for solutions to the land and water issues we each face. We hope that readers will not only obtain an understanding of aquatic and riparian ecology but also strategies for the sustainable restoration and management of both aquatic systems and riparian forests. The challenge is not to find the easy answer but to find the knowledge to read the land and read the river. Then we will have no fear of what you will do for them.

Meeting the Challenge for the 21st Century

Our challenge is first to understand current and healthy conditions for a wide range of riparian forest and water resources. This is the melding of management experience and the research that shows cause and effect along with the error of interpreting measured effectiveness. It is the base that defines the realm of possibilities. But the accumulation and synthesis of knowledge is only the beginning. "The acid test of our understanding is not whether we can take ecosystems to bits and pieces on paper, no matter how scientific, but whether we can put them together into practice and make them work (Bradshaw, 1983)." Equally important, or perhaps more important, is the common vision of what we want to see across our riparian landscapes. Deriving this vision, the socially derived, desired future condition will not be easy because of the very feature that makes it so powerful: all of us must participate.

Each of us realizes we must be competent in our own discipline, and each of us suspects we may lose competence as time moves on in spite of our experience. A training course to keep up is useful only when we apply our new knowledge with on-the-ground experience. Often, one course is not enough. We may need two, three, four . . . and more. We should not see disciplinary knowledge as the quiver of arrows carried into a consensus building session, but as our own base of confidence we can share with other disciplines and with other viewpoints. A greater challenge is to learn parts of other disciplines important to our own. Build not only your base of confidence but also your base of understanding.

When we walk beside streams and through forests, we sometimes are proud of what we see. We are sometimes discouraged by what we see. Sometimes we see evidence of stewardship and integrated management. At other times, we see a landscape (or pieces of a landscape) dominated by a single use. Is the integrated vision by chance? Or, did someone understand what it meant to do integrated management? Our purpose in this book is to reduce the element of chance. Our purpose is to give each of us the eyes of the other, to help develop the common understanding, appreciation, and vision to manage riparian areas in the Continental Eastern United States.

A municipal trail comes to the water's edge in West Duluth, Minnesota.

Sandy Verry

Ana Marie Becker

Fine sand from a native material road crossing (above) fills the spawning gravel and cobble of a brook trout stream.

Catch and release! Secret Muskie waters. It's the fishing!

Sandy Verry

Bob Leibfried

Canoeing the Au Sable River in the Huron-Manistee National Forest, Michigan.

Headcuts in the Nemadji River basin. Part of the Lake Superior Clay Belt on the Minnesota, Wisconsin border. Glacial landrebound, and landuse changes cause channels to change.

Andy Dolloff

Thaw reveals the winter thalweg in an over-wide Virginia river.

Sandy Verry

A 5-inch storm washed an 8-foot channel around a 3-foot-wide culvert in a local access road. The brook trout stream above and below the culvert is 8 feet wide at the bankfull elevation.

Sandy Verry

Autumn in the Suomi Hills Semi-Primitive Recreation Area. Chippewa National Forest, Minnesota.

Chapter 2

Defining Riparian Areas

Bonnie L. Ilhardt, Elon S. Verry and Brian J. Palik

"The riparian corridor is the heart of the drainage basin since it may be the ecosystem level component most sensitive to environmental change."

Naiman et al. 1992b

What is a riparian area? Does it include the aquatic environment or only the transition between aquatic and terrestrial environments? Does the transition include only lands with saturated or seasonally saturated soils or all land that influences or is influenced by the aquatic environment? Definitions vary with the perspective of the author and user. Typically, they center on which components of the landscape are included, which characteristics of a component, which landscape scales are considered, or which legislative mandates for water quality best management practices (BMPs).

We review the many definitions available from agencies and from several disciplines and offer a definition based on the long-term sustainability of ecologic functions in riparian areas. An ecosystem function is a collection of processes that govern the flow of energy and materials (e.g., sunlight, leaf carbon, water, sand, and nutrients). Our definition includes the water body, riparian land, and parts of upland areas that have a strong linkage to the water. This definition is not associated with a management prescription or BMP, rather, it defines a *riparian area* by the strength of the ecosystem functions associated with it. Under this definition, the boundaries of riparian areas typically are less uniform than those associated with fixed-distance prescriptions for filter strips, buffer strips, or streamside management zones (SMZs). We delineate riparian areas by examining how the ecosystem function changes with distance from the water. Finally, we offer a field key to delineate riparian areas on the ground.

Many Definitions

Many states define riparian areas to regulate land disturbance activities, to protect water quality, and to comply with the Federal Clean Water Act. "Streamside management zones," "buffer zone" or "buffer strip" and "riparian management zones," are the terms most frequently used, and minimum widths are usually specified. The riparian management zone excludes the aquatic component and delineates the land and vegetation that buffers the surface water from land disturbance. In our conclusions from the following discussion, we include the aquatic component and use the broader term--riparian area.

Agency Perspectives

A review of state BMPs in the Eastern United States revealed that, while the terminology differs, the focus is the same--water-quality protection. New Hampshire has proposed revising 1972 BMPs to address maintenance and protection of the key riparian and wetland ecological functions. Minnesota revised its water-quality BMPs in 1995 to include protection of hydrologic function in wetlands. They were revised again in 1999 to include riparian management practices, site productivity, wildlife, and the cultural aspect of forest management.

Many federal agencies define riparian areas and identify riparian-area components, however, definitions are not consistent among agencies. The USDA Forest Service defined riparian area to include the aquatic ecosystem, the riparian ecosystem and wetlands (USDA Forest Service 1994). While this broadly defines riparian areas, a "riparian ecosystem" is restricted to those areas with soil characteristics or distinctive vegetation that requires free or unbound water (thus the stream, lake, or open-water wetland is not included).

The Eastern Region (Region 9) of the Forest Service recognizes the deficiencies in this definition and supports the following functional definition:

> "Riparian areas are composed of aquatic ecosystems, riparian ecosystems and wetlands. They have three dimensions: longitudinal extending up and down streams and along the shores; lateral to the estimated boundary of land with direct land-water interactions; and vertical from below the water table to above the canopy of mature site-potential trees (Parrott et al. 1997)."

In this definition, the "aquatic ecosystem" includes the stream channel, lake, estuary beds, water, biotic communities, and the habitat features that occur therein; the "riparian ecosystem" is defined as extending away from the bank or shore to include land with direct land-water interactions, and whose areal extent is variable based on its ability to perform ecologic functions. The wetland part of the definition is consistent with other federal agencies that have adopted the USDI Fish and Wildlife Service's wetland definition and includes "those areas that are inundated or saturated by surface or groundwater at a frequency sufficient to support, and under normal circumstances do support, a prevalence of vegetation typically adapted for life in saturated soil conditions (USDA Forest Service 1994)."

The Coastal Zone Management Act excludes the aquatic ecosystems and defines riparian areas as:

"...vegetated ecosystems along a water body through which energy, materials and water pass." They further characterize riparian areas as having a high water table, subject to periodic flooding and encompassing wetlands.

The USDI Bureau of Land Management definition excludes the aquatic component and defines riparian area as:

". . . a form of wetland transition between permanently saturated wetlands and upland areas (USDI, 1993)."

They further state that these areas contain vegetation or physical characteristics reflecting permanent surface or subsurface water influence. The BLM definition includes land along and adjacent to, or contiguous with perennially or intermittently flowing rivers and streams, glacial potholes and the shores of lakes and reservoirs but excludes ephemeral streams or washes that do not contain hydrophytic vegetation.

The Fish and Wildlife Service recently adopted a definition to guide their mapping of riparian areas in the Western United States where mean annual evaporation exceeds mean annual precipitation. They, too, recognize that many definitions are used by government agencies and the private sector, with some based on functionality and others on land use. Their definition, which follows, was developed to ensure consistency and uniformity in identification and mapping:

"Riparian areas are plant communities contiguous to and affected by surface and subsurface hydrologic features of perennial or intermittent lotic and lentic water bodies (rivers, streams, lakes, or drainage ways). Riparian areas have one or both of the following characteristics: (1) distinctly different vegetative species than adjacent areas, and (2) species similar to adjacent areas but exhibiting more vigorous or robust growth forms. Riparian areas are usually transitional between wetland and upland (USDI Fish and Wildlife Service 1997)."

Discipline Perspectives

The word "riparian" is drawn from the Latin word "riparious" meaning "bank" (of the stream) and simply refers to land adjacent to a body of water or life on the bank of a body of water. Following the Latin derivative, some authors exclude the aquatic component when defining riparian area and apply the root word literally, using only single factors such as soils, groundwater and surface water hydrology, or vegetative type (Karr and Schlosser 1978). Soil scientists prefer a definition based on water availability and define the riparian zone as:

> ". . . land, inclusive of hydrophytes, and/or with soil that is saturated by groundwater for at least part of the growing season within the rooting depth of potential native vegetation (Brosofske 1996)."

Riparian areas are also characterized by climate, geology, land forms, natural disturbances, soil, and vegetation (Swanson 1982). Naiman et al. (1993), offer this definition:

> "The riparian corridor encompasses the stream channel and that portion of the terrestrial landscape from the high water mark toward the uplands where vegetation may be influenced by the elevated water tables or flooding and the ability of the soils to hold water."

These concepts include the water body but exclude the high terraces and slopes that never flood, yet their falling trees and bank sediment can strongly affect channel habitats. These areas also regulate water temperature, fine and coarse organic matter input, bank stability, regulation of nutrient and sediment flows, and landscape corridor or habitat connectivity. The role of climate, geology, and land forms is widely recognized and forms the basis for hierarchical classification systems of terrestrial and aquatic ecosystems (USDA Forest Service 1993; Maxwell et al. 1995; see Chapter 3).

The presence of water and its flow pattern or regime is *the* most distinguishing characteristic of riparian areas. The delivery and routing of water, along with the transport of sediment and woody debris, are responsible for defining riparian area boundaries (Naiman et al. 1993) and are used to classify natural stream reaches (Rosgen 1996). Riparian areas, because of their landscape position, are subject both to more surface flows (runoff and flooding) and more subsurface flows (saturated soil horizons near the surface or true groundwater). Riparian areas generally have more water available to plants and animals than adjacent uplands (Gregory et al. 1991).

Soil properties in riparian areas are quite variable, ranging from saturated to well-drained over short distances (Gregory 1991), so decomposition rates are also variable. Where the soils are saturated or seasonally saturated, decomposition rates are lower than in adjacent upland soils. Riparian areas with well-drained soils have decomposition rates similar to adjacent upland soils (Gregory et al. 1991; Naiman et al.1993; Brosofske 1996). The nutrient content of riparian soils also varies. It is generally higher due to the deposition of sediment and nutrients from floods and nutrients from upland runoff (Brosofske 1996). Activities occurring in the watershed, such as farming or other land disturbances, may also influence the soil nutrient content (Lowerance et al.1984; Welsch 1991; Palone and Todd 1997).

Geologic land form, soil moisture regime, and depth to the water table are the conditions that vegetation adapts to. These interactions give rise to landscape vegetation patterns that repeat in the riparian corridor. Additionally, natural disturbances (those occurring within and away from the aquatic environment) affect riparian area plant composition, structure, and successional development. These include landslides, debris-torrents, fire, wind, and flooding (Swanson et al. 1982; Gregory et al. 1991).

Defining riparian areas more broadly is also supported by Swanson et al. (1982), Gregory et al. (1991), and Verry (1992). Gregory's (1997) recent definition of riparian areas includes the aquatic ecosystem and that portion of the terrestrial ecosystem, beyond the influence of elevated water tables, that has a functional connection to the water. Hupp and Osterkamp (1996) endorse the ecological context for defining riparian areas because doing so recognizes the importance of fluvial geomorphic processes in shaping the character of the riparian zone. See Brosofske (1996) and Palone and Todd (1997) for recent reviews of riparian area definitions.

An Evolutionary Perspective

It is not obvious how all these views of riparian-area definition should be reconciled. Yet on close examination, we see them as a reflection of how the professions dealing with riparian areas have evolved in the last three decades. First, there was a tendency to separate ecosystems by disciplines (i.e., aquatic versus terrestrial) and to separate management alternatives by disciplines (i.e., forestry or agricultural BMPs applied to land versus fisheries habitat management applied to water). Those not directly concerned with management tend to combine descriptors of riparian condition using a multidisciplinary approach. The multidiscipline ecosystem approach uses soil condition (soil science), citing saturated soils for at least part of the year combined with soil mottling. This system also combines the presence of hydrophytes (plant science) and the presence of the water table within the rooting zone as a measure of hydroperiod (water science). These same three factors are also used in defining wetlands. The emphasis in these definitions is to give each discipline an equal footing, and to separate management from science; that is, the management or multidiscipline definitions can stand alone and, perhaps most importantly, the definitions describe the state or condition of the ecosystem.

Descriptions of the state or condition of the riparian area can cause administrative confusion and can heighten disciplinary turf battles when ecosystems are mapped (see Chapter 4). How is a riparian area physically mapped, using both aquatic valley segments and terrestrial land type phases, and how are agencies or landowners to use these designations to track their resources? Both the aquatic and terrestrial ecologic classification systems (ECS) map ecosystems that are defined by climate, geology, soils, and biologic community. Concerns about mapping separate professionals seeking agreement on riparian-area definition. The most recent definitions deviate from considering only soils, plants, and water to include geologic and landscape setting and the geomorphology of streams and lakes. They focus not on ecosystem state, but on ecosystem function.

Our *functional definition of riparian area* that follows differs from those based on static state variables by using the flow of energy and materials (an ecosystem function) as the basis. Hence, movement of things is the basis for the definition rather than a fixed map unit. It uses water movement to define a stream as having a bank and bed since water must form these features almost annually regardless of whether the stream is perennial, intermittent, or ephemeral. It enhances the linkage concept between terrestrial and aquatic ecosystems universally accepted as characteristic of riparian areas. It allows state variables to be mapped using both aquatic and terrestrial ECS systems and allows overlays of riparian information on top of exclusive ecosystem map units. The management corollary is the design of roads that move vehicles, people, things, and water across both terrestrial and aquatic mapping units. It asks the questions: what linkages are important and where are they important enough to be included in a functional definition of riparian area? It includes aquatic, classical Greek riparious-riparian, and parts of terrestrial ecosystems. We consider the functional definition as an *interdisciplinary* approach that recognizes ecosystem functions developed and applied from many professional disciplines in a common landscape rather than the equal grouping of soil water and plant variables (*multidiscipline*). While we will offer guides to many aspects, professional judgment should be used as needed from site to site.

A Functional Definition

Despite these differences in riparian components and their character of components from an ecological perspective, there is agreement that a riparian area:

- Includes the water or feature that contains or transports water for a portion of the year
- Is an *ecotone* of interaction between the aquatic and terrestrial ecosystem
- Has highly variable widths or boundaries

Accepting these concepts, we offer the following functional definition for "riparian area":

> **Riparian areas are three-dimensional ecotones of interaction that include terrestrial and aquatic ecosystems, that extend down into the groundwater, up above the canopy, outward across the floodplain, up the near-slopes that drain to the water, laterally into the terrestrial ecosystem, and along the water course at a variable width.**

This definition is appropriate for natural resource management because it recognizes riparian areas by the ecological functions that occur at various scales. These regulate water temperature, fine and coarse organic matter input, bank stability, regulation of nutrient and sediment flows, and landscape corridor or habitat connection. Riparian areas are more than just buffers, and this functional definition recognizes this. As discussed by Swanson et al. (1982), a definition that incorporates scale, and recognizes the interactions that occur at each scale, enables us to address all the functions critical to maintaining healthy riparian areas (Table 2.1).

Adopting a functional definition means recognizing that the riparian boundary does not stop at an arbitrary, uniform distance away from the channel or bank, but varies in width and shape. While riparian areas can be mapped, a functional approach to delineating their boundaries is preferable to applying a uniform width. This approach is addressed in the following sections.

Moving from Definition to Delineation of Riparian Areas

Delineating riparian areas requires that we see them in a landscape perspective. Understanding the geology and landscape of the areas, and knowing their length and width, will help us define riparian areas on the ground.

Geomorphology is the Basis for Delineating Riparian Areas

Hunter (1990) considered "scale" in answering the question, " Just what is a riparian zone?" The riparian zone, at the smallest scale, is the immediate water's edge, where some aquatic plants and animals form a distinct community. At the next scale, the riparian zone includes those areas periodically inundated by high water. At the largest scale (and in forested regions), the riparian zone is "the band of forest that has a significant influence or conversely

Table 2.1 Functions of riparian vegetation with respect to aquatic ecosystems. Adapted from Swanson et al. (1982) and Nilsson et al. (1991).

Location	Component	Function
Above the ground or channel	Canopy and stems	Shade controls channel temperature and in stream primary production Source of large and fine plant detritus Wildlife habitat
In the channel	Large debris derived from riparian vegetation	Control routing of water and sediment Shape habitat — pools, riffles, cover Substrate for biological activity
Stream banks	Roots	Increased bank stability Create over-hanging bank cover Nutrient uptake from groundwater and stream water
Floodplain	Ground vegetation	Spawning for fish and for insects moving up channel Source of detritus Retards nutrient loss Reduces runoff through evapotranspiration

is significantly influenced by the stream." (In this context, Hunter uses riparian zones in a functional context (as we have used riparian areas) to distinguish from a riparian *management* zone (RMZ) often associated with Best Management Practice buffers). Landscape geomorphology constrains the stream valley, lake basin, and vegetation type (Swanson et al. 1982; Goebel et al. 1996; Naiman et al. 1997). It is in this context that we examine the physical controls on riparian functions.

Longitudinal Geomorphic Controls on Riparian Areas

The river Adour in France does not have a continuous floristic corridor because of interruptions caused by environmental factors such as climate boundaries (with elevation change or nearness to large water bodies) and the confluence of tributaries (Tabacchi et al.1990). This confirms the work of Buchholz (1981) in New Jersey where the intersection of tributaries with main-stem floodplains strongly affected the woody species distribution of the riparian area. In northeastern Minnesota, Ericsson and Schimpf (1986) show that steep streams in bedrock consistently were lined with conifer forests, while flat, glacial-till areas with lower stream slopes consistently were associated with mixed forest types. Nilsson et al. (1991) took this geology control concept further and considered whether vascular plant species richness along rivers was similar for small and large rivers in Sweden. Species richness was similar for rivers differing by a factor of tenfold in their discharge, but more importantly, greatest species richness for all rivers occurred where the channel was moderately steep and the soil moderate in texture.

The sequence of vegetation and geology interaction in a downstream direction may be reflected in a similar sequence of natural stream types (tens to hundreds of meters long) that are also based on geology (see Chapter 6). For instance, the Natural Stream Type sequence, B-C-B-C, in the Appalachians corresponds to the dominance of relatively narrow-valley, high-slope forests (B Stream Type where there is no developed floodplain) with intervening sections of wide-valley and either low- or high-slope forests adjacent (C Stream Type with a developed floodplain). Similarly the Stream Type sequence, E-G-E or E-B-E corresponds to mixed floodplain forests (E Type) versus conifer forests on steep side slopes (G, or not quite as steep, B Types) in the northern Lakes States and Maine. Forest type change may not be reflected in stream type change. Rather, only a change in site indexes or changes in shrub and herbaceous plants may occur. Riparian area width varies not only as stream width increases but also as the valley type changes along the river corridor.

Characterization of riparian forests is facilitated by incorporating the hierarchical structure of landscapes into a classification system in the same manner that terrestrial ecosystems can be arrayed and understood hierarchically (Bailey 1996). While this process is the focus of Chapter 3, it is discussed here briefly as it relates to delineation of riparian areas. In glacial landscapes, riparian and wetland forest characteristics differ among different landforms such as outwash plain and moraine (Zogg and Barnes 1995). Valley segment classifications (based on segments that are many kilometers long) account for groundwater sources, stream chemistry, and stream biology that reflects geologic position in the broader landscape (Chapter 3). The landscape setting constrains the development of stream valley or lake basin characteristics (Goebel et al. 1996). For streams, the constraining features of both the terrestrial landscape and the aquatic system on riparian area development may vary in predictable ways, from headwater streams to confluences at the bottom of watersheds, i.e., longitudinally (*sensu* Malanson 1993). Lake riparian characteristics may vary in predictable ways as well, both within individual basins (e.g., between the opposite steep and flat sides of ice-carved kettle lakes in glacial landscapes) and among different lake basins whose depth may or may not intercept aquifers with a strong groundwater flow.

In 1994, Rosgen developed field measures that quantified the relationship of a valley to a stream in terms of the energy of flowing water and used these relationships to plan the restoration of riparian vegetation (as it relates to the water table) and stream channel morphology (Rosgen 1996). Two measures are most important. First is the entrenchment ratio that is measured in transverse cross-section, and is taken as the ratio of valley width (at the elevation of the approximate 50-year flood) to the stream width (measured at the bankfull elevation). Second is the belt width ratio sometimes measured on an aerial photo or map. This is the distance between opposing meander bends over a stream section compared again to bankfull stream width. The belt width ratio is usually the larger of the two and defines how the entire stream moves across the floodplain over time. Together, the two measures allow for a distinction to be made between vegetation on the floodplain and vegetation on terraces (higher, abandoned floodplains). They define the volume of the valley filled by large floods (in the channel and across the floodplain). Why use the 50-year flood elevation rather than the 100-year flood elevation? Simply put, the 50-year elevation nearly always intersects a terrace slope or other up-sloping surface, while the 100-year flood elevation can be so high as to carry beyond glacial or climate change caused terrace elevations (Rosgen 1996).

Lateral Geomorphic Controls on Riparian Areas

Hack and Goodlett (1960) showed that riparian forests in the headwaters of Appalachia's Shenandoah River in Virginia were best correlated with slope position (reflecting how much water came to and passed through the site), soil texture (that controlled water availability above the water table), and depth to the water table. These plant community observations later evolved into the hydrologic concept of wet, riparian areas (variable in size over time) that generated storm flow in streams (see Chapter 6). Hupp and Osterkamp (1985) expanded the Virginia work at Passage Creek with an emphasis on flood plain forests, and Harris (1985, 1988) broadened the perspective to include valley and stream types as they influenced riparian vegetation in the Cottonwood Creek basin of California's eastern Sierra Nevada range.

While vegetation was described in terms of size, amounts, and diversity, geomorphic valley and stream classifications were restricted to conceptual cross sections depicting U- or V-shaped valleys, alluvial fans, and depositional flats (floodplains). Similarly, Kovalchik and Chitwood (1990) added an indeterminate depth to the water table in these conceptual cross sections to differentiate tree, shrub, and herbaceous communities. Verry (1997) quantified depth to water table using depth-duration curves and related the 50th percentile of the daily elevation of the growing season water table to maximum vegetation height development (see Chapter 5).

The geomorphic ecotone of a riparian area constrains the development of soil and plant communities (Geobel et al. 1996; Naiman et al. 1997). The mechanisms behind soil and plant development include geomorphic influences on water tables and disturbance regimes. Water table depth often greatly affects vegetation composition along aquatic to upland gradients in riparian areas (Frye and Quinn 1979; Van Cleve et al. 1993; Malanson 1993; Viereck et al. 1993). This effect may be direct through influence on water availability to riparian plants, particularly during establishment (Dawson and Ehleringer 1991). It may also be indirect through influence on soil development. As an example of the latter, a persistent shallow water table in boreal and cool temperate environments inhibits plant decomposition, resulting in organic soil formation and development of characteristic peatland plant communities (Spurr and Barnes 1980).

Disturbances affecting riparian plant communities also are related to geomorphic features of stream valleys and lake basins. Flooding controls vegetation development in some riparian areas by influencing both the establishment and the mortality of plants (Swanson et al. 1982; Harris 1987; Hupp 1988; Viereck et al. 1993). Valley floor landforms (e.g., active floodplain vs. the terrace slope) regulate the specific response of vegetation to flooding by controlling inundation depths and duration (Hupp 1988; Viereck et al. 1993). Landforms may influence fire frequency in riparian ecosystems and vegetation composition and structure (Malanson 1993).

Roots and Canopies at the Bank and on the Floodplain

Ikeda and Izumi (1990) show how strong riparian vegetation roots will deepen and narrow gravel-bed rivers in North America, Europe, and Japan. When stream shoot cutoffs occur at meander bends, a strong living, pre-existing root system, several channel widths away from the original channel will ensure that new channels are shaped to the modal width/depth ratios for natural streams and that they will not become overwide and more shallow. Similarly, Parker (1978a,b) explained the concept that allows the coexistence of mobile streambeds and immobile bank regions in gravel-bed and sand-silt rivers by examining the relationship of sediment shear stress in the streambed and shear stress on streambanks. It is common for streambeds to scour up to several feet each year during bankfull flows and redistribute themselves with the riffle, pool pattern and sediment size distribution that existed before the bankfull flow.

In contrast to these two reviews that consider the entire floodplain and the movement of meandering stream channels over time, many studies and reviews have focused only on site-specific aspects of bank stability and conclude that streambank stability involves trees mostly within one-half tree-height of the bank (O'Laughlin and Belt 1995). Many studies and reviews also address large woody debris recruitment as a streambank phenomenon and suggest the source for most coarse woody debris recruited into a stream is often the riparian forest within one tree height of the bank; however, many of these have been done on entrenched stream systems with small or nonexistent floodplains (O'Laughlin and Belt 1995) (see also Chapter 7). Floodplain streams (with low valley entrenchment) recruit woody debris from the entire floodplain. Tree stems on the floodplain regulate the flow of woody debris to the stream channel in these stream types by catching floating wood just as wood in the stream channel regulates the flow of leaf litter. Thus, maintenance of living root systems and tree stems on the floodplain will allow channel cutoffs to produce new channels with normal channel dimensions and moderate the occurrence of log jams in the channel. Where water tables are high enough to exclude trees, grasses and sedges protect streambanks equally well if exposed banks contain sufficient root mass (Pfankuch 1975, Rosgen 1996).

Dick (1989) used spatial modeling of stream temperatures to show that the shade effect is derived partly from riparian elevation and slope and partly from vegetation canopies. Similarly, Sinokrot and Stefan (1993) showed that stream bottom color and the slowing of wind by the riparian canopy (tree, shrub, and herbaceous) also influence stream temperature (about 2/3 shade related and 1/3 related to the slowing of hot winds).

Using Riparian Functions to Delineate Riparian Areas

Defining riparian areas functionally (by the movement of material and energy between the land and water) avoids problems associated with assessing whether a terrestrial setting is part of the riparian area based solely on soils, or vegetation, or frequency of flooding. A functional definition implies several important things to remember when developing riparian area management prescriptions. The width of riparian functions:

- Are greater than the area associated directly with floodplain or wetland indicators

- Will vary with the function being considered

- Will be difficult to determine with certainty for some functions

The extent of a riparian area into the terrestrial setting varies with the strength of each function rather than at a fixed distance from the water. The number of functions contributing to riparian and aquatic ecosystem processes decreases with distance from the water ecosystem (Figure 2.1). But the distance at which a particular function is no longer important may be difficult to determine with certainty. This is why professional judgment should be used as needed from site to site.

Finally, a functionally delineated riparian area may not, and likely will not, translate directly into a riparian management zone designed to "buffer" the stream, but is usually larger. Riparian management buffers are designated where special silvicultural and operational considerations are needed. The size of the management buffer will depend on specific aquatic and terrestrial conditions; on the proposed practices and resultant impact of the management practices; and on the consensus reached in a particular organization or government division.

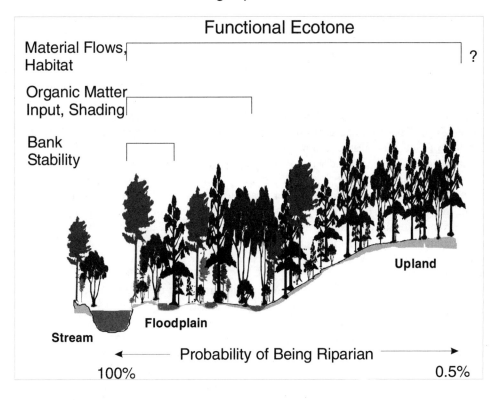

Figure 2.1 The probability of a function being riparian varies with each function across the riparian ecotone.

On the ground assessments of riparian area width based on functionality would be costly and complex, if not impossible, for natural resource managers. Indeed, research on riparian functional delineation has lagged behind the need managers have to locate and delineate riparian areas. Geomorphology is a likely surrogate for some functions that can be easily measured, either directly on the ground or from existing data. Few studies of riparian forest characteristics have been conducted with the transverse geomorphic structure of the riparian area in mind. Like riparian management zones, most riparian studies have been conducted with data plots at fixed distances from the water rather than by geomorphic position (floodplain, terrace, low terrace, tributary junction, etc.). This transverse structure reflects constraining geomorphic influences of both the upslope (terrestrial) and downslope (aquatic) environment. Understanding it will facilitate delineating riparian areas to capture the true extent of their functionality. Stream valley or lake basin geomorphology is the basis for delineating the riparian ecotone. Modeling relationships between condition variables and riparian functions will assist riparian area delineation based on land features rather than those based on lines at a fixed distance. Until such models are widely available, understanding the geomorphic ecotone of riparian areas is a practical approach to on-the-ground, riparian-area delineation.

In stream settings, recognizable and repeatable valley floor landforms reflect fluvial processes of deposition and erosion. Similar sequences of landforms define the riparian ecotone of lacustrine systems, although the processes that form them are different. Two points need to be remembered when identifying riparian geomorphic ecotones in the field. First, not all landforms occur in all riparian systems, and some landforms may repeat themselves within the same system. For example, an easily definable floodplain does not occur in steep-sided stream valleys, and multiple terraces occur where river or lake levels have dropped repeatedly, i.e., forming abandoned floodplains and lake beds. Second, the geomorphic features of the upland end of the riparian ecotone are often considered terrestrial in origin but may have been formed by the action of water, such as through glacial or marine deposition.

Quantifying the ecotonal structure of riparian areas and the interrelationships among geomorphology, soil, and vegetation along this gradient is necessary for understanding natural variability in structure and function among riparian areas for developing functional delineation criteria of riparian area width; and for developing options for managing riparian areas. With these goals in mind, a practical approach to delineation that inherently captures the geomorphic structure of riparian ecotones is outlined in the final section. Take time to consider how the scale of a riparian area affects your judgment of its extent (See Figure 2.2 for an example of scale effects).

Identifying Functional Riparian Areas on the Ground

Stream channels have a defined bank and a scoured bed whether they have water in them or not. Many are associated with a floodplain where the channel water spills at the flood stage (the bankfull elevation). Floodplains are deposition areas and regularly receive the sediment of the stream's valley. Terraces are abandoned floodplains and occur at a higher elevation than the active floodplain but may have been deformed over time by slumping. Abandoned stream channels on active floodplains or terraces are common.

Terraces more than 3 to 6 ft above the floodplain were likely caused by channel down-cutting as the result of climate change, glacial action, or regional changes in the base level of rivers (e.g., the glacial Great Lakes escaping to the Atlantic Ocean, or tectonic uplift). Lower terraces can be caused by land-use changes; in severely eroded landscapes, land-use change

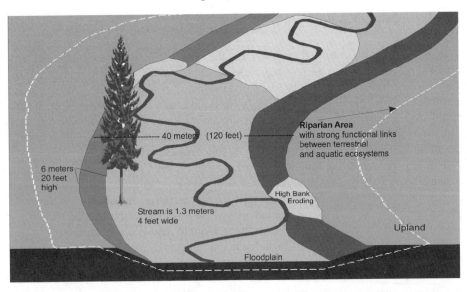

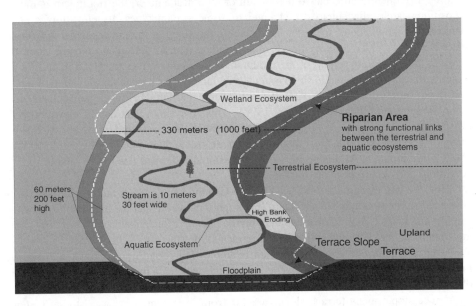

Figure 2.2 This diagram shows how the scale of the stream and its valley impacts the delineation of a riparian area. The relative scale of the stream width and its valley are the same in each frame.

can fill existing floodplains with 6 ft of new sediment (Allen 1965; Knox 1977; Trimble 1983).

Identifying the floodplain, the terrace, and the slope between the floodplain and terrace (terrace slope) in the field is the best way to define riparian areas. "Learn to read the land, and when you do, I have no fear of what you will do with it..." (Leopold 1949). Always look for these features first. However, in some areas, the flood simply flows between steep mountain or gully slopes, carrying the sediment to a floodplain at a lower elevation; in others the slopes away from the stream or lake are so slight that only a continuum is seen. In these cases, the rules of thumb given below can help delineate functional riparian areas.

This key for identifying functional riparian areas on the ground are guides, and should be tempered by your professional judgment as to the importance of a particular ecosystem function (a process that moves material between the terrestrial and aquatic portions of the riparian area) (Figure 2.3). Some examples of where this judgment is needed will help you read the land (river) and allow you flexibility to adapt to many field situations.

Let's consider the components included in the functional riparian area. First, the water body is always included (stream, lake, pond, wetland). Second, the floodplain of streams is always included (and the highwater area of lakes), as well as the wetlands within the floodplain frequently adjacent to the stream or lake. Remember, the floodplain is the flat depositional area adjacent to the channel of some stream types. Its elevation is easily identified on the *inside* of meander bends where the slope of the point bar rising from the water flattens. Since 50-year floods also inundate part of the upland terrace slope that confines the floodplain or part of the mountain slope that confines the flood in entrenched systems, this part of the upland slope is always included. To estimate the elevation of the 50-year flood, simply double the maximum channel depth measured below the bankfull elevation at a riffle section of the river (Rosgen, 1996). This "two times max depth" elevation is the 50-year flood elevation (+ or - 10 years). This information is tempered by the knowledge that many functions reach 60-80% of their importance close to the water, usually within 100 ft or about a tree length. At this point, your professional judgment needs to evaluate specific conditions and occurrences in your area.

First, we will discuss the first two On-The-Ground key items (A and AA/B) describing where terrace or mountain slopes occur. It is important to consider the high bank adjacent to streams and lakes. The high bank occurs on the *outside* of meander bends, but only where the stream channel is flowing against the terrace slope (the top of this terrace bank is well above the floodplain elevation). A common example of this condition is shown in Figure 2.2. From this figure, it is easy to see that sediment and trees from the high bank produce much of the stream sediment and large woody debris. But not all meander bends occur at the terrace slope; many terrace slopes are far from the channel in large stream systems. Remember that all floods will flow against all terrace slopes but that channel bank erosion will be concentrated where the meander bend flows against the terrace slope. You can decide whether to include these high banks cut into the terrace slope and a tree length at their top.

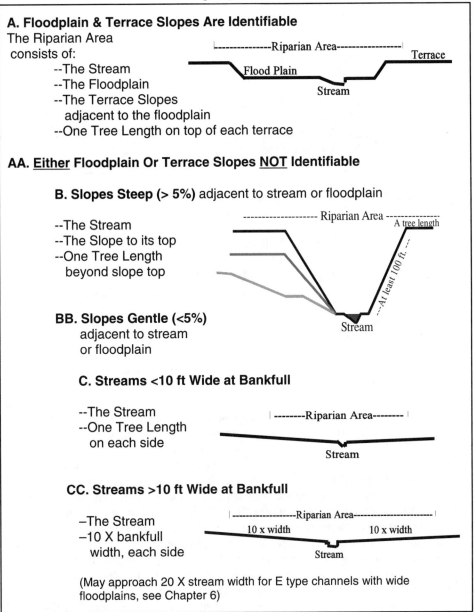

A. Floodplain & Terrace Slopes Are Identifiable

The Riparian Area
consists of:
  --The Stream
  --The Floodplain
  --The Terrace Slopes
   adjacent to the floodplain
  --One Tree Length on top of each terrace

AA. Either Floodplain Or Terrace Slopes NOT Identifiable

 B. Slopes Steep (> 5%) adjacent to stream or floodplain

  --The Stream
  --The Slope to its top
  --One Tree Length
   beyond slope top

 BB. Slopes Gentle (<5%)
  adjacent to stream
  or floodplain

  C. Streams <10 ft Wide at Bankfull

  --The Stream
  --One Tree Length
   on each side

  CC. Streams >10 ft Wide at Bankfull

  --The Stream
  --10 X bankfull
   width, each side

  (May approach 20 X stream width for E type channels with wide
  floodplains, see Chapter 6)

Figure 2.3 Field key to define riparian areas for streams.

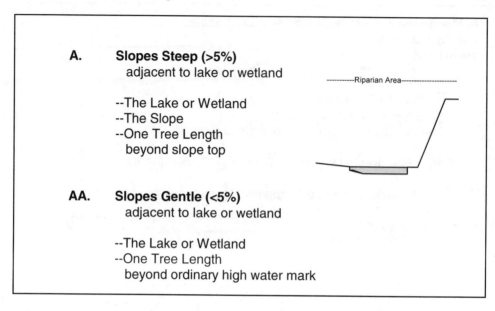

A.   **Slopes Steep (>5%)**
        adjacent to lake or wetland

        --The Lake or Wetland
        --The Slope
        --One Tree Length
          beyond slope top

AA.   **Slopes Gentle (<5%)**
        adjacent to lake or wetland

        --The Lake or Wetland
        --One Tree Length
          beyond ordinary high water mark

Figure 2.3 continued. Field key to define riparian areas for lakes and wetlands.

Whether you include all terrace slopes and tops or only part of them depends on the size of the valley. For small streams and valleys (say, streams 10 ft and valleys 300 ft wide), you may want to include all terrace slopes and terrace tops because leaf recruitment, for example, and many other functions are important to an arbitrary tree length of at least 100 ft. Where the stream and its valley are wider, you may want to exclude those upper terrace slopes above the 50-year flood elevation where the stream itself is far away from them.

Professional judgment is also needed to predict how changes occur over time. Meander-bends, for example, move down-valley with time. In many situations this happens so slowly that their location against the terrace wall appears static, even though they are moving downstream. In certain situations where land-use changes are rapid, or where reservoir operation impact reaches downstream, the meander migration rates may accelerate, resulting in significant changes within a decade. Know your situation.

Landslides and slumps can occur at these high bank locations in floodplain stream systems and frequently occur in mountain stream systems without floodplains where adjacent mountain slopes are steep and soils unstable. These slumps can extend 300 ft above the stream. Thus, how far you extend the functional riparian area up steep, high slopes will depend on the occurrence of slumping in your area. Extend the functional riparian area at least 100 ft vertically, regardless of their slope. Similarly, high slopes may be interrupted by a series of old glacial or climatic terraces. You must decide how many, if any, of the terraces are included in your functional riparian area. Similar conditions occur adjacent to lakes, frequently one side of a lake will have a steep slope formed by ice pushing up a tall moraine, and the other side will have a gentle slope formed by ice scraping the landscape nearly level.

For water bodies with gentle adjacent slopes (< 5%; key level AA/BB/C or CC), a simple rule of thumb is applied to estimate the meander-belt width that may not be obvious. Meander belt ratios can vary from 2 to 80 (Rosgen 1996); however, the lowest values are associated with highly entrenched systems where mountain slopes constrain the stream system, and the highest values are associated with some of the E stream types that have wide floodplains. Many of these have grasses, sedges, or shrubs; but where water tables are low enough, they are forested. Since the key should be used in series (first item first, etc.), the lower meander belt ratios are already considered in the first two key items (A and A/BB). This leaves only moderate- to high-meander belt ratios to consider for the water bodies with gentle adjacent slopes.

Judgment should be used when using the gentle-slope rule of thumb. First, remember that the meander-belt ratio accounts for the entire meander belt — it includes both sides of the stream. And in referring to Figure 2.2 it is obvious that the stream does not always (actually rarely) flow in the middle of the meander belt. Where you *cannot identify* the actual meander belt (floodplain and terraces), *define the functional riparian area as 10 x Bankfull Width on each side of the stream.*

This rule of thumb is half the meander belt ratio of 20 which includes all C stream types and the low end of E stream types. E channels are typical in low-slope eastern forests as are C stream types. Also, realize that we have already used some professional judgment in making the division between key level C and CC. If we assume a tree length is 100 ft and that many of the shade, woody debris recruitment, bank stability, and litter fall functions approach 60 to 80% of their maximums within 100 ft, then the rule of thumb is overridden by these considerations when streams are small (10 ft wide or less). For larger streams (> 10 ft wide at the bankfull elevation), the importance of the flooded area should also be considered. Lake and wetland guides are identical to stream guides, but gentle or steep adjacent slopes may, and often do, occur on opposite sides of the same water body. In time, we have no doubt you will be able to read the land and the river, and we have no fear of what you will do for them.

Chippewa National Forest

James O. Sneddon

Sandy Verry

Jim Hornbeck

Riparian areas are diverse. A lake bordered by sedges and tamarack in northern Minnesota (upper left), a peatland stream bordered by sedges and black spruce (upper right) in northern Minnesota, a road and mountain sides bordering a stream in the Fernow Experimental Forest in West Virginia (lower left), and the stream and its mountain slopes on the White Mountain National Forest in New Hampshire (lower right).

Chapter 3

Diversity in Riparian Landscapes

Thomas R. Crow, Matthew E. Baker
and Burton V. Barnes

"We will not save the riverine forests without protecting the floodplains, nor will the orchids be preserved without preserving the marshes. Our own fate is linked to the limits we set on the domestication of the world around us and to the offsetting effort we devote to maintaining the life-blood of the Home Place, the natural beauty and health of the creative, sustaining, enveloping Ecosphere."
Stan Rowe, from Arks Can't Save Aardvarks, in *Home Place: Essays on Ecology*

Biological diversity (biodiversity) is the number of organisms and their distribution within the ecosphere (Earth). Conserving biodiversity has become a major issue in the conservation and management of Earth's natural resources (Noss and Cooperrider 1994). However, studies of biodiversity have focused primarily on the variety of species within a given area. In total, biodiversity depends on the diversity of ecosystems within a landscape as well. Thus, both organism diversity and landscape diversity are needed for a holistic view of biodiversity (Rowe 1992). Landscape ecosystems are volumetric, structured segments of the Earth. The ecosphere is the largest ecosystem we know, and the Earth can be subdivided into a hierarchical series of ecosystems from large to small — from global to local (Rowe 1992; Bailey 1996; Barnes et al. 1998, 34-40). A perceptible ecosystem is a topographic unit, a volume of land and air plus organic contents, extending over a particular part of the Earth's surface for a certain time (Rowe 1961).

Therefore, in this chapter we focus on ecosystem diversity, defined as the number, kind, and pattern of landscape and waterscape ecosystems in a specified area and the ecological processes that are associated with these patterns (Lapin and Barnes 1995). One can then characterize ecosystems as to their composition, structure, and function — the attributes of diversity (Crow et al. 1994). Our objectives are to: (1) provide an example of a landscape ecosystem approach to characterizing ecosystem diversity in riparian areas by presenting a case study, (2) consider the importance of riparian areas to regional ecosystem diversity, and (3) examine and summarize management practices that conserve diversity in riparian areas.

43

Riparian Ecosystems and Their Diversity

Ideas and perceptions about the relation between land and water have changed dramatically during recent decades, and these changes in thinking have reshaped recommendations for managing riparian areas. Above all, there is greater appreciation for the interconnections between the aquatic and terrestrial ecosystems and the importance of scale when considering these interconnections (Swanson et al. 1988; Roth et al. 1996; Allan et al. 1997; Allan and Johnson 1997; Ilhardt et. al. Chapter 2; and Hornbeck and Kochenderfer Chapter 5). As defined by McCormick (1979), riparian wetlands are "lowland terrestrial ecotones which derive their high water tables and alluvial soils from drainage and erosion of adjacent uplands on the one side or from periodic flooding from aquatic ecosystems on the other." Implicit in this definition is the recognition that riparian areas are an integral part of a larger landscape, and therefore, riparian habitats are influenced by factors operating at various spatial and temporal scales (Odum 1979; Crow 1991; Nilsson 1992; Richards et al. 1996). At the regional scale, geomorphology, climate, and vegetation affect stream hydrology, sedimentation, nutrient inputs, and channel morphology. At the local scale, land use and related alterations to stream habitats can significantly influence the biota of streams (Richards and Host 1994; Roth et al. 1996).

If ecosystems are considered to be multi-scale, volumetric units of the Earth, then riparian areas should be defined as three-dimensional ecosystems directly interacting between terrestrial and aquatic ecosystems. Given this perspective, riparian areas are a collection of ecosystems that extend outward from the water to include the floodplain and some distance landward onto the terrace slope as defined in Ilhardt et al. (Chapter 2), downward through the soil profile, and upward into the canopy of the streamside or lakeside vegetation. Using this volumetric model, riparian ecosystems exist laterally between terrestrial and aquatic systems as well as vertically from the soil profile to the forest floor, through the canopy of vegetation, and throughout the air layers within and surrounding the vegetation.

Riparian areas have unique characteristics common to their physical environment as well as their diverse biota and may share characteristics with the adjacent upland and aquatic ecosystems. Both the abundance and richness of species tend to be greater in riparian ecosystems than in adjacent uplands (Odum 1979). For example, in a study of riparian and upland habitats used by small mammals in the Cascade Range of Oregon, Doyle (1990) found their abundance and richness to be greater in riparian than in upland forests based on mark and recapture sampling techniques. Further, he found riparian habitats often acted as a species source and upland areas as a dispersal sink for small mammals. Likewise, when comparing the richness of bird communities in different ecosystems along the river Garonne in southwest France, Décamps et al. (1987) found the average number of species based on point surveys in riparian woodlands > terrace woodlands > slope woodlands > *Populus* plantations. This pattern of bird richness followed a similar pattern in the richness of vegetation composition and structure in these communities: riparian woodlands, the most varied, *Populus* plantations, the most homogeneous. Although many species are specifically adapted to riparian conditions, many others, such as large mammals (including most common game species), require periodic access to stream margins or lake margins for survival, even if most of their time is spent elsewhere.

Riparian ecosystems often have elongated shapes with high edge-to-area ratios. Their edges can be very open, with large energy and nutrient fluxes and biotic interchanges occurring in the aquatic ecosystem on the inner margin and the upland terrestrial ecosystem on the outer margin. Sharp physical gradients characterize riparian areas, and the related differences in the composition and structure of vegetation that occurs along these gradients create diversity. Brosofske et al. (1997) measured soil and air temperature, relative humidity, short-wave solar radiation, and wind velocity along a transect running perpendicular from five streams across the riparian zone and into the uplands in western Washington. They found temperature and humidity in the transitional riparian zone to be generally intermediate between those above the stream and within the upland forest.

In addition to spatial variation, temporal variation also creates diversity. Riparian ecosystems are one of the most dynamic portions of the landscape (Swanson et al. 1988). They are characterized by frequent disturbances related to inundation, transport of sediments, and the abrasive and erosive forces of water and ice movement that, in turn, create habitat complexity and variability in time as well as in space, resulting in ecologically diverse communities (Brinson 1990; Gregory et al. 1991). A third factor, the exceptional fertility and productivity of riparian areas, is also responsible for great diversity (Hunter 1990; Sparks 1995). Many plants and animals have adapted to take advantage of this fertility and to survive the periodic flooding that indirectly creates the fertility in the form of deposited organic and inorganic materials. An abundance of empirical evidence supports the contention that riverine forests produce more biomass than upland forests in similar geographic locations (Brinson 1990).

At least two conclusions about the diversity of riparian ecosystems seem justified. First, riparian ecosystems are relatively productive areas with great diversity in their physical environments and their biological components. Second, because of their richness and their spatial distribution, the relative contribution of riparian ecosystems to total compositional diversity far exceeds the proportion of the landscape they occupy.

Relating the Physical Environment to the Diversity of Riparian Ecosystems

To understand variation in riparian forests from one stream to another, or even between sections of the same stream, we first need to understand the physical environment in which the stream exists. At continental scales, climate produces geographic variation in the composition and structure of riparian vegetation. Lindsey et al. (1961) used phytosociological analyses in the floodplain forest to show a continuum along a 230-mile latitudinal gradient, identifying regional climate as a controlling factor. Floristic variety in lowland forests diminishes rapidly northward and also westward from the Mississippi River.

In cold climates, spring snowmelt and ice can significantly affect both the hydroperiod and the vegetation of floodplains. When flooding occurs in winter or early spring, ice flows will often damage riparian vegetation severely (Lindsey et al. 1961). For vegetation on floodplains in cool, humid climates, anatomical and metabolic adaptations are important for surviving extended periods of saturated soils and anaerobic conditions during the growing season.

When rivers flow through flat regions, riparian habitats can cover large areas. Along low gradient rivers, diminished floodwater velocity results in the deposition of progressively smaller alluvium from the suspended sediment load. The physiography in large floodplains includes these features: bottoms, natural levees and alluvial terraces adjacent to the river channel, meander scrolls with ridge-and-swale topography and relict meander bends or oxbow lakes, and point bars (Brinson 1990). Each of these features produces characteristic vegetation. Levees support gallery forests that are adapted to frequent flooding but also to the rapid drying of the soil when water levels subside. Oxbow lakes and depressions are the most hydric of the floodplain features, so they support species that are adapted to long periods of flooding and anaerobic soil conditions. Early successional and colonizing species are maintained on point bars through periodic disturbances and by rapid deposition rates.

In contrast to large river-floodplains on flat land, steep gradients produce narrow valleys, more restricted floodplains, and reduced duration of floods (Swanson et al. 1988). Colluvium, or material transported from valley sides, can be an important source of material for floodplain deposits in steep, narrow valleys. Even where the floodplain is restricted, species composition and the general structure of the vegetation are often distinctly different from those in the adjacent uplands (Nilsson 1992). Differences among riparian ecosystems can be also related to the size of the catchments. Small catchments generally have shorter hydroperiods than large catchments, thus reducing the time floodwaters interact with the floodplain (Brinson 1990). In addition, the amount of material available for alluvial deposition decreases as the size of the catchment diminishes (Allen 1965).

Topographic, hydrologic, and edaphic features collectively and interactively create a highly heterogeneous environment for plant establishment and growth in riparian areas (Pautou and Décamps 1985; Brinson 1990; Malanson 1993). These factors create differences in soil moisture and aeration, and for riverine systems, they also reflect differences in the frequency and duration of exposure to the force of flowing water, and in the north, to the force of ice movement. At the wettest end of the moisture spectrum, riparian forests are limited either by the force of water currents or by soil aeration that is inadequate for tree establishment and growth. At the opposite extreme, reduced soil moisture and sandy soil can create areas with scattered trees, open canopies, and low basal areas that are savanna-like in their structure.

Where rich alluvial soils are adequately drained, however, dense stands with closed canopies and high basal areas characterize the riparian forest.

Composition of Riparian Forests

The term "bottomland hardwood forests" applies to the extensive forests that occupy the floodplains along streams and rivers in the Southern United States (Sharitz and Mitsch 1993). Küchler (1964) describes the major plant communities of this type as consisting of medium and tall forest of deciduous and evergreen trees and shrubs. Sharitz and Mitsch (1993), Kellison et al. (1998), and Hodges (1998) offer detailed descriptions of the vegetation common to southern bottomland hardwood forests. On sandbars and along the margins of rivers, pioneer species such as *Salix nigra* and *Populus deltoides* commonly occur. In the backswamp behind the levee, *Carya aquatica* and *Quercus lyrata* are likely to be found. With slightly higher ground and better drainage, *Quercus laurifolia, Q. nuttallii, Ulmus americana, Gleditsia aquatica, Acer rubrum, Fraxinus pennsylvanica*, and *Celtis laevigata* increase in abundance. *Nyssa aquatica* and *Taxodium distichum* are the common species associated with the wettest sites that are subject to frequent and prolonged flooding (e.g., oxbows, backswamp depressions, and swales between relict levees). Water in these deepwater swamps comes from runoff from surrounding uplands and overflow from flooding rivers. Anaerobic conditions result from the flooding and these conditions can persist for long periods. Tree species common to deepwater swamps have developed morphological adaptations such as the buttressing on the lower portion of the tree trunk that give tall trees stability where rooting is shallow.

Although bottomland forests are typically nearly level, changes in elevation of only a few inches can produce different hydrologic conditions, soils, and plant communities. On the higher floodplain beyond the first embankment, a shorter hydroperiod and better drainage allows oak (*Quercus alba, Q. phellos, Q. laurifolia, Q. nigra, Q. michauxii, and Q. falcata*) to dominate along with *Carya ovata* and *Liquidambar styraciflua*. And finally, *Pinus taeda* and *Quercus virginiana* are found at the highest levels in the bottomland hardwood forest.

Bottomland hardwood forests extend northward up the Mississippi River valley and farther up the Ohio River, resulting in a mixture of northern and southern species in the riparian forests of this region. Johnson and Bell (1976) studied the composition and distribution of biomass among tree species along the Sangamon River in central Illinois. In the floodplain forest where the probability of annual flooding ranged from 3 to 25%, *Acer saccharinum* dominated the forest with *Gleditsia triacanthos, Fraxinus pennsylvanica*, and *Platanus occidentalis*. In the transition zone between the upland and the floodplain, *Acer saccharinum* and *Quercus imbricaria* shared dominance. Other species present in the transition zone included: *Euonymus atropurpureus, Carya cordiformis, Prunus serotina, Ulmus rubra*, and *U. americana*. In the upland forest, with a slight probability of flooding, *Quercus alba* dominated. Forests on the floodplains of the Wabash and Tippecanoe Rivers in Indiana are dominated by *Populus* spp., *Salix nigra,* and *Ulmus americana* on the "first bottoms" (the lowest elevation along the river) and *Acer saccharum, Aesculus glabra, Cercis canadensis, Fagus grandifolia,* and *Ulmus americana* in the slightly higher "second bottoms" (Lindsey et al. 1961).

In northern regions of the Midwest and Eastern United States, cool temperatures, glaciated terrain, and abundant water combine to create a variety of riparian and related wetland ecosystems. As observed in northwestern Lower Michigan (Baker 1995; Baker and Barnes 1998), *Acer saccharinum* is a major dominant of many alluvial wetlands. *Fraxinus nigra* is a co-dominant overstory species on poorly drained sites, while *Populus deltoides* and many riparian *Salix* spp. are characteristic of well drained, but frequently flooded sites.

Other studies from the Northern United States bear out this general relationship. In their comparing floodplain and basin wetlands in southeastern Wisconsin, Dunn and Stearns (1987) found floodplain wetlands dominated by *Acer saccharinum* and by *Fraxinus pennsylvanica, Salix* spp., and *Quercus bicolor.* Although *Acer saccharinum* dominated wetland basins where low soil pH and standing water resulted in organic matter accumulation, *Betula alleghaniensis, Fraxinus nigra,* and *Acer rubrum* were also important associates. Pierce (1980) related the composition of an Allegheny River floodplain in southern New York to flood frequency and geomorphology. Once again, alluvial flats were dominated by canopies of *Acer saccharinum,* and despite a relatively sparse shrub layer, other species included *Crataegus* spp., *Juglans cinerea, Ulmus rubra,* and *Platanus occidentalis.* The forests of poorly drained sloughs were distinguished from alluvial flats by the presence of many understory shrubs, as well as *Quercus bicolor* and *Fraxinus nigra.* On an excessively well drained, sandy island in western Wisconsin, Barnes (1985) found a gradient from *Populus deltoides, Salix* spp., and *Betula nigra* on the frequently flooded edge, to *Acer saccharinum, Ulmus americana,* and *Fraxinus pennsylvanica* on the higher, drier interior. Detailed examples of northern riparian ecosystems are included as part of the case study presented later in this chapter.

Exotic Species and Diversity

A discussion about the composition and structure of vegetation in riparian and related aquatic communities would be incomplete without considering the occurrence of exotic (non-native) species. The introduction of nonnative organisms to river, lake, wetland, and riparian ecosystems in North America is so pervasive that few natural communities remain unaffected (Hedgpeth 1993, Noss and Cooperrider 1994, Lövel 1997). In studying spatial patterns of nonnative plants on the Olympic Peninsula, WA, DeFerrari and Naiman (1994) found nonnative species richness was about 1/3 greater in riparian zones than on uplands, and the mean number and the cover of nonnative plant species were more than 50% greater in riparian zones than in uplands. The list of nonnative plants and animals common to aquatic and riparian ecosystems in the East is extensive and their threat to native species is serious. Among the most noxious of the introduced plants are purple loosestrife (*Lythrum salicaria*), Eurasian milfoil (*Myriophyllum spicatum*), and curlyleaf pondweed (*Potamogeton crippus*). An introduced animal causing great concern is the zebra mussel (*Dresissena polymorpha* and *Dresissena bugensis*). First introduced to the Great Lakes in ballast water about 1985, the zebra mussel has since spread quickly to other freshwater bodies in the Eastern United States. Once established, these alien guests successfully compete against native species and severely reduce the biological diversity of the aquatic or terrestrial system. In addition, exotic pathogens and insects are profoundly affecting riparian communities, e.g., mortality of *Ulmus americana* caused by Dutch elm disease. A network of riparian corridors facilitates the movement of organisms through a landscape, although there is little evidence that riparian corridors act as sources of nonnative plants in undisturbed uplands (DeFerrari and Naiman 1994).

Diversity of Landscape Ecosystems in River Valleys of the Huron-Manistee National Forests, Northern Lower Michigan

This case study was designed to examine the full range of diversity of river valley ecosystems occurring in areas of the Huron-Manistee National Forests in Lower Michigan. It illustrates the complexity of ecosystem diversity and physiographically mediated differences in landscape ecosystems at regional and local scales.

Our research on ecosystem diversity begins at the regional scale using three classification levels: Region, District, and Subdistrict (Albert et al. 1986). An ecological classification is an attempt to simplify the tremendous diversity found in Nature. It is a grouping of ecosystems based upon their similar physical and biological characteristics and their spatial relationships with one another. It is also useful for examining the patterns of ecosystems and the plants and animals associated with them. The case study was conducted in Districts and Subdistricts of Region II of the regional landscape ecosystem classification of Albert et al. (1986) as illustrated in Figure 3.1. Ecosystems below the regional level are distinguished hierarchically as (1) Physiographic Systems (Outwash Plain, Moraine, Ice-Contact Terrain, Lake Plain, etc.), (2) Landforms (e.g., kettle, kame, and esker landforms within Ice-Contact Terrain), and (3) local ecosystem groups or types within landforms. At the finest scale, ecosystem types may range in size from less than 2 acres to more than 60 acres.

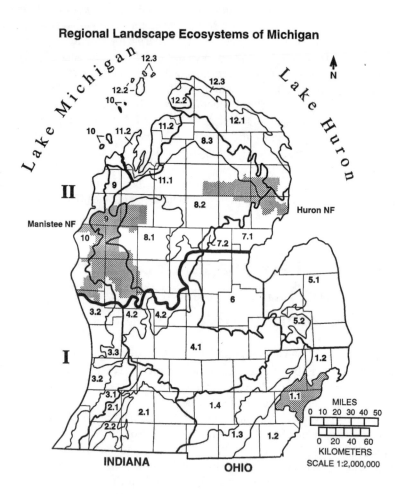

Regional Landscape Ecosystems of Michigan

Figure 3.1 Map of the regional landscape ecosystems of Lower Michigan (Albert et al. 1986. With permission.) Roman numerals I and II indicate landscape ecosystem regions divided by the thick dark line across the center of the state. Within each region, hierarchical landscape ecosystem districts and subdistricts are indicated by thinner lines, as well as different integers and decimals, respectively. The Huron-Manistee National Forests are shown as the shaded areas within Region II. Most of the Manistee National Forest occurs in District 9 (Newaygo), whereas the Huron National Forest occurs primarily in both Subdistrict 8.2 (Grayling) and Subdistrict 7.1 (Standish).

This local ecosystem hierarchy is similar to the Landtype Association, Landtype, and Landtype Phase of the national hierarchical framework of ecological units used by the USDA Forest Service (Avers et al. 1994; Bailey 1996; Barnes et al. 1998). In the sections that follow, we present examples that illustrate riparian diversity at several spatial scales using a classification framework.

Regional Landscape

This first level of classification represents a broad landscape unit distinguished primarily on the basis of gross physiography and macroclimate (Albert et al. 1986). Such factors mediate the movement of water, the formation of soil, and the distribution of plants and animals. The greatest difference in the kinds and patterns of ecosystems found in this classification is between regional landscape units. For example, a certain physiography (low-level outwash plain and coarse-textured moraine) and climate (relatively moist, lake-moderated) characterize District 9 (Newaygo), whereas others (high-level outwash, ice-contact terrain, fine-textured moraine and a drier, more extreme climate) characterize Subdistrict 8.2 (Grayling: Albert et al. 1986). The significance of regional ecosystems for riparian landscapes is twofold: (1) the same factors that distinguish regional landscapes also drive the hydrology of river systems and (2) these factors shape the local landscapes that, in combination with river systems, contribute to riparian ecosystem diversity.

The Manistee and AuSable Rivers originate near large, ice-contact features in the Grayling Subdistrict (8.2), north-central Lower Michigan. Both rivers have large portions of their catchments in the coarse sand and gravel of the Grayling Subdistrict, and both are known for their hydrologic and thermal stability (Bent 1971; Richards 1990; Wiley et al. 1997). However, from its source, the Manistee River flows south and west through the Manistee National Forest in the Newaygo District 9, whereas the AuSable flows east and south through the Huron National Forest in Subdistricts 8.2 and 7.1 (Figure 3.1). The glacial landforms and macroclimate of each regional ecosystem produce vastly different streamside wetlands.

The Manistee basin in the Newaygo District is somewhat larger with a lower gradient than the AuSable basin in the Grayling Subdistrict. The Manistee River flows along the Port Huron moraine initially, but most of its length in District 9 lies in a broad, flat, outwash plain. The floodplain is periodically wet during the growing season due to uniformly low topography and seasonal inundation from the river and a high water table. Its silty soil supports large expanses of silver maple swamps, black ash backswamps, and spring-fed, northern white-cedar meander-scar swamps (Figure 3.2A). The species composition of *Acer saccharinum, Fraxinus pennsylvanica,* and *Fraxinus nigra* in the Manistee River floodplain (Figure 3.2A) is similar to that reported in more southerly floodplains of the Northern United States (Pierce 1980; Dunn and Stearns 1987; Brinson 1990).

A. Manistee River Valley at Sergants Bayou, District 9

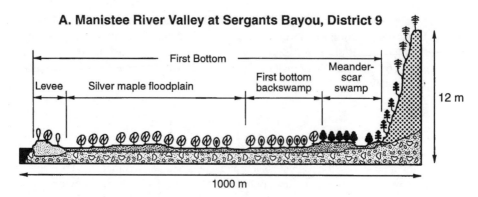

B. AuSable River Valley near Comins Flats, Subdistrict 8.2

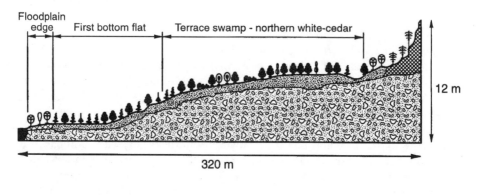

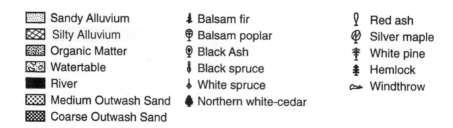

Sandy Alluvium	Balsam fir	Red ash
Silty Alluvium	Balsam poplar	Silver maple
Organic Matter	Black Ash	White pine
Watertable	Black spruce	Hemlock
River	White spruce	Windthrow
Medium Outwash Sand	Northern white-cedar	
Coarse Outwash Sand		

Figure 3.2 Representative river valley cross sections from the Manistee River in District 9 (Newago) and the AuSable River in Subdistrict 8.2 (Grayling), Huron-Manistee National Forest, northern Lower Michigan.

In contrast, the AuSable River in Subdistrict 8.2 flows past several large ice-contact hills before encountering the West Branch and Glennie moraines. Its riparian areas are more narrow (Figure 3.2B) with a species composition similar to northern floodplains described by Nanson and Beach (1977) and Brinson (1990). Unlike these floodplains, the AuSable riparian area does not appear to experience frequent over-bank flooding. Instead, its water source is primarily groundwater rather than streamflow. As a result, the alluvial terraces along the river receive substantial groundwater, have large accumulations of organic soil, and are dominated by northern white-cedar swamps. At the river's edge, sandy loam soil and an overstory of *Populus balsamifera*, *Fraxinus pennsylvanica*, and *Abies balsamea* characterize this riparian ecosystem (Figure 3.2B).

These differences in riparian ecosystem composition reflect the distinct combinations of gross physiography and macroclimate that characterize the Newaygo District and the Grayling Subdistrict. However, within each regional unit there is much more variation.

Physiographic Systems

In the next level in our classification, we recognize the influence of landform on the structure of the riparian zone. We distinguish glacial landforms as *physiographic systems* — distinct groups of ecosystems that repeat in a landscape mosaic within each regional unit; these include outwash plain, moraine, ice-contact terrain, and lake plain. Different physiographic features produce differences in both form and type of soil parent material encountered by rivers in the landscape, so physiographic features are often associated with changes in gradient, channel pattern, local hydrology, or fluvial landforms (Bent 1971; Schumm 1977; Hupp 1982; Kalliola and Puhakka 1988).

Just as channel boundary sediments are known to affect channel cross-sectional shape, the materials in different physiographic systems can result in markedly different valleys with distinct patterns of fluvial landforms (Swanson et al. 1988). In the Manistee National Forest, most rivers occur in nonpitted outwash plains (Baker 1995). However, where these rivers occur adjacent to moraines, their valleys often encounter the underlying glacial till. For example, along the Big South Branch of the Pere Marquette River, valleys in the outwash plain have broad floodplains with uniform topography and few terraces (Figure 3.3A). Ecosystems in these valleys are relatively contiguous with proportionately more interior area than the edge. Valleys in outwash-plain-over-moraine have narrow floodplains with diverse topography and many terraces (Figure 3.3B). Ecosystems in these valleys are relatively discontinuous with a large ratio of edge-length to interior-area. As the Pere Marquette flows away from a moraine and into an outwash plain, its valley and riparian landscape rapidly change from one to the other (Figure 3.3C).

In addition to influencing the spatial patterns of fluvial landforms, physiographic systems also contribute to riparian ecosystem diversity by affecting the kinds of ecosystems that occur

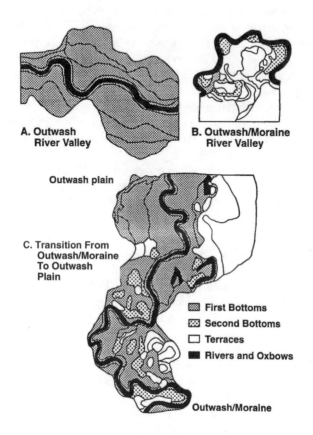

A. Outwash River Valley

B. Outwash/Moraine River Valley

Outwash plain

C. Transition From Outwash/Moraine To Outwash Plain

☒ First Bottoms
▨ Second Bottoms
☐ Terraces
■ Rivers and Oxbows

Outwash/Moraine

Figure 3.3 Fluvial landforms within non-pitted outwash plain of the Big South Branch of the Pere Marquette River in District 9 (Newaygo), Manistee National Forest, Lower Michigan. Outwash plain valleys (A) have several ecosystem types (thin black lines) on broad, continuous first-bottom floodplains (dark shaded areas). Outwash-over-moraine (outwash/moraine) valleys (B) have narrow and discontinuous first bottoms, large second bottoms (light shaded areas), and many nonflooded terraces (unshaded). At the transition from outwash/moraine to outwash plain (C), the narrow, multi-level, discontinuous outwash/moraine floodplain becomes gradually more uniform in elevation and broadly continuous as it passes into outwash plain. (From Baker and Barnes 1998. With permission.)

on each landform. For example, natural levees in outwash plain (Ecosystem Type 1, levee-silver maple-red ash, in Table 3.1) are lower and wider than those in outwash plain over underlying moraine (Type 13). Outwash-plain first bottoms contain more ecosystem types (Types 2 to 6), and these are lower, wider, and wetter than the single ecosystem type where outwash occurs over a moraine (Type 14).   In addition, a marked dominance of *Acer saccharinum* rather than *Fraxinus pennsylvanica*, and the unique occurrence of *Larix laricina*, *Platanus occidentalis*, *Juglans cinerea*, and *Quercus bicolor,* distinguish these ecosystems (Table 3.2). The pattern of fluvial landforms in a river valley is the physiographic expression of variation in the relationship between a river and the local landscape. Such differences in valley geomorphology result in distinct segments of riparian ecosystem along a river valley.

Fluvial Landforms

Fluvial landforms (levee, first-bottom floodplain, and terrace in Table 3.1) form the next hierarchical level within the ecosystem classification of river valleys. Within a physiographic system, the diversity of riparian ecosystems is closely linked to fluvial landform. Hupp and Osterkamp (1985) and Osterkamp and Hupp (1984) reported that different fluvial landforms had distinct soil as well as vegetation. On these landforms, as distance from and elevation above the river channel increase, flood frequency and duration decrease.

In District 9 of the Manistee National Forest, this pattern is also quite clear. Along these rivers, floodplain inundation occurs due to both over-the-bank flooding and water table fluctuation (Baker 1995). Natural levees (Types 1 and 13 in Table 3.1) occur near the river channel but are generally higher and drier than adjacent first bottoms (Types 2 to 6 and 14). In both outwash and outwash-over-moraine floodplains, the levee has few tree species (Table 3.2). The first-bottom, a flat low-lying surface beyond the levee, is generally lower and has poorer drainage than either the levee or second-bottom. In both outwash plain and outwash-over-moraine, the first and second bottoms are distinguished by their relative elevation. Such physiographic differences affect soil drainage and pH, depth of plant rooting, and plant composition (Table 3.2).

Differences in species composition can also be related to the position of terraces. High terraces, formed early in the development of the river valley, often develop deep soil profiles. Because of their origin, glacio-fluvial terraces in outwash-over-moraine valleys typically have coarse sand soil and vegetation such as *Quercus velutina*, *Q. alba*, and *Pinus strobus* (Type 18 in Table 3.1). Younger soil profiles, more silt and clay, and species of the northern hardwood forest characterize lower alluvial terraces (Type 17). Thus, the fluvial landforms in a river valley represent different kinds of both floodplain and terrace ecosystems.

Table 3.1 A partial classification of river valley landscape ecosystems in District 9 of the Manistee National Forest, and Subdistricts 8.2 and 7.1 of the Huron National Forest, northern Lower Michigan. This table shows only the detailed hierarchical structure for outwash plains in District 9 and Subdistrict 8.2. An abbreviated name for the ecosystem type emphasizes the local physiography and vegetation. (From Baker and Barnes 1998 w/p.)

District 9 (Newago)
- A. Outwash Plain
  - 1. Non-pitted outwash plain (reworked by water)
    - River valley segments associated with outwash plain
      - Floodplain
        - Levee
          - 1. Levee–silver maple–red ash
        - First Bottom
          - 2. First bottom flat –silver maple forest
          - 3. First bottom flat–alder–willow thicket
          - 4. First bottom flat–cattail–iris marsh
          - 5. Backswamp–black ash–silver maple–northern white cedar
          - 6. Meander scar swamp–northern white cedar–hemlock–tamarack
        - Second Bottom
          - 7. Second bottom–sugar maple–northern red oak–swamp white oak
      - Terrace
        - Well-drained
          - 8. Terrace plateau–hemlock–white pine–beech
          - 9. Terrace flat–white pine–hemlock–white oak–black oak
          - 10. Terrace riser–hemlock–white oak
        - Poorly-drained
          - 11. Terrace swamp–black ash–hemlock
          - 12. Terrace seep–white pine–hemlock
    - River valley segment associated with moraine
      - Floodplain
        - Levee
          - 13. Levee–red ash–alder–American elm
        - First Bottom
          - 14. Swale–red ash–willow–black ash
        - Second Bottom
          - 15. Well drained ridge–sugar maple–basswood–northern red oak
          - 16. Poorly drained backswamp–black ash–hemlock–silver maple
      - Terrace
        - Well-drained
          - 17. Low terrace flat–hemlock–northern red oak–sugar maple
          - 18. High terrace flat–white oak–hemlock–white pine–red pine
          - 19. Terrace riser–white pine–hemlock
        - Poorly-drained
          - 20. Terrace swamp–northern white cedar–white pine–yellow birch
          - 21. Terrace seep–yellow birch–northern white cedar
  - 2. Pitted outwash plain (not reworked by water)
      - Floodplain
        - 22. Levee–red ash–balsam poplar–alder
        - First Bottom
          - 23. First bottom–balsam poplar–northern white cedar
          - 24. Backswamp–black ash–alder–American elm
        - Second Bottom
          - 25. Second bottom–white pine
          - 26. Backswamp–northern white cedar
      - Terrace
        - Well-drained
          - 27. Terrace flat–white oak–white pine
          - 28. Terrace riser–white oak–white pine
- B. Ice-Contact Terrain (kettle-kame topography)
    - Floodplain
      - First Bottom
        - 29. First bottom flat–red ash–balsam poplar
        - 30. Backswamp–northern white cedar
      - Second Bottom
        - 31. Second bottom–sugar maple–northern white cedar–basswood
    - Terrace
      - Well-drained
        - 32. Low terrace flat–sugar maple–hemlock
        - 33. High terrace flat–white pine–red pine–white oak–north. pin oak
        - 34. Terrace riser–white pine–white oak
      - Poorly-drained
        - 35. Terrace meander scar–northern white cedar
- C. Lake Plain[1]
- D. Moraine
    - Floodplain
      - Levee
        - 36. Levee-red ash–willow–alder
      - First Bottom
        - 37. First bottom–red ash–basswood
      - Second Bottom
        - 38. Well-drained second bottom flat–white pine–basswood
        - 39. Poorly-drained second bottom swamp–northern white cedar
    - Terrace
      - Well-drained
        - 40. Low terrace flat–northern red oak–hemlock
        - 41. High terrace flat–white oak–white pine
        - 42. Terrace riser–white oak–hemlock
      - Poorly-drained
        - 43. Terrace swamp–hemlock–black ash

Subdistrict 8.2 (Grayling)
A. Outwash Plain
   1. Non-pitted outwash plain
      River valley segments associated with outwash plain
        Floodplain   44. Floodplain edge–balsam poplar–red ash–paper birch
                             45. First bottom flat–northern white cedar–white spruce
                             46. First bottom backswamp–black ash–American elm
        Terrace
          Well-drained   47. Low terrace flat–white pine–balsam fire
                             48. High terrace flat–jack pine–red pine–northern pin oak
                             49. Terrace riser–red pine–white pine–northern pin oak
          Poorly-drained   50. Terrace swamp–northern white cedar–black spruce–balsam
                                 poplar–balsam fir
                             51. Terrace seep--white pine–northern white cedar

      River valley segments associated with moraine
        Floodplain
          First Bottom   52. First bottom–red ash–basswood–balsam poplar
          Second Bottom   53. Well drained flat–sugar maple–balsam fir–white pine
                             54. Poorly drained swamp–northern white cedar–black ash
        Terrace
          Well-drained   55. Terrace flat–northern pin oak–white pine
                             56. Terrace riser–white oak–northern pin oak
          Poorly-drained   57. Terrace swamp–northern white cedar–balsam fir–black spruce
                             58. Terrace seep–white pine–northern white cedar
   2. Pitted outwash plain[1]

  B. Ice-Contact Terrain[1]
  C. Lake Plain[1]
  D. Moraine[1]

Subdistrict 7.1 (Standish)
A. Outwash Plain (non-pitted outwash plain over lacustrine clay)
        Floodplain
          First Bottom   59. First bottom–willow–ninebark
          Second Bottom   60. Second bottom–red ash–basswood
          Third Bottom   61. Well drained ridge–sugar maple–basswood--white ash
                             62. Poorly drained depression–northern white cedar–black ash
        Terrace
          Well-drained   63. Terrace flat–northern hardwoods
                             64. Terrace riser–white pine–white oak–northern pin oak
          Poorly-drained
            Shallow organic (< 2 ft of organic soil)
                           65. Wet-mesic swamp–hemlock–red maple
                           66. Wet swamp–silver maple–black ash
            Deep organic   (> 2 ft of organic soil)
                           67. Terrace swamp–northern white cedar–black spruce

  B. Moraine[1]

---

[1] These units occur in the landscape but were not studied.

Ecosystem Types

Within a single fluvial landform, floodplain ecosystem types are typically distinguished along lateral gradients perpendicular to the river channel. These gradients include decreasing flow velocity during floods, decreasing particle size, and increasing soil saturation (Bell and Johnson 1974; Schumm 1977). Flushing from periodic floods combined with regular soil aeration prevent peat accumulation close to the river (Bell and Sipp 1975; Brinson 1990). Farther from the river, flow velocity decreases and with it the ability of the floodwaters to retain their suspended load. Fine particle deposition away from the river often results in poor drainage and the formation of backswamps. On broad, fluvial features, these backswamps

Table 3.2 Comparison of selected overstory tree species (% density (De) or % dominance (Do)) on floodplain landforms in outwash plain and outwash over moraine, District 9, Manistee National Forest, northwestern Lower Michigan.

Physiographic System	Outwash Plain on						Outwash over Moraine					
Landform	Levee		First Bottom		Second Bottom		Levee		First Bottom		Second Bottom	
Density or Dominance	De	Do	De	Do	De	Do	De	Do	De	Do	De	Do
Tree species/No.	12	12	66	66	14	14	3	3	8	8	8	8
Acer rubrum			0.1	0.1	2.9	2.9					1.3	1.7
Acer saccharum			0.2	0.1	20	13			1.3	1.1	26	22
Acer saccharinum	41	49	60	71	7.4	13	11	5.0	6.9	6.4	4.1	6.9
Carpinus caroliniana	0.6	0.2	1.3	0.2	2.1	0.3			4.5	2.3	5.4	0.1
Fagus grandifolia					5.1	8.8					12	15
Fraxinus americana					12	16					3.6	2.4
Fraxinus nigra	1.3	0.5	8.3	5.2	14	11	11	4.0	10	10	3.4	3.4
Fraxinus	25	24	11	7.7	2.1	2.0	42	41	33	37	4.3	2.8
Juglans cinerea					0.9	0.7						
Larix laricina			0.7	0.9								
Picea mariana			0.2	0.2								
Platanus occidentalis			0.2	0.1	0.9	1.0						
Populus balsamifera			0.1	0.2					6.3	6.0		
Prunus serotina					1.2	0.8					2.2	3.5
Quercus bicolor					2.7	4.2			1.3	2.0		
Quercus rubra					1.6	5.1			0.8	3.4	4.6	12
Salix spp.	11	9.1	1.0	1.3					4.6	6.3		
Thuja occidentalis			8.9	8.5	2.4	3.6			1.3	3.6		
Tilia americana	15	13	3.3	2.4	11	7.8			11	12	23	21
Tsuga canadensis			0.9	1.2	12	8.0			3.5	4.3	4.8	5.0
Ulmus americana	4.2	1.9	2.3	0.6			33	50			5.6	3.7

may also experience prolonged soil saturation from a high water table whose fluctuations are much more moderate than that of the open channel (Bell and Johnson 1974).

In the Huron-Manistee National Forests, we observed lateral gradients in both floodplains and terraces. For example, on a single fluvial landform, such as the first bottom in outwash plain, significantly different site factors and vegetation characterize the first-bottom flat (Ecosystem Type 2 in Table 3.1), backswamp (Type 5), and meander-scar swamp (Type 6). The first-bottom flat occurs closer to the river, is sandier, and has less soil organic matter than either the backswamp or the meander-scar swamp. In addition, there is a slight but significant elevation difference between the first-bottom flat and the meander-scar swamp. These abiotic characteristics are also reflected in overstory vegetation. The first-bottom flat, backswamp, and meander-scar swamp are dominated by *Acer saccharinum*, *Fraxinus nigra*, and *Thuja occidentalis*, respectively (Baker and Barnes 1998).

In these valleys, occurrence of the meander-scar swamp (Ecosystem Type 6 in Table 3.1) appeared to be closely related to its location at the foot of the valley wall where groundwater seeps saturate the soil. A similar pattern was also observed on higher terraces in many valleys, particularly those in outwash-over-moraine (Types 20 and 21). On these fluvial features, the step-like physiography of sandy terraces and glacial till at the valley margin produces groundwater-fed wetlands at the foot of large terrace slopes. These wetlands often experience slumping as upslope soils, heavy with water, slide down and collect at the base of the slope. *Thuja occidentalis* and *Betula alleghaniensis* dominate the overstory vegetation of this ecosystem group. Although not directly related to the river channel, these wetlands may moderate the amount of groundwater reaching the stream and certainly provide a sharp contrast to the dry oak-pine forests (Type 18) of surrounding terrace ecosystems.

Summary

Results of our research, together with those of others (e.g., Host and Pregitzer 1992; Host et al. 1988) illustrate that (1) markedly different landform-based patterns of ecosystem diversity occur at both regional and local levels and that (2) many more ecosystems are associated with rivers and streams than with adjacent upland areas. For example, eight ecosystems were mapped adjacent to the Pine River in the Huron Mountains, Marquette County, MI, compared to one in uplands adjacent to the river (Simpson et al. 1990; Lapin and Barnes 1995). In the Cyrus H. McCormick Experimental Forest in Upper Michigan, twice as many ecosystems (primarily wetlands) were found along the Yellow Dog River as in adjacent upland transects of the same length. Other "hot spots" of ecosystem diversity were found along the Maple River, Carp Creek, and Van Creek of the 10,000 acre University of Michigan Biological Station, Emmet and Cheboygan Co., northern Lower Michigan (Pearsall 1995). Detailed studies of ecosystem diversity at the University of Michigan Biological Station by Pearsall (1995) demonstrated that, in fact, biodiversity was markedly greater in areas of greater ecosystem diversity.

Great diversity characterizes riparian ecosystems at different spatial scales within the Huron-Manistee National Forests. At a broad scale, we found marked differences between riparian ecosystems in different regional ecosystems. These differences reflect the effect of physiography and macroclimate on each river system. At local scales within a given regional ecosystem, rivers in different physiographic systems have distinct patterns of fluvial landforms, and different riparian ecosystems are characteristic of each fluvial landform. Different fluvial landforms represent a gradient of edaphic characteristics and vegetation arranged laterally away from a river. Across a given fluvial landform, physiographic position results in distinct hydrologic conditions that enhance ecosystem diversity at fine scales.

By starting "from the top down" and examining differences in both the physical environment and vegetative communities at decreasing spatial scales (Region, District/Subdistrict, Physiographic System, Landform, etc.), we observe not only that the patterns of the ecosystem types differ among the scales, but that their composition, structure, and hydrology also differ. Although a landscape ecosystem approach has been applied in many areas of Michigan at the local level (Barnes et al. 1982; Pregitzer and Barnes 1984; Spies and Barnes 1985; Archambault et al. 1990; Simpson et al. 1990; Lapin and Barnes 1995; Pearsall 1995; Barnes 1996), rarely has the approach been applied across multiple scales. However, this case study illustrates one such application and the remarkable ecosystem diversity thereby revealed. Explicit management strategies are required to maintain the great ecosystem and biological diversity of riparian landscapes.

Managing with Diversity in Mind

Based on the general review of the literature and on the case study, recommendations are made for managing riparian ecosystem for multiple benefits, including maintaining and enhancing biological diversity. These are general recommendations or guiding principles that should be considered when making management decisions.

Principle 1
Know your ecosystems

Ecosystem classifications and ecosystem maps, such as those developed by Albert et al. (1986) for Michigan and Albert (1995) for Minnesota, Wisconsin, and Michigan, provide a useful framework of ecological units that are essential for "knowing your ecosystem." By taking time to develop local ecosystem classifications and mapping the classification within the broader context provided by Albert's or similar classifications, managers can gain an understanding of the ecological processes that create diversity within a given riparian landscape. In addition to being a powerful teacher, classification and mapping are also effective tools for sharing ecological information among people involved and interested in riparian management.

The case study presented in this chapter illustrates the utility of classification and mapping to assess regional and local differences in ecosystem diversity. An important lesson from the process of ecological classification is that ecosystem diversity, and hence biological diversity, at any given site may be controlled by many different factors operating at multiple spatial scales. Without question, the management of natural resources as entire ecosystems can be more effectively implemented and conducted using an integrated, multi-factor, multi-scale classification approach rather than the arbitrary separation of aquatic, riparian, and terrestrial ecosystems into categories based on single factors such as water, vegetation, or soil (Barnes 1985, 1996; Barnes et al. 1998).

Principle 2
Apply a landscape perspective to riparian management

The importance of linkages and interdependencies between upland and lowland ecosystems is a major theme in our studies of river valleys in northern Lower Michigan as well as in studies elsewhere. These linkages demand a broad perspective and an integrated approach to resource planning and management. The Huron-Manistee case study illustrates how these linkages operate at several different spatial scales. In particular, the valley segment scale is important because it is at this spatial scale that both river and riparian ecosystems change.

Bedford and Preston (1988) are correct in asserting that a sound scientific basis for managing riparian wetlands will not come solely from acquiring more information but also from the "recognition that a perceptual shift to larger temporal, spatial, and organizational scales is overdue." The cumulative impact of many local actions should be evaluated over entire regional ecosystems (Figure 3.1) and watersheds and over both short- and long-term time frames. Perspectives that include larger temporal, spatial, and organizational scales are beginning to be incorporated into Best Management Practices (BMPs). In developing BMPs for forested wetlands, for example, Welsch et al. (1995) present three underlying principles as the basis for specific recommendations, one of which is to "consider the relative importance of the wetland in relation to the total property to be managed."

Principle 3
Maintain or restore natural processes
that regulate riparian ecosystems

This principle deals with ecosystem dynamics. It is the combination of geomorphologic and hydrologic processes that create a diverse physical environment in riparian ecosystems that, in turn, fosters biological diversity. Geomorphic processes associated with flowing water create a complex array of landforms and create periodic fluctuations in the wetness and aeration of these landforms by over-the-bank flows or water level changes, or both these process result in a spatially and temporally diverse set of physical environments that support an incredible variety of plants and animals. Management actions aimed at controlling seasonal fluctuations of water will reduce the functional, structural, and compositional diversity of plants and animals in riparian areas (Poff et al. 1997).

Some variation in the frequency and intensity of flooding is probably needed to maximize ecological diversity (Sparks 1995). That is, no single pattern benefits all species. The cottonwood, for example, requires flooding and subsequent deposition of suspended materials to provide suitable substrates for seed germination, followed by several years of reduced flows that enable the tree seedlings to grow large enough so they are not destroyed by the next flood. More frequent and regular flood pulses favor establishment of herbaceous species. It is important to understand this variation and to understand how proposed management practices might change this variation (and the implications of these changes) *before* implementing a management practice.

Management should be directed at maintaining the geomorphologic and hydrologic processes that create diversity in the physical environment. Additionally, in some river valley segments, over-the-bank flooding and associated geomorphologic processes may be much less important than seasonal fluctuations in the water table and groundwater flow. In the Huron-Manistee National Forest case study, northern white-cedar swamps would not exist without local groundwater inputs. The stands of silver maple in outwash floodplains would almost certainly have a different species composition in the presence of rapid and frequent flood pulses rather than prolonged seasonal floods. However, this linkage to groundwater is not so pronounced in all river systems in the Huron-Manistee National Forests. Along the Pine River in the Standish Subdistrict (7.1 in Figure 3.1), where watershed hydrology is somewhat less stable and over-the-bank flooding does occur in response to storms, riparian ecosystems are markedly different from those along more stable rivers in District 9 and Subdistrict 8.2. Such hydrologic variations probably greatly influence the way riparian ecosystems and rivers respond to logging, damming, and development.

Maintaining geomorphological and hydrological processes is far cheaper than restoring them. However, human influences on rivers and their associated riparian areas (e.g., dams, diversions for irrigation, dikes, and levees) are so common and have often so reduced the compositional, structural, and functional diversity in these systems, that restoration may be the only option. Not only are these attempts to control the movement of water expensive, but they are often ineffective. In the East, it is the meandering path and broad riparian wetlands that provide the essential physical setting and biological function necessary for clean water and productive aquatic ecosystems. Large-scale riparian restoration projects are underway (e.g., the Des Plaines River Wetlands Demonstration Project in Illinois), and the technology and knowledge to support restoration at a landscape scale are growing rapidly.

Obviously, site-level restoration will be effective only if it is consistent with processes operating at the watershed scale. Before conducting site projects, first understand the likely impacts of ongoing or potential alternations (e.g., wholesale land transformation or major water diversions) occurring within the larger watershed. Starting with a landscape perspective (Principle #2) is helpful when conducting local management actions.

Principle 4
Favor native species

Non-native species pose a significant and increasing threat to conserving biological diversity in aquatic, riparian, and terrestrial ecosystems. What makes riparian ecosystems so susceptible to species introductions is the frequency of natural disturbances which allows invasive species to propagate and establish along with the mobility provided by flowing water and the connectivity provided by riparian corridors.

An important part of managing riparian ecosystems is developing strategies for dealing with the introduction and rapid colonization of exotic species. Managers should avoid introducing nonnative species to riparian habitats. And in some cases, measures to control or eradicate nonnative species may be necessary, and seeding or planting native species will be required where riparian vegetation has been severely degraded.

Principle 5
Buffer width? There are no pat answers

What minimum buffer width is needed to protect the riparian environment? The answer to this frequently asked question depends on many factors. These factors vary from place to place, but among the factors to consider are groundwater and flood hydrology, critical species habitat, the structural characteristics of the riparian forest, the gradients controlled by physiographic factors such as slope, and the degree of contrast between the riparian area and the adjacent landscape.

Buffer widths have been determined empirically from various ecological gradients and for various purposes. Based on air and soil temperature, Brosofske et al. (1997) recommend uncut buffers 300 ft wide for small streams (6 to 12 ft wide) to maintain unaltered microclimatic gradients near streams, but they caution that changes in microclimate associated with forest edges can extend up to 1000 ft for some variables. Maintaining wildlife habitat is another basis for establishing buffer widths. In their summary of wildlife buffers along wetland and surface waters for wildlife, Chase et al. (1997) recommend buffers 20 to 30 ft wide for small mammals, 10 to 300 ft for amphibians, 250 ft for pine marten, and 650 ft for the cavity-nesting wood duck. It is also important to assess factors that influence the relative rate of hydrologic transport (e.g., slope, infiltration rate, soil porosity) across the landscape. Because these factors affect the rate of transmission of upland activities to the river and because this rate affects the magnitude of impact on the river system, the relative rate of hydrologic transport can be used as guides for determining buffer width. In general, buffer width needs to increase as slope increases and as the infiltration rate and soil porosity decrease. Still another recommendation for buffer width is based on maintaining a supply of large woody debris and for providing shade to the water surface. BMPs provided by Welsch et al. (1995) call for a buffer width equal to one and one half tree heights in width, but they suggest that this will vary with climate, streamside slope, and stream direction.

Recommendations for buffer width also depend on the context in which riparian areas exist in the broader landscape. When contrast is low — for example, a riparian forest abuts an upland forest that is under uneven-aged or extended-rotation management — a narrow uncut buffer or streamside management zone one or two tree heights in width should protect aquatic systems from most activities in the uplands. When contrast is high — riparian forests embedded in a landscape matrix dominated by farmland or urban development — buffers that are hundreds of feet in width are needed for adequate protection.

There will never be a simple answer to the question about minimum buffer width. Too many variables need to be considered so the answer is always — "It depends." A more useful approach is to apply a "gradient of impact" where the impact or intensity of treatment declines as the distance to water increases. Further, buffer width may be less important than the continuity of the buffer strip along the stream (Weller et al. 1998). Gaps in riparian buffers are important points for the discharge of materials and nutrients from uplands to streams and so eliminating gaps in buffers should be a management priority. In forested landscapes, road stream crossings are the most prevalent gaps, and their number and significance as a fine sediment source may have a greater impact on stream quality than variation in buffer width along the stream. This may not be true, however, in agricultural and urban landscapes.

Principle 6
Timber production is a secondary benefit

Although timber production will continue to be an important objective in managing riparian forests, it may be a secondary benefit when applying silvicultural treatments to guide stand development. The primary objective in riparian management is to maintain or restore riparian habitats and ecological processes. A suitable management regime for producing timber in riparian forests is one that does not degrade or seriously disrupt ecosystem processes. Through the input of organic litter, including leaf litter and other organic detritus, riparian forests provide sources of energy for aquatic organisms. Shade from streamside vegetation prevents excessive warming of water during summer months and thus helps moderate temperature regimes in aquatic systems. Woody debris from riparian vegetation provides habitat for aquatic organisms, and it influences the development of channel morphology. Through the regulation of overland flow of water, riparian vegetation also affects sediment transport and, thus, reduces terrestrial inputs of nutrients to aquatic systems from agricultural and urban sources.

All of these interactions between the terrestrial vegetation and the aquatic ecosystem suggest the need for caution when it comes to timber management. Harvest operations that cause severe reductions in canopy coverage and stocking levels, or cause significant rutting and compaction of the soil, should be avoided. There are some riparian forests in which timber production is clearly undesirable. In their guide for managing black ash in the Lake States, Erdmann et al. (1987) recommend concentrating silvicultural treatments on better sites where seedling and sprout regeneration can be expected. They do not recommend timber production on wet sites where the site index is less than 45 ft at age 50, but instead they recommend management that focuses on maintaining wildlife habitat and protecting water quality.

It should also be recognized that the environmental heterogeneity common to riparian ecosytems makes it difficult to apply universal guides and management prescriptions. The suitability of silvicultural treatments depends on the condition of the forest and the physical environment. Each can change dramatically over short times and small spaces in riparian forests. For the most part, we lack guides for managing timber that are specific to riparian forests. When available, the recommendations often sound similar to those commonly proposed for upland forests. In these cases, proceed with caution. New and innovative silvicultural techniques are needed for managing riparian forests (see Chapter 14).

Art Elling

The difference between silviculture and riparian silviculture is that riparian silviculture benefits the water as well as the forest.

Landscape position determines vegetation types on the low and high terraces of the AuSable in Michigan (upper). Depth to water table determines vegetation structure near this small stream in the Suomi Hills, MN.

Chapter 4

Classifying Aquatic Ecosystems and Mapping Riparian Areas

Harry Parrott, Clayton Edwards and Dale Higgins

"In any system of classification, groups about which the greatest number, most precise, and most important statements can be made for the objective serve the purpose best."   M.G. Cline 1949

The natural world is so complex and large that we cannot avoid using classification in the management of land and natural resources. By grouping similar objects into classes, we are able to make management generalizations and reduce the number of named objects so we can comprehend and remember them. The problem is to evaluate the resource and land-management problems and specify the information we need (Bailey et al. 1978).

The *Hierarchical Framework for Aquatic Ecological Units in North America* (Maxwell et al. 1995) presents a continental view of aquatic classification with progressive divisions of the Nearctic Zone (North America) into smaller and nested watersheds. The divisions are predicated on the similarity or dis-similarity of aquatic animal communities within the watershed divisions. At low levels it focuses on within-channel (and within-lake) processes (including physical, chemical, and biotic processes), but they are not directly transferable to adjacent ecosystems in riparian areas. The terrestrial *Ecological units of the Eastern United States* (Keys et al. 1995) also presents a continental view of classification using the progressive division of geologic units. It uses geologic, climatic, soil, and potential vegetation criteria; however, it also does not specifically address riparian areas. Rather than develop a new riparian-area classification, we recommend that:

• Riparian area map units are drawn using the strong functional relationships between terrestrial and aquatic ecosystems (Chapter 2).

- Since parts of an existing terrestrial unit may be either within or outside a riparian area, we recommend amending the existing terrestrial map labels to include an R where that unit is within the riparian area and that both terrestrial and aquatic unit descriptions be amended to document functional relationships important to a broad range of management options.

The aquatic and terrestrial classification systems are compared in Table 4.1. It should be pointed out that the relative sizes of map units in the two systems are similar, and more importantly, the shaded classes show where functional linkages between the two systems (on-the-ground) may exist. We intend to show how classification systems are structured and how these systems are useful to land and water management at a variety of spatial scales. Later, we will return to this comparison of the aquatic and terrestrial systems to show how field mapping can be done.

Like most contemporary management issues, the objectives of ecological classification have become complex. For example, 23 years after the USDI Fish and Wildlife Service began the first national wetland classification (Shaw and Fredine 1956) with the single purpose of assessing the amounts and types of waterfowl habitat, the objectives expanded to provide additional information for the multiple-use management of wetlands and deep water habitats (Cowardin et al. 1979). Issues have expanded from single-factor issues (e.g., water quality, fish habitat, riparian management, Parrott et al. 1989) to multiple factor issues (e.g., distribution and viability of species, hydrologic function, ecosystem integrity, restoration of fisheries, sustainable forestry, and collaboration of managers, Paustian et al. 1997; PCSD 1996). Concurrently, classification has evolved from the domain of single disciplines (e.g., hydrologists, biologists, foresters) to multi-discipline solutions that combine individual resources.

There is a rich history of ways to classify streams and lakes, but they are generally limited by use of a single-factor classification (e.g., intermittent or perennial). A consensus is emerging for classifying stream habitat units (Bisson et al. 1982; Hawkins et al. 1993), morphological stream reaches and physical and biologic valley segments (Montgomery and Buffington 1993; Higgins and Reinecke 1995; and Rosgen 1996), and aquatic communities in watersheds and larger river basins (Maxwell et al. 1995) (see Table 4.1 for scales). The importance of watersheds as an aquatic classification criteria is broadly recognized (Frissel et al. 1986; Parrott et al. 1989; Naiman et al. 1992; Bayley and Li 1992; Montgomery and Buffington 1993; Maxwell et al. 1995; and Imhoff et al. 1996). The hierarchal classification and mapping framework proposed by Maxwell et al. (1995) begins with a division of the Nearctic Zone using the continental divides and their major drainage basins. This is defined by groups of aquatic species within the major basins that have less than 20% overlap in taxa. Finer divisions are made with smaller watersheds, geoclimatic zones, valley types (lake types), stream reaches (lake zones), and channel units (lake sites).

Table 4.1 The general framework of aquatic and terrestrial ecologic classification systems. Proposed by Maxwell et al. (1995) and Keys et al. (1995), respectively. Primary functional linkages between the two systems are shown where shading occurs. Aquatic subbasins are subscribed within river basins using the lines drawn for terrestrial classes, in this case, terrestrial subsections (Figure 4.3). The functional linkages between the valley segment and stream reach groups, and landtype and landtype phase groups are strong in riparian areas and allow for user-defined mapping that displays their objectives.

NORTH AMERICA (Nearctic World Zone)		
AQUATIC CLASSIFICATION Aquatic Taxa, Geol., Climate, Water Qual. & Hydrologic Unit	**Mapping Scale**	**TERRESTRIAL CLASSIFICATION** Geology, Climate, Soil Taxa & Potential Vegetation
Macro-Ecosystem Level		
SUBZONES	**1:2,000,000⁺**	**DOMAINS**
REGIONS	**1:2,000,000⁺**	**DIVISIONS**
SUBREGIONS	**1:2,000,000⁺**	**PROVINCES**
Meso-Ecosystem Level		
RIVER BASINS	**1:2,000,000**	**SECTIONS**
SUBBASINS	**1:2,000,000**	**SUBSECTIONS**
(1,000s of m) **WATERSHEDS**	**1:100,000**	
SUBWATERSHEDS	**1:100,000**	**LANDTYPE ASSOCIATIONS**
Ecosystem Level		
(100s of m) **Valley Segments /Whole Lakes**	**1:24,000**	**Landtypes**
(10s of m) **Stream Reaches /Lake Zones**	**1:12,000**	**Landtype Phases**
(meters) **Channel Units /Lake Sites**	**1:1,000**	

Classification, Inventory, and Mapping

Classification objectives are meant to: assess conditions, transfer knowledge, identify opportunities or outcomes, and facilitate communication (Frayer et al. 1978). Cline (1949) suggested four principles of classification:

- A class is tied by bounds about the modal individual
- Individuals are placed by their similarity to the class mode
- Classes are defined with characteristics that differentiate among classes
- Accessory characteristics will co-vary with the class-defining characteristics

Classification in its strictest sense means simply ordering or arranging objects in groups on the basis of their similarities among the objects (e.g., genus *Populus*) or on the basis of relationships among the objects (invertebrate shredders, scrapers, gatherers). Placing an object into a group (class) is called identification (Sokal 1974), and subsequent delineation over an area of land of similarly identified objects is called mapping. While classification is used generally as including the terms classification, identification, and mapping, the science of classification is called taxonomy (Bailey et al. 1978).

Gilmore (1951) promoted the science of taxonomy by stating:

- Classification is prerequisite to conceptual thought
- We can make inductive generalizations about the classes we form
- Particular classes are formed for a particular purpose
- We need a different classification system for each discipline we want to work in
- Some classifications have a more general use; they are called natural (e.g., soil series)
- Some classifications have a more specific use, they are called artificial or technical (e.g., site index or index of biologic integrity)

Classification Perspectives

Bailey et al. (1978) discuss three formats for classifying a population of individuals into classes (Table 4.2). The single-factor and multi-factor formats are distinguished by the methods used to construct classes and how they are presented for use by others. For example the measure of tree productivity — site-index (tree height at age 50 or 100) is a single-factor classification system. It is an artificial or technical classification. The multi-factor classification of soil types is based on climate, texture, structure, chemistry, etc., and is used for several purposes (e.g., evaluating plant productivity, soil moisture regime, and building foundation stability).

Table 4.2 The three formats of classification schemes

Item	Classification Formats	
Number of Factors	Single	Multiple
Number of Levels	Single	Hierarchical
Direction of Grouping	Aggregation	Division

The classification systems are described as single (tree, site-index) or hierarchical (soil types, habitat types, and plant taxonomy), where classes are nested in ever-larger classes. In the hierarchical classification, each class is mutually exclusive of all others, and each higher level includes only the classes in the level immediately beneath it. If a class is known at any level, you automatically know all other classes above it (e.g., species, genus, family, order, division, kingdom). Thus, you know where a white oak, a brook trout, or a Landtype Phase fits in the plant, animal, or land classification systems because each system is hierarchical. In the single-level system, a site-index-100 individual could be in many classes (sugar maple, red spruce, white oak, etc.). Knowing only the site index of an individual tree would not tell you which species it is.

The third system describes how individuals are formed into groups. Aggregation begins with all the individuals and puts them into similar groups (bottom-up approach). Division starts with the entire population and divides it into similar groups (top-down approach). In practice, most classifications are developed using both the aggregation and division at the same time, constantly testing if the lower groups form concise divisions of the entire population (note that populations may be plants, animals, or land).

Using Maps to Understand Riparian-Area Capability

Forest stand maps and forest cover maps are applications of taxonomic classification, and land type (or aquatic reach) and land type association (or aquatic valley segment) maps are examples of regionalization. Taxonomic forest stand maps are independent of forest cover maps that cover a larger area, and the classification of individual stands is independent of the classification of adjacent stands. By contrast, a landtype association (LTA) is the parent of the land types contained within it. The ecologic potentials for smaller map units are directly affected by the dominant environmental factors that occur in a larger area, just as a child's potential depends partly on their parents' genetic make up. Drawing a regionalized map requires a knowledge of functional ecologic processes at both the larger and smaller scales (Bailey et al. 1978). Regionalized map units are both scale specific and spatially dependent.

Taxonomic classification and mapping have served us well in dealing with individual resources. In fact, Gilmore (1951) recommends that each resource have its own taxonomic classification; resources are too different and their interactions too complex to be combined in the same classification. Bailey et al. (1978) state that once resources have been classified independently, different resources can then be compared objectively to study interactions. The traditional application of the taxonomic approach to mapping is to prepare several acetate maps to overlay aerial photos: one for soil types, one for forest types, one for roads, streams, etc. The owner, manager or scientist interprets the interaction implied by seeing through all layers. Sometimes we prefer to use each acetate map separately and consider our management one resource at a time without integrating through the stack of acetates. Realization that this may be narrow-sighted has led to the second mapping approach — regionalization.

Regionalization is a classification of *landscapes*. As a mapping approach, it is generally subdivisive, but most applications use a combination of the aggregation (bottom-up) and division (top-down) methods. *At every level of its hierarchy, regionalization always deals with geographically associated objects.* A major contribution of regionalization maps is that they display spatial patterns. It is through this spatial pattern that people use a different form of resource integration in their minds. With regionalization maps, we see resources as components of the landscape because we believe that the relationship between landscape components and physical or biologic processes is nearly always through spatial pattern and structure rather than through composition (the taxonomic map) alone.

How an owner, manager, or scientist interprets the capability of a riparian area depends on whether he/she uses a taxonomic map or a regionalization map; many times we use a combination of the two. When we use taxonomic maps, we are concerned about the characteristics of the individual resource; when we use regionalization maps, we realize that resources on-site, above the site, and below the site are interrelated. We realize that broader considerations (climate, land form, etc.) control the site we are managing. We realize that managing a site in a riparian area depends on the conditions above the site and that how we manage a given riparian site will influence sites downstream.

Regionalization

In *The Hierarchical Framework for Aquatic Ecological Units in North America* (Maxwell et al. 1995), the North American continent is one of six zoogeographic *Zones* on Earth (Darlington 1957). The Continental Divide and southern limits of the Colorado and Rio Grande River basins subdivide the continent into three *Subzones* that have aquatic species lists with less than 20% overlap. This process is repeated on successively smaller hydrologic units whose boundaries define significant barriers to aquatic organisms. These units, named *Aquatic Regions*, *Aquatic Subregions*, and *River Basins*, are currently mapped using native aquatic specie groups that overlap less than 45%, 70% and 95% respectively.

River Basins (in the 4,000 to 40,000 square mile range) may be subdivided further into *Subbasins that no longer use a smaller watershed boundary to define them.* Instead, the aquatic subbasins are delineated by terrestrial ecologic units. Terrestrial subsection lines are used in the three examples shown in Figure 4.1.

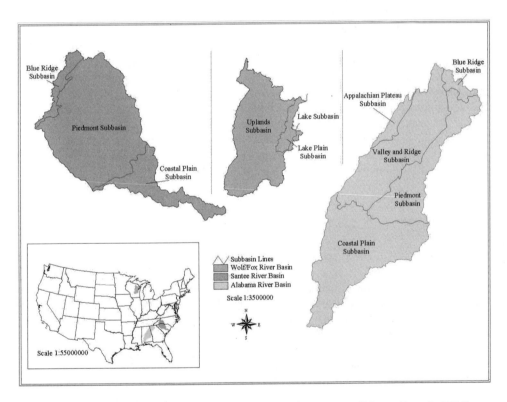

Figure 4.1 Basins (shaded) in the aquatic classification system (Maxwell et al. 1995) are subdivided into subbasins on the basis of subsection lines used in the terrestrial classification system (Keys et al. 1995).

By design, aquatic communities of different river basins will be dissimilar, but this dissimilarity may be predicated on specie varieties, commonly known as genetic stocks in fisheries management. While the species may be the same, they may be of different stocks. A New York and Pennsylvania brook trout *(Salvelinus fontinalis)* study demonstrated that 22.5% of genetic diversity was due to river basins (Perkins and Krueger 1993). Subbasins are further divided into watersheds and subwatersheds whose properties can be often stratified by additional physiographic features (e.g. landtype and landtype phase) and species distributions.

The importance of biology as a factor (Amundson and Jenny 1997) for defining ecosystems and hence ecological units, is exemplified in aquatic environments where some physical, chemical, and thermal boundaries are biological barriers for the flux of species and genes. This condition underlies the debate on the use of watershed (used generically and scale-independent, inclusive of basin) in delineating ecological units (Omernik and Bailey 1997). The issue is resolved if we look at the objectives about which we want to make precise statements. To make statements about aquatic ecosystem capability and biodiversity, from the community to genetic levels, we must be able to predict species assemblages and intra-species characteristics. Ecological units that integrate watersheds that have distinctive faunal histories and geoclimatic regions provide a zoogeographic foundation that enables people to make statements about both species assemblages and genetic diversity. Classification and mapping of the physical environment (e.g., climate, physiography, watersheds), without use of faunal history as a boundary criterion diminishes, our ability to make statements about zoogeography (Mayr 1976).

Streams

Streams are conceptualized as a continuum with longitudinal physical, chemical, and biological gradients (Vannote et al. 1980; Minshall et al. 1985) and lateral dimensions associated with over-bank flooding (Junk et al. 1989). However, they may be subdivided based on patterns of structural and functional "patches (Stanley et al. 1997)." Single-factor valley segment classifications that are used both independently and in combination in multi-factor valley (and reach) classifications are sediment source, transport and deposition zones (Schumm 1981; Montgomery and Buffington 1993), channel pattern (Leopold and Wolman 1957; Rosgen 1996), and biological zonation (Shelford 1911; Cummins 1977; Church 1992).

Hydrology, water temperature, substrate, and land type are important variables for regional aquatic classifications (Poff and Ward 1990; Bayley and Li 1992). Streamflow variability, flood regime, and intermittency were used to classify 78 streams from across the United States into nine types and each was hypothesized to have different community processes and patterns (Poff and Ward 1989). At 34 midwestern sites, flow stability was used to predict functional species assemblages with an 85% accuracy (Poff and Allan 1995).

Substrate and channel control by debris were used to describe a lateral succession of channel forms that are characteristic of the seven stream regions hypothesized by Brussock et al. (1985) for the conterminous United States. For the Glaciated Igneous and Eastern Mountain regions, they describe longitudinal succession from debris-regulated streams, to gravel and debris regulated streams, and to sand and gravel regulated streams.

Maxwell et al. (1995) use the top-down hierarchy of river basin, watershed, valley segment, stream reach, and channel unit to map a space-dependent, nested series of classification units for stream systems (see Figure 4.2 for relative sizes and disturbance interval times).

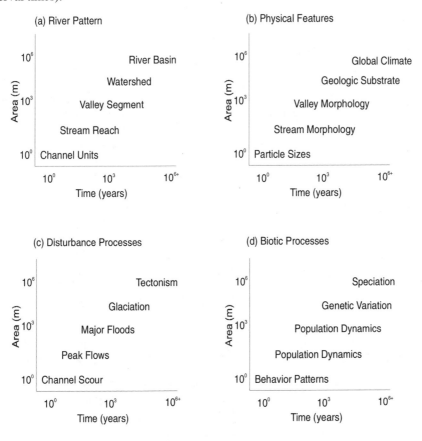

Figure 4.2 Space- and time-scaled patterns of river systems (a). These patterns are created by physical features (b) and disturbance processes (c) and directly influence the biotic processes (d) that occur in river systems. Space and time scales will vary with region (Maxwell et al. 1995).

We have found valley segments, stream reaches, and channel units most useful for an on-the-ground dialogue about a spatial pattern of stream conditions and to management. Management actions that are enhanced by these classifications include communication to resource professionals in different disciplines; communication to partners and clients in other organizations; and guides that offer a quick evaluation of resource degradation or enhancement caused by various land management activities. How and what they are used for varies with the landscape scale and classification unit size.

Valley Segments

"Valley segments stratify the stream network into major functional components that define broad similarities in fluvial processes, sediment transport regimes, and riparian interactions" (Maxwell et al. 1995). In the Continental Eastern United States, valley segment classification differentiates distinct biologic zones that co-vary with chemical and physical parameters. Three provisional valley segment classifications are: (1) The Nature Conservancy's (TNC) "Classification Framework for Freshwater Communities (Higgins et al. 1998)," (2) Michigan's Department of Natural Resources (MI-DNR) landscape based stream classification (Seelbach et al. 1997), and (3) application of the USDA Forest Service's, "A Hierarchical Framework of Aquatic Ecological Units in North America (Maxwell et al. 1995)" in Wisconsin.

Table 4.3 Defining and associated criteria for mapping valley segments and stream reaches as applied by national forests in Wisconsin and The Nature Conservancy in a pilot study (Poff and Ward 1989, P1; Poff and Ward 1990, P2; Bayley and Li 1992, B; Maxwell et al. 1995, M; Poff and Allan 1995, P3; Imhof et al. 1996. I; Rosgen 1996, R; Higgins and Reinecke 1995, F1; Stevens-Savery et al. 1998, F2; Jonathan Higgins, TNC, personal communication, T).

Hierarchical Level	Defining Criteria	Associated Criteria	Criteria as Applied
Valley Segment	Flow $_{M,P1,P2,P3,B,I}$ Network position $_{M,T}$ Water Source $_{M,P2,B}$	Fluvial processes Trophic function Aquatic community Water chemistry Water temperature	Stream width $_F$ Stream link $_T$ Alkalinity $_{F1}$ Troph./alkalin. class$_T$ Water temp. class $_{F1, T}$

Hierarchical Level	Defining Criteria	Associated Criteria	Criteria as Applied
	Valley slope $_{M,.R}$	Geomorphic processes	Gradient class $_T$
	Geology $_{M,R,B}$ Morphology $_{M,R,B}$	Flow characteristics quantity, duration, and quality	Surficial geology $_T$ Adjacent geology $_T$ Landform $_T$
	Confinement $_{t,\ M.R}$	Floodplain processes Sediment processes	Entrenchment $_{F1}$
Reach	Channel pattern $_R$	Fluvial processes	Stream gradient $_{F2}$
	Channel shape $_R$	Fluvial processes	D_{50}, pebble count $_{F2}$
	Entrenchment $_{M,R}$	Floodplain processes Flow energy	Entrenchment $_{F2}$
	Stream gradient $_{MR}$	Stream power Habitat pattern Bed form	Stream gradient $_{F2}$
	Channel materials $_{M,R}$	Hydraulic processes Habitat quality	D_{50}, pebble count
	Sinuosity $_{M,R}$	Channel unit pattern	Sinuosity $_{F2}$
	Bedform $_M$	Fluvial dynamics Habitat patterns	
	Width/Depth ratio $_{M,R}$	Fluvial processes	Width/depth ratio
	Riparian vegetation $_{M,I}$	Habitat structure Water chemistry Bank stability	Width/depth ratio $_{F2}$

Biologic alliances in The Nature Conservancy's version are defined as macrohabitats at a map scale of 1:100,000, and use stream map characteristics (network position and stream node level), trophic class, water temperature, geology, and valley gradient class (Table 4.3). Concurrently, and also at a map scale of 1:100,000, the Michigan DNR is building a river valley segment classification with 20 different valley types (Seelbach et al. 1997). In contrast to The Nature Conservancy and the Michigan DNR, the Forest Service has taken an intensive field-data-collection approach to identify important defining characteristics for valley segments mapped at the 1:24,000 scale. Using stream width, temperature, and alkalinity, Higgins and Reinecke (1995) defined 13 valley segment types with corresponding and representative fish and mussel communities.

These eastern projects use physical, chemical, and biological data for classification (Step 1) of valley segments. The intent is to delineate map units (Step 2) based on physical characteristics available remotely (maps, aerial photos, satellite images) or are easily collected in the field, with chemical and biological conditions as accessory (co-varying) characteristics. Water temperature and chemistry are functions of groundwater flow that can be an accessory characteristic of terrestrial classification. "Geofisheries" described by Dean et al. (1991) used terrestrial map units to estimate brook trout potential in Ontario.

A significant outcome of the The Nature Conservancy, the Michigan DNR and Forest Service studies is mapping the biological capability of streams at a spatial scale that can show the aggregation of valley types at two map scales (1:100,000 and 1:24,000). They use surrogates or actual measures of flow, thermal characteristics, and chemistry as defining characteristics. Additionally The Nature Conservancy and the Michigan DNR are using a connectivity variable that characterizes valley position. Osborne et al. (1992); Lyons (1992); and Lyons et al. (1996) have shown in the Midwest that the index of biotic integrity (IBI) is a function of location (headwaters, main-channel tributaries, large rivers or lakes) within a large watershed (river basin).

Stream Reaches

Stream reaches have highly uniform flow and channel morphology (Maxwell et al. 1995). They are intermediate in length between the larger valley segments and smaller channel units. Church (1992) described a reach as a homogeneous subdivision of a river within which "controlling factors" (e.g., volume and time distribution of water and sediment, sediment characteristics, channel materials and geology) are relatively uniform. The Rosgen (1996) system classifies stream reaches based on a similar morphology for both the channel and its confining valley. Inclusion of the adjacent floodplain or confining terrace slopes, along with normalized channel shapes, allows the system to describe how the channel handles the flow of water and sediment; does it keep it all within its channel (highly entrenched, high energy), or does it spread it readily on a floodplain (slightly entrenched, low energy).

The Rosgen method defines nine basic stream types, using channel entrenchment in the valley (three classes), channel shape (width/depth ratio in three classes), and channel sinuosity (in three classes) (Figure 6.6 in Chapter 6). All of these types can be further described by four channel slope classes and six channel substrate classes. With these descriptors, the nine stream types are found to occur in 94 subtypes. Rosgen (1996) hypothesizes that natural channels that meet the class criteria (although at least one of the criteria may vary +/- 20%) are stable, while those that do not are undergoing a transition from one type to another and are unstable. The system is a multi-(physical) factor, hierarchical classification, but it is not regionalized because it is only nested within the watershed valley. The stream types are useful for interpreting how easily stream reaches can become unstable through changes in land use and for understanding which in-channel fish habitat structures are likely to be successful in a given stream type (Rosgen 1996).

The Forest Service has demonstrated the applicability of Rosgen classification criteria to stream reaches in low-relief Lake States terrain (Stevens-Savery 1994). However, additional low channel slope ranges are recommended especially for the B stream types, and sinuosity is lower than modal values (Stevens et al. 1998 and Fig. 1.4).

Alhough not yet demonstrated to be applicable in the East, Montgomery and Buffington (1993) subdivided Pacific Northwest alluvial valleys into reaches based on bed morphology of single-thread streams: cascade, step-pool, plane-bed, pool-riffle, regime, and braided streams. These bed morphology classes were observed to match reach-level delineations based on Rosgen criteria, and to have potential applicability to biotic capability on the Monongahela National Forest in WV (Robert Ries, Forest Service, personal communication).

Channel Units

Channel units subdivide reaches and are the smallest geomorphic feature that we recommend mapping. Examples include pools, riffles, runs, steps, etc. These physical features are associated with characteristic categories of water velocity (Sullivan 1986), and Aadland (1993) has shown that micro-locations of water velocity ranges are critical to various fish life stages. Hawkins et al. (1993) proposed classifying channel units into 18 types: five in turbulent, fast water; two in nonturbulent fast water, six scour pools, and five dammed pools. While there is some consensus on this classification of channel units, Poole et al. (1997) found that "correct" identification varied from 29% to 80%. Inventory and monitoring should use repeatable, quantitative measures of physical, chemical, and biologic conditions (e.g., bankfull mean channel depths at the riffle or pool locations and measures of pool volume). Poole et al. (1997) were concerned that using frequency of descriptive channel units for monitoring would lead to direct installation of habitat structures without considering the habitat-forming biophysical processes in the riparian area or watershed.

Lakes

Lakes may be classified in whole, or part, like streams. Lake classification has been guided by limnologists, with an emphasis on biology and water quality, in contrast to the morphological basis of stream classification.

Classification by trophic status was suggested by Weber (1907) to describe the nutrient conditions that determined the flora of German peat bogs. Thienemann (1918) classified lakes as oligotrophic and eutrophic, depending on summer oxygen depletion and on the types of benthic organisms associated with oxygen conditions. Strom (1930) and Lundbeck (1934) incorporated depth into trophic level classification. They focused on the observation that the ratio of water volume to sediment surface was less in shallow lakes, which influenced the amount of nutrients that diffuse from sediments into the water column. Naumann (1919) classified Swedish lakes as oligotrophic or eutrophic based on the amount of plankton biomass.

Lake classification based on fish has met with variable success. Johnson et al. (1977) evaluated 3378 lakes in Ontario for combinations of terminal predators: (lake trout, northern pike, walleye, and smallmouth bass) associated with various physical and chemical measurements. The only clearcut association in the 15 groups they proposed was that lake trout were found in lakes with a Morphoedaphic Index (Ryder 1965) less than 5 (deep, softwater lakes). The number of groups expanded to 63 by adding brook trout and lake whitefish to the list (OMNR 1978). Tonn et al. (1983), using an approach similar to that of Johnson et al. (1977), classified Wisconsin lakes using northern pike, largemouth bass, and mudminnow. Their sample of 18 lakes smaller than 220 acres was used to develop a dichotomous key using lake area, watershed size, maximum depth, and pH (or conductivity). They used ordination and discriminant analyses techniques. Schupp (1992) evaluated 3029 Minnesota lakes and derived 44 lake classes based on principal component analysis of a suite of physical and chemical parameters. The resultant lake classes were then compared to fish community data; groupings of major and minor species occurrences were made to add a biological component to the process. A significant finding in this approach was that 19 of the 44 classes were located in the glacial scour zone, whereas the remainder were located in the glacial deposition zone of Minnesota. Results of each of these projects reflect complications when exotic species impact the community structure of native species.

Lake size is considered an important classification criterion, with the large lakes having higher biological diversity (Barbour and Brown 1974), especially when normalized by climate (Schlesinger and Regier 1982) and depth (Marshall and Ryan 1987). Another feature that is gaining prominence in the classification of lakes is its location within the watershed (Magnuson et al. 1998). This feature, when considered in the context of river and ground water linkages, has an important bearing on water-stage regime, mineral content, and nutrients. The importance of groundwater and water-residence time to the chemical, physical, and trophic state of lakes has been amply demonstrated (Born et al. 1974; Winter and Woo 1990; Winter 1992; Vollenweider 1975). Specific conductance has been used to classify perched, groundwater, and mixed lakes in Minnesota (Hawkinson and Verry 1975). The Carlson (1975) trophic state index and Vollenweider (1975) carrying capacity model were used by Garn and Parrott (1977) to classify lake condition, determine sensitivity to pollution, and predict impacts using lake retention time, chemistry, and nutrient inputs.

Maxwell et al. (1995) recommended classification and mapping of lakes based on two whole-lake criteria followed by subdivision into zones and further subdivision into sites analogous to channel units. Primary whole-lake criteria include genesis (Hutchinson 1957), physiography, hydrology, and morphometry. Secondary criteria include thermal regime and stratification, retention time, and other physical, chemical, and biological characteristics. Littoral, pelagic, and profundal zones may be subdivided into ecologically significant sites (e.g., springs, and spawning substrates). Similarly, Higgins et al. (1998) has proposed a whole-lake classification using criteria that includes connectedness, hydrologic regime, lake size, and shoreline complexity.

Wetlands

The Fish and Wildlife Service's *Classification of Wetlands and Deepwater Habitats of the United States* (Cowardin et al. 1979) is a taxonomic system for classifying streams, lakes, and wetlands using water regime, hydric soils, and vegetation, with emphasis on the last. Although hydrology is the driving variable for wetland function, and morphology may be used as an indicator of hydrology (Brinson 1993), these are relegated to modifier use in the Fish and Wildlife Service system. Since publication of *Classification of Wetlands and Deepwater Habitats of the United States*, Winter (1992) and Brinson (1993) have identified "hydrogeomorphic" criteria, including physiography, climate, water source, and hydrodynamics that can complement and add functionality to the Fish and Wildlife Service classification.

Riparian Areas

Riparian areas include aquatic ecosystems, wetland ecosystems, and elements of terrestrial ecosystems with strong functional links between land and water (see Chapter 2). Issues affecting riparian area management have expanded from water quality protection and fisheries management (Parrott et al. 1989) to restoring ecosystem integrity and resource sustainability (PCSD 1995). The function of classification has thus expanded from gathering information about delivery of pollutants to streams and lakes and managing fish species within affected waters, to characterizing ecological function and capability.

In the mid-1980s, some national forests numerically rated riparian conditions, but forests did not classify riparian capability (Parrott et al. 1989). A decade later, the Bureau of Land Management and the Forest Service implemented "proper functioning condition" ratings for riparian areas (USDI BLM 1993, only for western range lands and not eastern forests). Although each agency has on-going classification and mapping programs for aquatic and terrestrial resources, neither have implemented a riparian classification, inventory, and mapping system.

Traditional inventory and classification practice has included parts of riparian areas within soil inventories and currently (within the Forest Service) as components of terrestrial ecosystem classification and mapping. However, these inventories may not define distinct riparian map units. In addition, the national wetlands' inventory contains classifications and map units that are, somewhat, inclusive of riparian areas. Chapter 2 provides a functional definition and criteria for delineating the outer boundary of riparian areas.

The classification of resources leads naturally to resource mapping. Mapping can be simple or complex. We have shown how mapping within either the terrestrial or aquatic classification systems can be hierarchical from the continental to the Land Type Association or subwatershed scale in the respective systems (the lowest part of the meso-ecosystem scale, Table 4.1). Now we will suggest ways to map on the ground with a variety of resource information that uses criteria from several broad classification systems. In the past, resource managers have mapped with the aerial-photo/acetate-overlay method which works well at the site level. We suggest a similar approach that can still use acetate overlays, or it can use GIS procedures that combine aquatic and terrestrial classification system components at the ecosystem level. It requires knowing the interactions among physical, chemical, and biologic variables at the local level. As an example, we will use the frameworks provided by Maxwell et al. (1995) and Keys et al. (1995) for the classifying of aquatic and terrestrial ecological units, respectively. Tentative valley segments developed on the Chequamegon National Forest in north-central Wisconsin will be used for the aquatic example.

The aquatic portion of the riparian area along a stream can be subdivided into valley segments and reaches. Higgins and Reinecke (1995) used three defining criteria, each with three levels to classify valley segments that correspond to fish and mussel communities. We feel that each organization needs to do this type of aquatic classification before riparian-area mapping at the ecosystem level has meaning for the local manager. The three criteria used on the Chequamegon National Forest are shown in Table 4.4. Out of 27 possible combinations, they found 13 that related well to fish and mussel communities. For instance, the valley segment **nsw** (narrow, soft, warm) is associated with the major fish species: creek chub, brook stickleback, central mudminnow, finescale dace, and blacknose dace and the rare occurrence of one mussel (fat mucket). In time, these local associations and criteria may be combined among localities where appropriate.

Higgins and Reinecke used a set of local attributes to type valley segments: (1) an adaption from the Rosgen classification system (mean bankfull width), (2) a modification of the long-standing division between cold-, cool-, and warm-water fish, and (3) a chemical criterion (alkalinity) that relates well to groundwater inflow, ionic concentration, and total streamflow. The physical and chemical map unit criteria are used to type other valley segments, and to predict fish and mussel community make-up and speculate about the management impact on the resource. Other systems, such as Ohio's index of biotic integrity applied at the local scale, have found invertebrates to be useful in strongly contrasting forest and agricultural settings (see Chapter 17 on monitoring). Valley segments can be further subdivided into Rosgen stream types at the reach level to describe and understand the current and potential morphological characteristics of a stream.

Table 4.4 Valley segment criteria for the Chequamegon National Forest in north-western Wisconsin. Labels are derived by combining the bolded descriptor letters in a left-to-right order (e.g., an **mso** valley segment has a **m**oderately wide stream with a **s**oft water alkalinity, and **c**ool maximum stream temperatures).

Stream Width$_{bankfull}$		Alkalinity		Stream Temp.$_{max}$	
narrow	≤20 ft	**a**cid	≤6 ppm	**c**old	<72°F
moderate	>20 ft ≤40ft	**s**oft	>6≤20 ppm	**c**ool	≥72°F ≤80°F
wide	>40 ft	**m**od. hard	>20 ppm	**w**arm	>80°F

For the terrestrial portion of the riparian area, existing classification systems can be used to describe the current and ecological potential of the area. However, neither the terrestrial land type or valley segment systems include criteria to differentiate riparian areas. We propose using the riparian-area definition and the delineation criteria and keys found in chapter 2 to remedy this problem. For example, the national hierarchy of terrestrial ecological units (Keys et al. 1995) could be amended to include a riparian criteria so that map units would break at the riparian-area boundary. Another option is to map riparian areas using the riparian delineation criteria and keys, and then overlay the riparian areas with both the terrestrial and aquatic classification system classes (preferably as part of a geographic information system database). The promise of object-component data models in GIS databases allows data to be associated with an object (within a riparian area) while having various other associations as well (landtype phase, compartment or stand designation) (Maguire 1998). The classification systems could include landtype phases, existing vegetation, wetland, soils, or other resource characteristics.

The top panel in Figure 4.3 is an example of how the latter system might be displayed, while the lower panel in Figure 4.3 displays how landtype phases might be split to match riparian-area boundaries. Note that, for large stream and valley systems (see Figure 2.2, lower panel), the landtype phase modifier (-R) might be split along the terrace slope. In all cases, GIS, with an object-component data model, can be utilized.

In Figure 4.3, the highest aquatic classification level is valley segment defined as **nac** (**n**arrow, **a**cid, **c**old); it is divided into two Rosgen stream types at the reach level (C5c-, and E5). The C and E labels describe the major characteristics of entrenchment, width/depth ratio and sinuosity. The 5 designates the stream substrate as sand dominated, and the presence (or absence) of the small letter and modifier (c-) designates the channel slope. The terrestrial ecosystem is classified as Landtype 300, and is subdivided into Landtype Phases (LTPs): 310A, 310C, and 315A. The letter modifiers (A or C) are slope classes, and the ones and tens place of the number designate soil phases based on major phase characteristics of internal drainage, texture, and elevation, and geologic land form. In this example, the dashed line (riparian area) would serve as an additional delineation criterion for development of LTP map units and LTP 310A and 310C would each be split into two units to identify LTPs with direct riparian functions (hypothetically coded as LTPs 310A-R and 310C-R in the lower panel). LTPs 325A and 315A are on the floodplain (a component of the riparian area) and thus they too are identified as 315A-R and 325A-R.

Conclusions and Recommendations

General

- Classification solutions can transcend institutional and professional boundaries.

- The classification systems for aquatic and terrestrial ecosystems both recognize the importance of ecologic function (the movement of material and energy in time and space).

Streams

- Valley segments are a collection of stream reaches with different physical stream characteristics, yet have the same characteristic fish and mussel community. They are coarse scale (1:24,000 or larger) delineations of hydrologic condition and biologic potentials.

- Valley segment criteria include its flow regime, stream chemistry, thermal regime, and network position.

- Reaches characterize parts of a stream valley with the same flow characteristics and similar morphology.

- Reach criteria, in addition to flow, include channel cross section, sinuosity, slope, and sediment sizes, as well as the ratios of channel width to flood-prone width and channel width to channel depth. Measurements of these physical values denote one of several reach types which are useful for evaluating habitat, channel stability (condition relative to modal values), choice and placement of in-stream structures, channel and floodplain restoration, and interpreting resistance to change caused by the placement of roads, land use, or natural debris flows, etc.

- Channel units and habitat niches are formed by water volume, velocity, depth, and turbulence, and lead to pools and riffles. They are useful for monitoring if they are measured physically, but less useful if they are simply counted and used in frequency ratios for habitat goals or placement of in-channel structures.

Lakes

- A whole lake classification system is proposed that further subdivides a lake into zones and sites that share similar functions.

- Defining criteria include physical, chemical, and biologic characteristics, such as geology, genesis, hydrology, chemistry, and trophic condition.

Wetlands

- The Classification of Wetlands and Deepwater Habitats of the United States is widely used and provides national map unit coverage, however, it does not use hydrology or physiography as driving variables.

- Integration of hydrogeomorphic criteria into wetland classification will increase ability to infer cause-effect relationships and to predict changes.

Regionalization and Aquatic Classification

- The distribution of fishes and intraspecies genetic diversity, in conjunction with watershed boundaries, are used as criteria to delineate aquatic classes.

- This approach allows the conservation of species and genetic diversity of aquatic species to be addressed in management.

Riparian Areas

- An alternative to developing a new classification for riparian areas is to map them with both aquatic and terrestrial class units and the on-the-ground the key in Chapter 2.

- Riparian functional linkages that are evident from mapping may be used to amend map unit descriptions and modify map codes.

Chippewa National Forest

Mapping the riparian area of the Boy River in north central Minnesota requires information on both the aquatic and terrestrial resources.

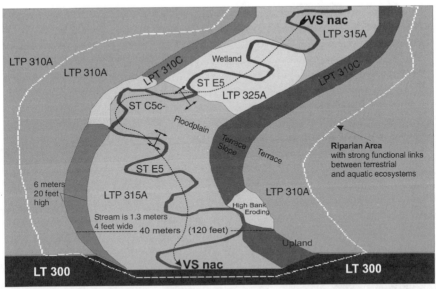

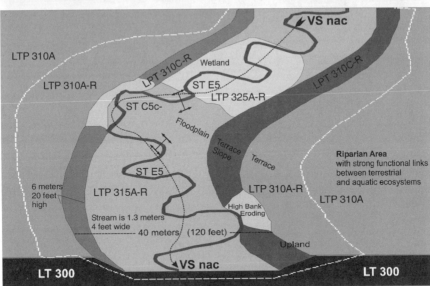

Figure 4.3 A depiction of aquatic classes (valley segments and stream types) and
terrestrial classes (land types and phases) in a riparian landscape. The upper panel uses
conventional landtype phase labels. The lower panel suggests adding -R to the phases
within the riparian area to ease analysis or riparian area data in a GIS data base.

Classification of aquatic and terrestrial ecosystems in riparian areas is based on defining criteria of fish community structure, geomorphology, geology, soils, and potential vegetation. Associated criteria can include resource condition and the R label for classes within a riparian area.

Chapter 5

Linkages Between Forests and Streams: A Perspective In Time

James W. Hornbeck and James N. Kochenderfer

...a tree planted by the water, that extends its roots by a stream...will not be anxious in a year of drought.   Jeremiah 17:8

Over the past half century, knowledge about linkages between forests and streams has been gathered through watershed management and watershed ecosystem research (Hornbeck and Swank 1992). These studies, usually conducted on small, experimental watersheds, have shown how contributions of water, sediment, nutrients, heat, and organic matter from forests to streams change as forests undergo succession or experience natural and human-related disturbances. In this chapter, we review this information and discuss how it can be blended with management practices to protect and enhance aquatic systems.

Disturbances and Forest Succession

The contributions from forests to streams are continuous and ever changing, depending on forest age and extent of disturbances. Table 5.1 summarizes how variables affecting streams from small watersheds (usually 100 acres) might change during five primary stages of succession. The table best describes a continental eastern hardwood forest subjected to a major disturbance such as fire or intensive logging. In the initiation stage (years 0 to 10), early vegetation is characterized by various naturally occurring herb, shrub, and tree species that undergo rapid fluctuations in numbers and dominance. However, tree species, including pioneers, become dominant by year 10 and canopy closure is nearly complete (Martin and Hornbeck 1989). As woody plants become dominant, biomass increases rapidly. For example, aboveground biomass on an herbicided watershed in West Virginia increased from 0.9 tons/acre at the end of the first growing season to 14.6 tons/acre 9 years later, with 77% being produced in the last 3 years of the period (Kochenderfer and Wendel 1983). During the organization stage (years 10 to 30), the forest takes on a more permanent identity as

89

remaining shrubs and pioneer species are overtopped or they die and the major species of a forest type assume dominance. Rapid increases in biomass continue during this stage. At the West Virginia site, biomass had increased to 37 tons/acre by age 21.

The aggradation stage (years 30 to 80 or longer) is characterized by a gradually increasing accumulation of tree-biomass components and forest-floor organic horizons (Bormann and Likens 1979) and by occasional mortality from windthrow, lightning, or insects and diseases. During the maturation stage (years 80 to 125), growth rates level off and the forest assumes a steady state in terms of biomass accumulation and production of litter and woody debris. Aboveground biomass in mature eastern hardwood stands might be expected to reach 80 to 165 tons/acre. The old-growth stage (years 125 to 250 and beyond) is characterized by increasing mortality, changing species composition, size, and age classes (a shift from old, even-aged to mixed-aged) as gaps from dead trees are filled. In hardwood stands, this stage may be accompanied by an increasing abundance of shade-tolerant species, including conifers. Disturbances such as harvesting, fire, or blowdown invariably occur at unpredictable times during any of the described stages. Depending on severity of disturbance, succession will usually revert to an earlier stage, perhaps with a new species mix. As Table 5.1 and the following discussion show, each successional stage, along with any reverting caused by disturbances, has implications for the linkages between forest and streams.

Table 5.1 Contributions from forests to streams during successional stages

Contributions from forests	Successional Stage				
	Initiation (Yr 0-10)	Organization (Yr 10-30)	Aggradation (Yr 30-80)	Mature (Yr 80-125)	Old Growth (Yr 125-250⁺)
Water yield and peak flows	Increased	Stable ⟶				Decreased
Erosion and Sedimentation	Increased	Variable ⟶				Decreased
Nutrient Leaching	Increased	Minimal ⟶				Gradual Increase
Water Temperature	Increased	Stable ⟶				Decreased
Organic Matter	Greatly Reduced	Gradual Increase, Changing Quality ⟶				Steady State

Contributions from Forests

Water Yield and Peak Flows

During the forest initiation stage, transpiration and evaporation losses are low because of low leaf area. Soils are wet and more water is available for streamflow or seepage to groundwater, resulting in increases in water yield and peak flows. In cases of complete canopy removal, annual water yield can be increased by as much as 40%, with most of the increase occurring in the growing season (Hornbeck et al. 1993). Peak flows may increase by about 20%, a small amount in terms of culvert design (Helvey and Kochenderfer 1988), but possibly contributing to localized erosion of streambanks. Increased water yield during the 10-year initiation stage can be beneficial in augmenting low flows, but it also can increase the potential for erosion and sedimentation.

Increases in water yield cease with the attainment of a closed canopy, and water yield usually returns to predisturbance levels before the end of the initiation stage. From then through the old-growth stage, annual evapotranspiration remains fairly uniform at around 20 inches/yr (Federer et al. 1990). Annual precipitation that exceeds this amount becomes streamflow or groundwater. Under similar climates, evapotranspiration is greater from conifers than hardwoods; any trend toward increasing conifer composition during the old-growth stage will result in small decreases in water yield and peak flows. Disturbances over the course of succession will increase water yield, but basal area must be reduced by at least 25% to produce detectable increases in yield ("...
forest are disturbed (on the order of square m..
during any stage of succession usually have m..

Erosion and Sedimentation

Erosion, resulting sedimentation, and turbidity...
Fortunately, the problem is less critical than i...
not as prone to erosion. Litter layers and orga...
snowmelt to rapidly infiltrate into the mineral s...
horizons are mostly well-drained, coarse-textur...
capacities. As a result, erosive overland flow s...
and sediment yields are among the lowest in N...
Although there is much year-to-year variatic...
undisturbed forests in the Northeast is about 27...
et al. 1984).

Martin and Hornbeck (1994) showed that annual sediment yields from undisturbed watersheds at the Hubbard Brook Experimental Forest, for years 1966 through 1990, ranged from .9 to 125 lbs/acre. The variability is not related to watershed size or amount of annual precipitation but to individual storms occurring during a given year (Bormann et al. 1974). For example, a large storm with a predicted recurrence interval of 50 years accounted for about 35% of an 11-year total sediment yield from a 96-acre forested watershed in West Virginia (Kochenderfer et al. 1997).

Depending on the type and extent of disturbance, erosion and sediment yields will be greatest at the beginning and during the initiation stage of forest succession (Table 5.1). The critical factor with harvesting disturbances is not the intensity of harvest, but the care taken during logging. Careless logging practices, particularly poorly designed truck roads and skid trails, cause much erosion and sedimentation (Patric 1976, 1978). Conversely, the use of best management practices (BMPs) that minimize compaction and other disturbances limits erosion and sedimentation to small amounts, mostly along the stream channel (Patric 1978). The regenerating forest lessens erosion from disturbed sources such as roads, trails, and stream crossings, usually within a few years, by providing protective cover and soil stabilization. Erosion along streambanks and channels may take longer to diminish, depending on the extent of disturbance in the riparian area and on whether debris dams, which help control sedimentation (Bilby 1981), have been washed out.

From the organization through the old-growth stages, erosion and sedimentation can be expected to fluctuate with the severity of storms that produce runoff (Martin and Hornbeck 1994) and with intervening natural or human-related disturbances. Contributions of sediment to streams should be minimal in the old-growth stage as organic horizons, soil stabilization, and contributions of organic matter and woody debris for instream dams are at maximums.

Nutrient Leaching

Nutrient leaching from forests to streams is affected by various factors including: mineral weathering, soil and hydrologic characteristics, vegetation, climate, biological processes, and natural and human disturbances. The balance among these factors determines the forest's ability to retain nutrients such as carbon, nitrogen, and base cations, and to control the mobilization of aluminum and heavy metals.

Immediately after a disturbance or early in the initiation stage, leaching of base cations and nitrogen in the form of nitrate usually is at a maximum. For example, commercial clearcutting of northern hardwoods in New Hampshire caused nitrate in streamwater to rise from < 5 ppm before harvest to about 25 ppm by the second year after harvest (Martin et al. 1986). Such increases are due to increased decomposition stimulated by warmer and wetter soils, and the absence of vegetation to sequester nutrients. In most eastern forests, the leaching losses are tempered quickly (usually within 5 years) as foliage production and plant uptake recover rapidly (Mou et al. 1993).

During the organization and aggradation stages, forest vegetation and soil organic horizons gain in nutrient content (Covington 1981) and nutrient leaching is at a minimum. However, during the maturation and old-growth stages, there may be a gradual loss of biotic control over nutrient uptake and accumulation, and nutrient leaching to streams may increase (Reiners 1981).

In general, streams draining forests are low in nutrient content; the only concerns about water quality and eutrophication are the increases in nutrient leaching immediately following disturbances. However, over the past three decades, atmospheric deposition and accompanying cation depletion via leaching have become major problems in some forests and streams of the Northeast. Cation depletion is accompanied by mobilization and leaching of inorganic aluminum to the extent that aluminum can become toxic to fish (Cronan and Schofield 1990). Cation depletion and aluminum mobilization associated with atmospheric deposition are a concern during all stages of forest succession (Shortle and Smith 1988; Federer et al. 1989).

Water Temperature

The temperature of aquatic systems greatly influences fish production, recreational use, and value for water supply. The principal source of heat for streams draining forests is solar energy striking directly on the surface of the stream (Brown 1980). Thus, the temperature of streams is controlled primarily by shade from overhanging vegetation.

During the initiation stage, water temperature is high if stream shading has been severely reduced or eliminated. Maximum daily stream temperatures after intensive harvests can increase 9 to 18^0F (Lynch et al. 1975). Heat added to streams by the sun is not easily dissipated. Streams in exposed areas eventually cool when flowing through shaded areas, but the cooling is more from mixing with the cooler water in the shade than from transfer of heat from water to air (Brown 1980). Thus, stream heating can have an impact for long distances downstream.

Once the stream channel is shaded by shrubs and regrowing trees, stream temperatures decrease and exhibit fairly uniform annual and seasonal variations through the remaining successional stages, or until another disturbance reduces or eliminates streamside shade. An increase in shade-tolerant species, including conifers, during the old-growth stage of hardwood stands could increase stream shading and slightly decrease water temperatures.

Organic Matter

Contributions of organic matter from forests to aquatic systems are important with regard to energy source, dissolved oxygen supply, habitats for aquatic organisms, and stream channel stabilization. Organic matter consists of leaves, needles, bark, fruits, branch wood, and boles. These materials are added to streams as direct or windblown litterfall (including tree harvesting) from surrounding and upslope vegetation or by erosion or streamwater transport.

After severe disturbances, organic matter contributions to and accumulations in streams can take a century or more to recover. During the initiation stage of forest succession, contributions of organic matter are minimal, although materials may have been retained in or added to streams during the prior disturbance. There is controversy over whether logging slash should be retained in or added to streams. Arguments for leaving or adding slash include improving habitat, stabilizing the stream, and adding an energy source; arguments against include obstructing fish passage, plugging of culverts, blockage and rerouting of existing stream channels, and the possibility of decreasing dissolved oxygen.

Toward the end of the initiation stage and during the organization stage, contributions of litter begin a gradual increase that continues through aggradation and early maturation. Initially, foliar contributions are high compared to woody contributions, but natural mortality during the aggradation stage eventually increases the coarse woody debris (branches and boles) contribution. These larger materials are especially important to the formation of stream dams and aquatic habitats (Bilby 1981). During the maturation and old-growth stages, contributions of organic matter to the forest floor and streams are at or near a steady state (Table 5.1) that can exceed 0.4 tons/acre (Bormann and Likens 1979).

Blending Linkages with Management Practices

As indicated above, there are good general understandings and sources of information about the various linkages between forests and streams. The challenge is to incorporate this knowledge into management practices. This will be easier when pertinent information is available about rates of the various contributions from forests to streams. In the absence of site-specific information, computer models can be useful substitutes. The JABOWA family of forest succession models (Shugart and West 1977) simulates death and growth of individual trees and can be used to model litterfall and production of coarse woody debris (Bormann and Likens 1979). Nitrogen cycling that includes the effects of disturbances can be simulated with the models FORTNITE (Aber et al. 1982) and LINKAGES (Post and Pastor 1996). Water yield from forests, including increases due to disturbances, can be simulated on a daily basis with the hydrologic model BROOK90 (Federer 1995). Brown (1980) presented equations that can be used to calculate stream heating and cooling under various shade regimes.

In the absence of site-specific data and use of computer models, general guides discussed below can be followed to protect streams. These guides have two things in common: (1) they emphasize protection during disturbance and the stand initiation stage or the time when linkages between forest and streams are most vulnerable (Table 5.1); and (2) they stress careful management of the riparian area that links forests and streams.

Best Management Practices

Adhering to BMPs, including construction and restoration of roads, landings, and stream crossings, before, during, and after logging, is a must in terms of being practical, obeying laws, and protecting linkages between forests and streams. For the most part, managers and loggers rely on various BMP guides that have been published by state agencies or researchers (e.g., Kochenderfer 1970). These guides stress protecting against erosion and sedimentation and changes in water temperature, but seldom address nutrient leaching, organic matter contributions, or water yield and peak flows.

When these unaddressed variables are a concern, or in situations such as managing municipal watersheds or abused or exceptionally erosive sites, precautions beyond the usual BMPs may be required. In such cases, it is doubly important to have as much site-specific information as possible to develop protective management practices. The overriding goal should be the same as for any disturbance: limit impacts to the shortest possible time. In the case of erosion and sedimentation, this may mean special attention to disturbances such as landings, roads, and stream crossings that create chronic problems. One solution to erosion is to spread gravel (3-inch clean is ideal) at least four inches deep along approaches to the stream crossing sites, landings, and other locations that might create chronic problems. In the case of nutrient leaching or water yield, promotion of rapid regeneration through management techniques, seeding, or planting may be necessary to help the site through the vulnerable initiation stage (Table 5.1).

Managing Riparian Areas

Riparian areas have a variety of uses and values (Hornbeck et al. 1994), but protecting streams from upslope disturbances continues to be among the most important. A properly designed and managed riparian area can provide a variety of amenities and still protect against stream temperature changes (Brown 1980); assure a continuous supply of organic matter (Webster et al. 1990); absorb nutrients, sediment, and water from upslope (Welsch 1991); and maintain a diversity of species composition (Naiman et al. 1993). Management of riparian areas is discussed, more fully, in later chapters.

Managing Upslope Forests

The protection capabilities of riparian areas must be supported by careful management of forests upslope or outside the riparian area. In the case of harvesting disturbances, application of BMPs is a given, but beyond that it is helpful to think holistically in terms of forest-stream relationships. For example, will the harvesting disturbance affect streams? If yes, what steps are necessary to minimize harmful impacts or maximize beneficial impacts?

One approach is to consider management objectives, including those for streams, in terms of present and desired future conditions with regard to stand age, stocking, and species composition. This can lead to selecting an appropriate silvicultural method, including cutting intensity and configuration, as shown in the following examples.

Example #1

Forest type: Northern hardwoods.

Current condition: Even-aged stand that regenerated from a clearcutting in 1910; usual mixture of sugar maple, yellow birch, and American beech; widely visible site that supports several second- and third-order streams.

Desired future condition: Continuous, uneven-aged stand to meet recreational and esthetic demands and protect water quality (including a continuous supply of litter and coarse woody debris to stream channel).

Silvicultural system: Selection cuttings at approximately 20-year intervals to develop a balanced distribution of size and age classes, increase the number of shade-tolerant species including hemlock and red spruce, and improve stand quality.

BMPs: Follow BMPs for road construction, stream crossings, and restoration of disturbed soils. Leave a variable-width, unharvested buffer strip along all perennial streams to maintain sources of organic matter and coarse woody debris contributions to streams.

Example #2

Forest type: Spruce-fir.

Current condition: Old-growth, uneven-aged stand; substantial windthrow and insect and disease problems; the stand may have been high-graded around 1900; located on a flat, unexposed site with deep soils.

Desired future condition: A healthy, productive commercial forest supporting wildlife and protecting water quality.

Silvicultural system: Shelterwood in which the overstory is removed in two or more cuttings. This system provides a seed source, protection for regeneration, and rapid growth on residual trees while minimizing impacts on contributions from forests to streams.

BMPs: Follow known precautions for preventing sediment. Buffer strip along streams may be included in shelterwood harvest, but with special precautions to avoid soil disturbance.

Example #3

Forest type: Mixed Appalachian hardwoods.

Current condition: Even-aged stands that regenerated from heavy cutting in the early 1900s; composed primarily of valuable intolerant species such as yellow-poplar, black cherry, and northern red oak; occurs on mesic site that supports a dendritic system of first- and second-order streams.

Desired future condition: A healthy, productive, and diverse forest composed of a high percentage of intolerant species that provide timber and wildlife benefits while protecting riparian values.

Silvicultural system: Group selection with individual openings about 2 ½ acres in size. This size opening will result in reproduction of a wide range of species, including desirable intolerants, and have little impact on water yield and nutrient leaching.

BMPs: Follow known precautions to prevent sedimentation. Avoid harvesting streamside buffers in order to maintain contributions of organic matter and coarse woody debris.

Sandy Verry

Andy Dolloff

The linkages between forests and steams are measurable in the Knife River on Lake Superior's North Shore (above). The valley and its conditions shape the Maury River in Virginia (below).

Chapter 6

Water Flow in Soils and Streams: Sustaining Hydrologic Function

Elon S. Verry

"We may conclude then that in every respect the valley rules the stream. . ."
Hynes, 1975

This chapter is concerned with how we manage soils, floodplains, and streams and their watersheds. Management should ensure that the water flows without undo restriction, and carries nutrients, detritus, sediment and water to plants and animals without adverse consequences to plant and fish productivity. Gregory (1997), in considering riparian management for the 21st century, calls for a landscape perspective of river networks. He suggests that the question of how to manage forest ecosystems at the scale of landscapes has remained largely unasked, let alone answered. This chapter is a partial answer. It draws on site-specific studies, paired watershed experiments, deductive reasoning, and a new stream classification to seek a more comprehensive approach to watershed and riparian management.

Water Flow in Soils

Sixty-five years ago, Horton (1933) showed that when the soil infiltration rate was exceeded, overland flow to streams occurred (e.g., surface runoff from plowed and often compacted cropland). However, Hewlett (1961) showed that the uniform, basin-wide, generation of runoff in forests with an intact forest floor was not explained by the generation of surface, overland flow.

Hewlett explained that streamflow in small watersheds in North Carolina came from saturated areas near slope bottoms and near channels. The extent of these saturated areas varies over time, so his idea became known as the variable-source-area theory. Whipkey (1965) in Ohio suggested the water supplied to these saturated, variable-source areas came from subsurface flow rather than overland flow. Hewlett and Hibbert (1967) described water movement within the soil matrix as "translatory flow." They perceived that water films

99

around soil particles thickened and moved water downslope by flowing in the films to saturated areas near the bottom (more recently see Liu 1987; Yeakley et al. 1994). Pearce et al. (1986), using oxygen-18 studies on steep basins in New Zealand, confirmed that old water contributed to stream water more than new, large-pore, water from the current storm

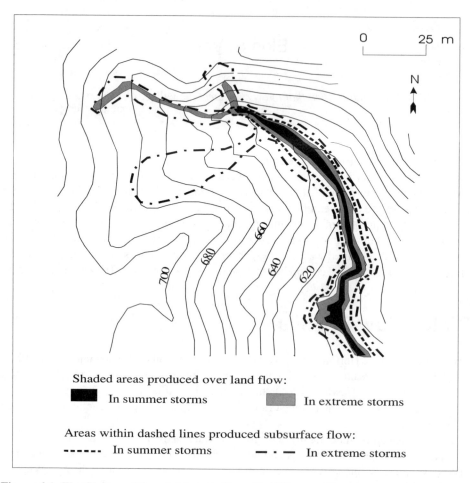

Figure 6.1 The Sleepers River basin near Danville, Vermont illustrates areas of saturated overland flow (shaded areas) and subsurface flow (dashed areas) (adapted from Dunne and Black, 1970). During 50- to 100-year rain-on-snow events, subsurface areas can become saturated and produce overland flow (Yarnal et al. 1997). With permission.

See Pearce et al. 1986 for a history of the variable-source-area concept. Most recently Bonell (1998) thoroughly reviewed the field studies and modeling efforts aimed at the processes that generate runoff in forested headwater basins.

Dunne and Black (1970) used runoff plots and stream hydrographs in Vermont to measure the contributions to channel flow and confirmed that overland flow in the expanding areas of saturation (variable-source areas) was a significant contributor. However, "the remainder of the watershed acts mainly as a reservoir during storms, and between storms it supplies base flow and maintains the wet areas that produce stormflow (Dunne and Black 1970)." DeWalle, et al. (1997) also used oxygen-18 to look specifically at baseflow in small West Virginia streams and a large Pennsylvania stream. Travel time was less than a year in the small systems and up to 5 years in the large system. Dunne and Black's "Sleeper Rivers" diagram illustrates the variable source area concept well (Figure 6.1). Later Dunne (1983) acknowledged that, in humid regions, overland flow dominates the stream hydrograph in basins with gentle slopes, concave lower slopes, and wide valley bottoms. In areas with steep, straight, or convex slopes, incised channels, narrow valley bottoms, and permeable soils, subsurface flow (whether derived from old or new water) dominates the stream hydrograph. Recently, Yarnal et al. (1997) showed that the 1996 rain-on-snow floods of the Susquehanna River in Pennsylvania produced overland flow for storms with a 50- to 100-year recurrence interval in areas away from the floodplain.

In the flat lands of the Lake States and parts of Maine especially, peatlands (making up 20 to 30% of a basin) commonly act as the variable-source-area of saturated flow to stream channels. Recently, Verry (1997) showed the relative importance of upland subsurface flow (entering a peatland) and peatland surface-outflow to a stream draining a 24-acre watershed in Minnesota. Subsurface flow in the mineral soil occurs as saturated flow within the A horizon (over the clayey, impeding B horizon). It flows rapidly, responding during the snowmelt or rain event through the large pores of the surface soil horizon. Horizons below the A/B interface may be unsaturated. This subsurface flow intersects a *Sphagnum* peatland (10 acre) with a saturated soil. Figure 6.2 (following page) clearly shows the importance of the saturated peatland to the April snowmelt peak flow. However, the mineral soil subsurface flow is nearly as important in April. In July, large evapotranspiration demands on the mineral soil deplete the limited soil water and diminish the subsurface flow. Streamflow from the peatland typically stops in midsummer.

Clearly, soil, seep, wetland, stream, and lake hydrographs are strongly linked in the riparian setting. Maintaining large soil pores is key to maintaining a free draining, well-aerated soil, and a soil that allows the full development of vegetation. Table 6.1 lists soil bulk densities that will support maximum vegetation development by being well drained and aerated.

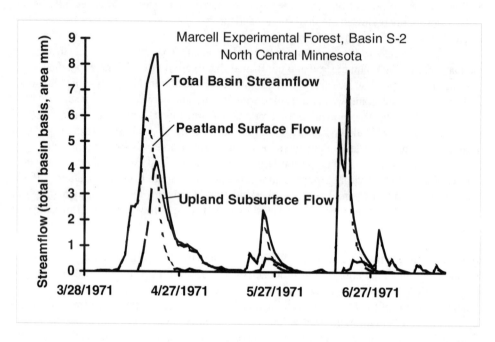

Figure 6.2   Variable-source-area flow partitioned between upland mineral soil subsurface flow and saturated peatland surface flow that gives rise to a first-order stream in Minnesota.

Table 6.1 Soil bulk density for maximum vegetation development. In soils with high rock content, subtract the volume and weight of the rock before calculating bulk density of the rooting medium.

Horizon	Mineral Soils	Organic Soils
0-6 inches	≤ 1.25	≤ 0.10
6-12 inches	≤ 1.50	≤ 0.10

Water Table Position and Soil Water Regime

Aerated soil depth above the water table determines maximum vegetation height. In wetlands or wet mineral soils next to lakes, seasonal pools, streams, and rivers, the average depth to the water table during the growing season controls the height of vegetation (Verry 1997). Figure 6.3 illustrates the rapid growth change associated with minor changes in depth to water. This gradation in riparian vegetation height causes a vertical and horizontal diversity of plant edges and increases habitat niches.

The 18-inch depth is significant from many viewpoints. Soils will begin to support heavy equipment when the water table is more than 18 inches below the surface (e.g., agricultural, field, tile placed 18 inches below the surface for tractor operation). Similarly, the use of

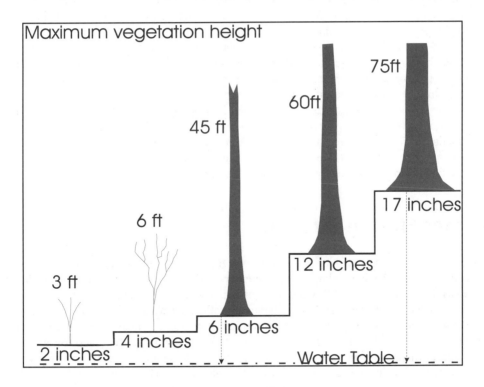

Figure 6.3   Depth to the water table greatly affects maximum vegetation height, and in the context of the topography of riparian areas, this gives rise to edge patterns that increase plant and animal diversity.

skidding and forwarding logging equipment becomes possible when water tables are more than 18 inches below the surface. Soil moisture within the soil matrix is the major control of vegetation development when water tables are lower. Maximum vegetation development then occurs as the number of growing season days increases with soil water content at 10 Kpa of water tension (field capacity). This tension value reflects the gravity-drained condition of the soil about three days after a heavy rain.

Operating heavy equipment on wet soil will compact the soil, leave it wet and cold too long in the spring, and thus reduce tree productivity. Aspen productivity in the first five years following logging on a Minnesota loamy sand was less than half that on an non-compacted site (Stone and Elioff 1998). High soil density after 7 years of winter and spring frost were unchanged (D. Stone, pers. comm.). Compacting surface soils significantly impairs site productivity. One indication that compaction is occurring is when equipment operation ruts 6 inches deep over distances of 300 feet or more. Additionally, invertebrates and amphibians living in the macropores of mineral soils are decreased (deMaynadier and Hunter 1995).

The use of walk-on-slash harvesting and forwarding equipment, and sequencing harvest from the back of the sale to the front (road-access side) will reduce soil compaction and reduce repeated skidding on soils made wetter by tree harvest. Compaction can also be reduced by harvesting when soils are frozen or dry. After large rains of 1½ inches or more, allow the soil matrix to drain for an additional 2 days. This will bring the soil to field capacity and avoid turning surface horizons to mud. The sustainability of optimum wood production over time is directly related to the condition of soil pores and their ability to transport water and air. Likewise, the sustainability of optimum invertebrate and fish production in streams is directly related to the condition of the channel system.

Water Flow in Streams

The central function of running water is to form the stream channel and much of what we consider as stream habitat for birds, fish, fur bearers, and invertebrates. If we understand how streams handle the energy of flowing water and sediment, we will understand how to place culverts, to locate road rights-of-way in the stream valley, select and build durable fish habitat structures, and evaluate watershed condition as a function of stream-channel condition. Understanding the dynamics of stable and unstable streams is the first step toward the confident management of watersheds, riparian areas, and streams. For a thorough review of how channels change over time, see the review volume edited by Gurnell and Petts (1995). It is a tribute to the English geomorphologist Ken Gregory, but it includes much of the work done in the United States, including the defining body of work done by Luna Leopold and his contemporaries in the U.S. Geological Survey.

Bankfull Flow is the Key

A stable stream has a normal dimension (channel cross-section), a normal pattern (plan-view sinuosity), and a normal profile (longitudinal pool/riffle or step/pool transect) (Figure 6.4). Leopold (1994), in his review of river dynamics, records the essential question he first asked in the mid fifties: "Why is the channel created and maintained by the normal flow of water relatively small?" Why isn't the channel large enough to handle some floods? Wolman and Miller (1960) answered Leopold's question when they considered both the magnitude of channel discharge and the frequency at which a given magnitude occurs (see also: Harvey 1969, 1975; Dunne and Leopold 1978; and Andrews 1980). The streamflow most effective in producing and carrying sediment is the flow that occurs when stream water is at the *bankfull elevation.*

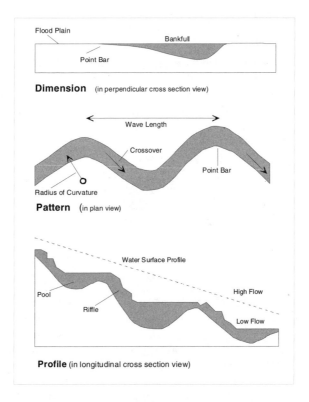

Figure 6.4 The dimension, pattern, and profile of stream systems (adapted from Leopold 1994. With permission.)

In short, water flowing at the bankfull elevation has the highest velocity (channel velocities are generally two or three times those on the floodplain). The bankfull velocities are thus better able to carry large amounts of sediment. A second part of Wolman and Miller's answer dealt with the frequency of various streamflow discharges. For instance, why don't large floods carry the most sediment since the discharge rate (cubic feet/second) is several times higher than bankfull discharge, in spite of lower velocities? Simply put, over a matter of decades, the large floods (say once in 50 years) occur so infrequently that the amount of sediment is small compared to the sediment carried roughly each year (over 50 years) at the bankfull elevation. These concepts were developed for stream and valley systems having floodplains, in steep, mountainous channels without floodplains, flood flows may be more important for carrying sediment and shaping channels.

How do you calculate it?

Bankfull flow and its elevation is a fundamental geomorphic feature of streams and their valleys. Flows at bankfull elevation shape the channel and define the elevation of the floodplain! This flow is also called the channel forming discharge, the dominant discharge, or the most probable annual "flood" (Dury 1969). The bankfull elevation is at the elevation of the depositional flat, immediately adjacent to the channel (Leopold 1994). At the bankfull elevation there are strong relationships between the channel characteristics: width, depth, cross section area, stream discharge, and the size of the contributing basin (Leopold and Maddock 1953; Leopold et al. 1964).

The frequency of any streamflow level is calculated using the largest peak value in each of at least 10 years. Typically, the bankfull discharge occurs as the 1.5-year event, but may vary between a frequency of 1.2 to 2.5 years in North America (Annable 1995, Rosgen 1996). Conceptually, it is the average annual peak discharge, but it does not occur every year because, in the long stretch of years, we must allow for the drought years and the large flood years. Usually it occurs two out of three years (the 1.5-year frequency).

When does it occur?

Only a single peak value for each year is used in the mathematical derivation of streamflow frequency; however, bankfull water stages actually occur in the Eastern United States 2 or 3 days a year. If we consider a plus or minus 10% variation in stage, then this stage occurs for 2 or 3 weeks each year. In snow-dominated areas, bankfull stage may last for several days or even a week or two when snowpacks are deep and the melt is slow. Summer rains of 3 to 5 inches will also bring channels to bankfull stage for a day or two. Spatially, the bankfull wave occurs within the basin on different days depending on the sequence of snow melt or the location of large summer rains. This nearly-annual surge travels along the stream shaping the channel.

Summaries of long-term measurements at USGS gaging stations are used to index the range of stream discharges and mean stream water depth to the bankfull discharge and its mean depth. At these locations, detailed cross-section data and flow measurements are available and mean channel depth is always calculated. These relationships define the relative size and depth of flows (Figure 6.5, following page).

The floodplain elevation in this Virginia valley is the bankfull elevation. It is easily seen as the mid-channel island elevation, the field elevation and the point bar elevation (right, center). Know the elevation of bankfull flow, and know your stream types (a C stream type above) so you can interpret stream condition, and manage the stream and its valley more effectively.

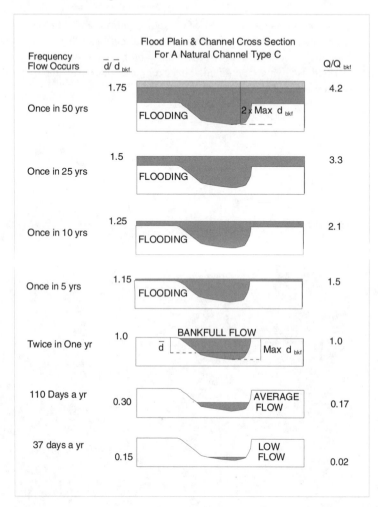

Figure 6.5 Empirical ratios of *mean* depth (d) to bankfull *mean* depth (d_{bkf}), and streamflow discharge (Q) to bankfull discharge (Q_{bkf}) at seven flow frequencies for natural channel type C in the Eastern United States (adapted from Dunne and Leopold 1978; Leopold 1994; 1997). *Max* d_{bkf} and 2 x *Max* d_{bkf} are practical field application tools used to classify stream channels (Rosgen 1996) (Note that depth is exaggerated for the C channel type). For consistency and minimum variation, channel width and depth are measured midway between meander bends on riffles or at the narrowest pool width in step/pool systems. With permission.

A Geomorphic System of Channel Classification

Natural channel types
Recently, eight natural channel types were defined using channel cross section shape, and the width of the 50-year flood (Rosgen 1994, 1996). Rosgen added the measure of *maximum* (as opposed to *mean*) channel depth to allow the quick field determination of channel entrenchment (see below). *Maximum* channel depth is measured at the midpoint of the riffle or at the base of steps for consistency. Figure 6.4 shows the maximum channel depth and twice *maximum* channel depth (the approximate elevation of the 50-year flood) to contrast a practical field tool with the detailed mathematical relationships developed only at long-term gaging sites. Table 6.2 shows field index values of stream discharge and flow depth using *maximum* channel depth that can be easily remembered to assess current and potential flow conditions. The geomorphic channel type classification has been extensively tested on many substrates in the Lake States, Ontario, Maryland, Virginia, Delaware, Pennsylvania, and the Western United States and is universally applicable. Many other methods of stream and valley classification that preceded the geomorphic system are discussed in Rosgen (1996). The valley-segment classification system is detailed in Chapter 3 and expands on the geomorphic classification system by including temperature, chemical, and biologic factors.

Table 6.2 Field indices of waterflow (Q) and depth (d) using maximum channel depth (adapted from Leopold 1994 and Rosgen 1996). With permission.

Recurrence Interval (yrs.)	d/d_{max}	Q/Q_{max}
50	2	4
25	1 ¾	3
10	1 ½	2
Bankfull~1	1	1
Ave. Flow	¼	.2

Reduce your variation
The geomorphic stream classification allows interpretation of current channel condition within a broadly applicable framework. It stratifies a continuum of variation in cells having

modal tendencies. The conclusions reached in many studies comparing streams or basins before the use of this classification framework do not partition variation caused by different stream types. Valley landform, bank condition, and in-stream structure evaluation, plus watershed assessment methods developed concurrently with the geomorphic stream classification, use the classification to reduce variation caused by major differences in the way streams handle the energy of flowing water and sediment (Rosgen 1996). Certainly improvements will be made, but it should be used as part of deductive reasoning in watershed assessment and as a rigorous application of stream geomorphology interpretation and design. Note that a given stream or river will usually have several channel types as it flows through different geologic formations.

The geomorphic channel classification uses three measurements to place channels into one of eight fundamental types (Figure 6.6).

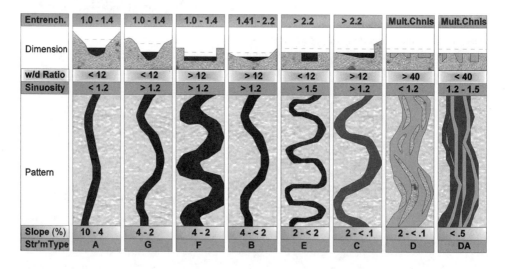

Figure 6.6 Natural channel types showing classification criteria of entrenchment, w/d ratio, sinuosity, and slope range. Dark areas shown in the channel cross section are bankfull stages, dashed lines are the approx. 50-year flood stage (adapted from Rosgen 1994). With permission.

These measures include the entrenchment ratio, the width/depth ratio, and channel sinuosity. Note the difference between single-thread and multiple-thread channels (stream width for multiple-thread channels includes the total distance between outside channels). The central bars of stream Type D are constantly shifting and not vegetated, while those of stream

Type DA (anastomose) are stable and vegetated (Rosgen 1996; Nanson and Knighton (1996) further divide what Rosgen considers as anastomose into several types under the heading of anabranching rivers).

Entrenchment

Entrenchment describes the relationship of the river to its valley. Kellerhals et al. 1972 defined entrenchment qualitatively as the vertical containment of a river and the degree to which it is incised in the valley floor. Highly entrenched rivers do not have much room to spread out flood waters, but slightly entrenched rivers do. Rosgen (1994) quantified the entrenchment ratio as the width of the valley at the approximate 50-year flood elevation (2 x *maximum* bankfull depth) divided by stream width at the bankfull elevation. He called this valley width the flood-prone area to distinguish it from floodplain that occurs in some channel types, but not all.

W/D Ratio

The width of the channel (at bankfull elevation) divided by its *mean* depth is the second classification factor. Measuring mean depth of a channel allows the cross-section area to be calculated. In unfamiliar areas, regional curves that relate basin area to channel cross section can be used to evaluate channel condition (Rosgen 1996).

Sinuosity

Channel sinuosity completes the classification criteria for the eight basic types. Channel sinuosity is the length of the channel (measured along the thalweg over at least two meander lengths) divided by the straight line length of the stream valley.

These above three measures are evaluated in the order given: entrenchment ratio, width/depth ratio and sinuosity. However, a close examination of Figure 6.6 shows that sinuosity is not required to type a channel if single-thread and multiple-thread channels are recognized. Frequently, channel sinuosity is less than the modal values used to type natural channels. This deviation from the modal type value has been caused by physical alteration of the stream channel and by cumulative land-use changes that increase bankfull discharge (see Chapter 1, Figure 1.5). Each of the eight geomorphic channel types is further subdivided using descriptors of stream slope and sediment size (see Rosgen 1996 for all 42 possibilities).

Using Stream Type to Understand Stability of the Stream and its Floodplain

A fundamental use of natural channel types is the evaluation of current stream condition and how it deviates from the modal values of the channel type. As stream dimension, pattern, and profile values become more distant from the mode, the stream becomes unstable, and floodplain dynamics may also change if water-table position or sediment deposition patterns are altered. Streams are unstable when they shift among the eight basic channel types. This shift may take decades to accomplish, or occur overnight. As a general guide, Table 6.3 shows modal values of width/depth ratio and sinuosity associated with each stream type in the Eastern United States. Values within a + ⅓ (w/d ratio) or a - ⅓ (sinuosity for E and F channel types) can be used to judge if a channel is near the mode (is stable) or deviating from it (unstable). Undisturbed or recovered sites in your region should be used to develop local modal values. A stream is stable when it maintains its dimension, pattern, and profile over time, so it neither aggrades nor degrades and transports, without adverse consequence, the flows and detritus of its watershed (Rosgen 1996). Do not confuse stream stability with a lack of sediment movement. Stable streams will typically scour a foot or more deep each year and redistribute their sediment into a normal pool/riffle sequence. Most of this scour occurs on the descending leg of the hydrograph. Thus, the use of scour chains and channel cross-sections with toe pins is recommended.

Table 6.3 Modal stream dimension values for the Eastern United States (estimated from Annable 1995; Rosgen 1996; and Stevens-Savory et al. 1998). With permission.

Natural Stream Type	A	G	F	B	E	C	D	DA
W/d ratio	7	7	20	20	8	24	50	50
Sinuosity	1.1	1.1	1.5	1.2	2.0	1.3	1.1	1.2

A stable stream keeps its channel type classification. Unstable streams can only transport their water and detritus (sediment and woody debris) by changing their dimension, pattern, or profile. They are characterized by channels with large w/d ratios (for their type), eroding banks on the straight parts of the channel and at meander bends (normal sites of erosion), rapid rates of either channel downcutting or channel aggradation (filling of pools and riffles with sediment), and by the formation of channel cutoffs that reduce stream sinuosity over time. Entrenchment and channel shape are perhaps the two most important measures for evaluating the degradation of channel habitat downstream of road crossings. Downstream degradation may occur if culverts are undersized, or if road fills across the valley do not allow flood passage, resulting in the channel culvert and the road bed washing out.

Seeing Bankfull in the Field

Stream or channel typing is a useful tool, but the entire system depends on identifying the bankfull elevation. Just where is this elevation, and how does it vary with stream type? Leopold (1994) defines the bankfull elevation as the depositional flat immediately adjacent to the channel, but in some channel types (A and G) there is no depositional flat (see Figure 6.6 for typical cross sections).

Highly entrenched
Type A and Type G channels are the steepest (A 10 to 4% slope, G 2 to 4% slope), and the channels are narrow so that sediment is carried through the A or G type to deposit in channel types with a wider flood-prone area downstream. In both A and G channel types, bankfull elevation is typically at the base of trees. In these types, the bankfull velocity is so high that trees are scoured away below this elevation.

In F channels, the power of the bankfull water and sediment is also contained within nearly vertical terrace walls (even the 50-year flood cannot get out of the high banks), and while the stream slope is relatively high (2 to 4%), the channel is wide (width/depth ratio about 20) causing sediment to be deposited inside the channel at many elevations. The *highest* mid-channel bar elevation inside of the F channel is the bankfull elevation. Trees, shrubs, or grass may or may not be growing on top of the highest mid-channel bar.

Moderately entrenched
The moderately entrenched B channel type will spread flood waters, but woody debris within the flood-prone area is still readily carried to the stream channel because entrenchment is still moderate. Trees will typically grow down to the bankfull elevation and along the length of the channel. Small, foot-size depositional flats typically occur at the bankfull elevation.

Poorly entrenched
The C, E, D, and DA channel types are poorly entrenched and all have extensive flood-plains. The floodplain elevation is the bankfull elevation. Meandering systems with point bars on the inside of bends are excellent places to see the bankfull elevation. At these point bars, the bars slope upward toward the floodplain on the inside of the bend. Where these point bars flatten is the elevation of the bankfull stage. Between bends on these channel types in the Eastern United States (where humid conditions produce rapid vegetation growth), typically

the channel side, rises rapidly from the thalweg, but nearly always will "round over or flare out" before reaching the floodplain flat. That is, there is nearly a 45-degree slope to the channel side right at the top of the channel before it flattens to the floodplain. In cross section, many channels tend to flare out at the top.

Vegetation - maybe yes, maybe no

Grasses, sedges, and near-stream trees always grow below the bankfull elevation in the Eastern United States as the growing season progresses. Willows always grow below the bankfull elevation in the Eastern United States. Trees growing on the rounded channel side nearly always grow below bankfull elevation in the Eastern United States. Except where banks are slumping, alder grows down to the bankfull elevation and typically (along with willow) covers the floodplain elevation, particularly in "beaver meadow" E channel types where the water table adjacent to the channel is high enough to exclude trees. Where trees grow below bankfull elevation, a change in lichen color on the tree trunk may confirm the depositional flat elevation. This change in lichen color can also occur on rock in any of the channel types. Be aware of natural levees in the Eastern United States and don't include them in determining bankfull elevation. Natural levees are not continuous, thus channel water will spread on the floodplain behind the levees through natural breaks. Use the floodplain elevation as the bankfull elevation.

Recognizing Change in Channel Systems

All eight stream types represent natural channels and all can be "stable" channels. When streams shift between stream types, they are unstable.

The stable condition depends on the cause

For the multiple-channel systems (D and DA), some explanation may help us understand how these types are sometimes either stable or unstable. The D type has constantly shifting in-channel bars that yield multiple channels. When the D type occurs on alluvial fans receiving large amounts of sediment from high energy streams above (As or Gs), the D type is the natural response to the recurring high levels of sediment and high flows. Heavy sediment loads may also occur in a C channel because of excessive grazing and loss of bank stability, and conversion to a D stream type may occur. However, the causes are not climatic/geologic and can be reversed. Thus, in the latter case, the D type is unstable in a setting where the normal geologic type would be a single-thread C channel.

DA (anastomose) stream types may also occur in a geologic setting where the intervening bars are naturally stabilized with vegetation in low-slope valleys (some of Alaska's north-slope rivers, for example), but they may also be induced by increasing the magnitude of the near-annual peak flow.   Two examples are "peaking" power-generation schedules at reservoirs and repeatedly dynamiting tall beaver dams. Again, the difference is whether the cause is the climatic/geologic or anthropogenic. The stream type is accurate in either case.

Unstable streams don't like their type

Streams generally do not change from one type to another quickly. The criteria used to classify streams (entrenchment, w/d ratio, sinuosity, and channel slope) do not change over time, but their values do change over time in an unstable setting. The values given in Figure 6.6 are classification or boundary values. The values, in fact, have a wide range with a modal distribution — a distribution commonly skewed in one direction (see Rosgen 1996 and Stevens-Savory et al. 1998). Values on the skewed tail of the distribution indicate a stream that is unstable and in the process of changing to a different type. In general, the entrenchment ratio will change rapidly when the 2 x Max. depth $_{bkf}$ elevation becomes either highly or slightly entrenched. The w/d ratios tend to get larger on the skewed end of the distribution tail, the sinuosity tends to become less; and the channel slope becomes higher. Channels near the mode of the distribution indicate a Proper Functioning Condition in the geologic sense, while Current Functioning Condition will include any value in the total distribution. Working toward a Desired Future Condition is a conscientious decision to reach a dimension, pattern, and profile agreed to or implied by a consensus of social values, frequently it is not the stream with perfect sinuosity or a perfect width/depth ratio, but one that is achievable within the geologic and social context of the river and land uses desired.

Knowing modal values of the w/d ratio and of sinuosity will help observers interpret measured values. The best sources of modal values are measurements within a similar land type and near the river of interest. As a poor substitute, use the values in Table 6.3. Sinuosity is the most variable. Sinuosity tends to increase as sediment size decreases (i.e. cobble to clay) (see Schumm 1963; and Ferguson 1987) and where vegetation roots bind bank soil. Organic sediment found in beaver meadows can be classed as clay since the organic particles are colloidal and similar in size to clay particles (Malterer et al. 1992).

A progression toward large w/d ratios (a wide and relatively shallow stream) is a common indicator of an unstable stream. Streambank erosion on the outside of bends is natural, but if bank erosion on either side of the channel in straight sections is occurring, the stream is experiencing an advanced rate of in-channel erosion most often caused by increases in the near-annual peak flow. The heavy sediment loads from in-channel erosion may be seen as an increase in mid-channel or cross-channel bars. The rate of bank erosion depends on the bank material (non-swelling clays are more stable compared to uniform sand that erodes easily). Bank erosion less than 3 inches per year is considered low, while bank erosion rates of more than 12 inches per year are considered excessive (Rosgen 1996). If bank sloughing wider than a foot is occurring, excessive in-channel sediment deposition may be happening if bankfull flows cannot spread it on the floodplain. Methods of measuring bank condition (part of the current condition) are detailed in Pfankuch (1975), and methods of measuring bank and channel erosion are detailed in Rosgen (1996).

Another indication of excessive in-channel erosion or high sediment loads from road crossings is a flattening of the pools. The pools should be about three times the depth of the riffle mean depth (distance below bankfull elevation). If not, excessive sediment is blurring the distinction between pools and riffles. One exception is in geologic settings where *only uniform sand occurs*; there is little tendency to form pools in this material (Leopold and Wolman 1957). In these uniform sand settings, large woody debris accumulation is especially important because it is the substrate for invertebrates, and the cause of random pool development. In most geologic settings, any *mixture* of pebble sizes should produce riffles and pools or steps and pools (in steeper gradients).

Seeing and Interpreting Causes of Unstable Channels

Causes of unstable channels

Head-cuts (eroding steps in the channel bottom that migrate upstream) indicate a change in the channel slope caused by: (1) down-cutting a larger order channel, (2) cutting off of meander bends to align culverts, (3) straightening channels to pass floods quicker, (4) impingement on the floodplain by locating road prisms that significantly increase channel entrenchment and decrease sinuosity, (5) excessive cleaning of large, woody debris from channels (trees that fall from the bank should be cut to only 1/3 the channel width to allow both pool development (Lisle 1986) and boat passage), (6) washout of large road fills at stream crossings, and (7) changes in land use that increase the near-annual peak flow. All these actions speed the flow of water. However, the stability of banks can also change without a change in the flow regime or because of secondary impacts on vegetation. Channel down-cutting may cause vegetation to dry out on sandy banks, causing a change from trees to shrubs then to grasses, and a loss of deep bank strength as root systems become shallow. Excessive grazing can both break down banks and cause a similar vegetation change by compacting soils leading to a change from shrubs to shallower-rooted grasses.

An important component of watershed analysis is the inventory of channel and bank stability, and derivation of landscape-scale conditions using inventories of land use, geology, soils, land types, climate, and road or drainage infrastructures (Montgomery et al. 1995). The Pfankuch method of estimating bank condition and propensity to erode is valid throughout North America (Pfankuch 1975); it was detailed in Rosgen (1996) along with extensive methods for measuring stream stability. Increases in the near-annual peak flow can put enough stream power in channels to erode banks and cause streams to change type. Harr et al. (1975) showed that road systems whose total right-of-way area makes up 15% of the basin will increase peak flow. The road ditches duplicate the stream system, which may make up only 10% of the basin.

Looking at watershed studies
Many watershed harvesting studies report an increase in streamflow peaks using the paired-watershed approach. In general, snow melt peaks are increased twofold to threefold with clear cutting, and the change may last up to 15 years in the Eastern United States (Verry et al. 1983). Rainfall peaks in the Eastern United States increased 1/4 to two times in magnitude but lasted only 5 to 10 years (Reinhart et al. 1963; Hewlet and Helvey 1970; Lynch et al. 1972; Hornbeck 1973; Verry et al. 1983). Other studies in small basins showed little or no change in peak flow with total or partial forest harvest (Hewlett and Hibbert 1967; Rothacher 1973; Harris 1973; Harr et al. 1975; Harr and McCorison 1979; Settegren et al. 1980; and Hornbeck et al. 1993). Burton (1997) suggested that even total harvests on small research basins may not increase peak flow, but partial cuts (25%) on large basins (> 5 000 acres) do (a 66% increase in annual daily maximum discharge).

Usually paired-watershed experiments cannot measure changes in the size of events at a given frequency-of-occurrence because the trees regrow before enough years pass to perform a sound frequency analysis. A substitute evaluation method combines streamflow models with long-term weather records to evaluate static land-use change or high rates of cutting that prolong young forest conditions on large basins. This was done in the Lake States by Lu (1994), who evaluated aspen clearcutting on a calibrated watershed in Minnesota using an 80-year weather record nearby. The results for snowmelt peaks (daily mean flows) illustrate an essential concept for stream stability: clearcutting does not significantly increase the 25- to 100-year flood peak (95% confidence) (Figure 6.7, following page). But, more important to stream stability, the 1.5-year snowmelt (the bankfull flow) is significantly increased by 150%. As Leopold pointed out, the bankfull flow shapes the channel!

Thus, the bankfull flows that are doubled will make a channel cross-section area twice its previous size, assuming similar channel roughness. Channels tend to adjust over time by widening with little change in channel depth. Permanent land-use change from forest to agriculture also causes this. An analysis of the 1.5-year peak flow for clay basins in Minnesota and Wisconsin near the western end of Lake Superior confirms this. On the same land type, peak flow was double from basins with more than 50% agriculture compared to basins dominated by forests (Figure 6.8, page 119).

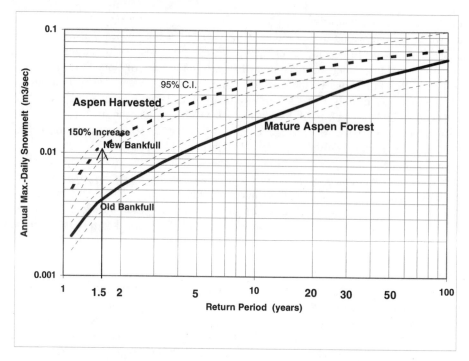

Figure 6.7 The change in the annual maximum daily snowmelt peak flow frequency from a mature aspen watershed to a recently harvested aspen watershed in Minnesota (adapted from Lu 1994).

   High rates of cutting on forested basins that result in 2/3 of the basin in young forest (0-15 years old) or in combination with farm land will also cause this (see Verry 1997 which shows a total range of *both decreased peak flow (at 25% clearing) and increased peak flow response* as the open area exceeds 2/3 of the total basin).   Just as road ditches increase peak flows, drained land (tiled and ditched agricultural or wet land) exceeding 1/3 of the basin will double bankfull flows (Verry 1988). When it comes to channel habitat, it is typically the every-year peak (bankfull flow) that shapes the channel, while floods, sometimes catastrophic, leave the channel and shape the valley. A catastrophic flood may relocate the channel and where mass wasting and debris flows occur in steep, mountain A or moderately steep G channels, mass wasting and debris flows can reshape a channel overnight. Still, the new channel width, depth, and cross-section will represent the bankfull condition.

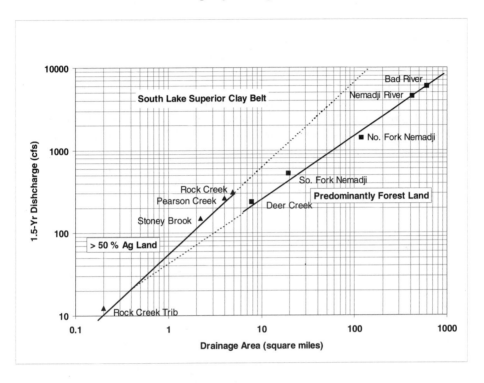

Figure 6.8 The 1.5-year return interval discharge (approximate bankfull discharge) for basins with more than 50% agriculture and basins dominated by forest land use on the south Lake Superior Clay Belt.

Combining Statistics, Deductive Reasoning and Classification

The observations above follow from detailed analyses of paired-watershed experiments and from experiments where levels of a single causal factor are controlled. They also emphasize anthropogenic causes of channel degradation. However, central to the concept of watershed analysis is the need to deduce the difference between climate/geology and land-use causes of channel change. Differences in geologic conditions, (e.g., bedrock basins versus outwash sand basins) or climate changes can foster differences in near-annual peak flow equal in size to land-use-caused changes (Warner 1987).

Richards (1987) offers insight to the use of deductive reasoning in physical geography methods (see also Haines-Young and Petch 1980, 1986; Popper 1959). It requires two equally true statements. The first, is a time- and space-independent natural law. For instance, alluvial rivers are braided if their stream power exceeds 5 watts/ft^2 and meandering if less. The second is a set of initial geographic or historical conditions that is bound by time and space. For instance, the South Platte River, braided before 1850 has experienced a decrease in flood magnitudes because of upstream reservoir development and stream power is now only 0.25 watts/ft^2. The deductive conclusion is: "The South Platte changed from braided to meandering because its stream power was reduced."

The framework is valid, but variation and complexity can make deductive reasoning difficult, yet not less powerful. The stream-power threshold has been shown to vary with sediment size and bank stability, thus a stream-power statement specific to the initial sediment sizes in the South Platte is needed. Further, an increase in irrigation water along the Platte Valley raised riparian water tables leading to more deeply rooted vegetation, more stable banks, and less bank erosion. The deduced conclusion is still valid but needs more information in both the set of conditions and the natural law. Pitty (1979) argued that environmental heterogeneity and complexity should form the basis for a geographical method of explanation. He contrasted the geographical method with the formal use of agricultural statistics where dependent factors are controlled (Fisher 1935). The geographical method uses the variability at the Earth's surface as a source of natural selectors and depends on the exploration for and discovery of localities naturally simplified by the marked presence or absence of controlling variables.

Watershed change analysis employs deductive reasoning through elimination of cause, isolation of one or several causes, observation of the repeated occurrence of both circumstances, and the presence of natural law explanations near threshold levels.

Managing the Basin and Riparian Areas for Stable Channels

Land-use thresholds
The impact of land use change on channel condition can be lessened by keeping young forest and open areas at less than 2/3 of the basin. Changes in channel width caused by peak flow have been observed on steep moraine areas (slopes 1 to 30%) in basins as small as one square mile and on flat outwash or lake plain basins (slopes 1 to 3%) as small as 10 square miles. Drainage networks should be less than 1/3 of the basin and ditched roads less than 15% of the basin. The development of new channels via shoot cut-offs or parallel anastomose channels can be kept within the normal w/d ratios by ensuring that the floodplain always has a vigorous, growing, and deep root system in place. Providing this resistance to channel widening is a fundamental reason for streamside buffers that keep vigorous forests in place.

This does not eliminate forest harvest but argues for establishment and thinning of riparian forests to maintain a vigorous forest cover.

Unpaved roads at stream crossings

Two aspects of road crossings at stream channels can significantly impact stream habitat. First, the use of road surfacing material that contains fine sand can degrade channels by filling in pools downstream of the crossing. This only occurs where the road slope approaching the channel is steep (usually more than 3%). The erosion of the sand is compounded by road maintenance when surface grading pushes the road surface into the stream. Immediately downstream of the road culvert, washed gravel beds that may serve as excellent redds occur. This is caused by the increased velocity at the culvert outlet. However, 100 to 200 feet below the culvert sand deposition begins. Pool-filling sand can occur for distances of ½ to 1 mile. Where road approaches to the culvert are long, the solution is to use road ditch turnouts that lead water onto the forest floor. On rock or gravel roads these turnouts are associated with broad-based dips every 150 ft. Additionally, the road will need to be surfaced with black top or crushed rock (passing a 1 ½ inch sieve, but caught on a ½ inch sieve) about 6 inches deep. Crushed limestone is ideal because fines produced by heavy load traffic are simply dissolved in rain and stream water, and the mixture of rock sizes packs well. In limestone-deficient areas, crushed igneous rock can be substituted. Clean rock with a mixture of sizes varying from the ½ inch to 1 ½ inch size is preferred. This mixture packs the surface better and reduces the road-rumble noise caused with a single, large rock size. Avoid crusher-run or pit-run rock or gravel because it contains too many fines less than ½ inch in size (see Chapter 16 for additional road considerations). Maintain vegetated shoulders with a brush hog rather than scrapping with a grader blade.

Wetland and stream culvert installations

Secondly, the installation of culverts tends to call for either a detailed engineering study or a one-size-fits-all operation for forest road or recreation trail construction. Some simple guides for culvert installation can save money in the long run and protect stream habitats below the culvert. In wetlands, culverts that simply allow for the normal waterflow of the wetland through the road bed (they are not at channel locations) should be at least 2 feet in diameter and must be buried about a foot below the soil surface. The bottom half of the culvert carries the *every-day* flow that runs through the porous, upper horizon of the organic soil (see Hewlett and Kochenderfer 1988; Welch et al., 1995; MN DNR 1999; and Copstead 1997 for details of installation).

At well-defined channels in mineral or organic soils, culverts need to meet three criteria to allow fish passage and maintain road bed integrity (Baker and Votapka 1990; Parola 1987). First, they should be laid at the channel slope and be buried 1/6 of their vertical depth to prevent vertical jumps that cannot be negotiated by fish. Second, they should be sized so that *in-culvert* water velocities do not exceed the short-term swimming speed of fish sustainable over the length of the culvert. Finally, they need to be sized so that design flows do not overtop the road and wash the culvert and road away. Where road surfaces are paved with black top, concrete, or 6 inches of 1½ inch (minus) crushed rock, and road embankments are well-vegetated, overtopping can be incorporated into culvert design. Squash culverts (arched pipe) are the most efficient for handling water flows and require less vertical space than round culverts. A squash culvert can be sized (for a wide range of channel slopes) by simply matching the average diameter (vertical plus horizontal/2) to the bankfull channel width. Similarly sized, round culverts will also handle the flow of small streams up to 2 feet wide, but for streams 3 to 6 feet wide, add one foot wider than the bankfull width to the culvert size. Table 6.4 details culvert sizing recommendations for both warm- and cold-water stream fisheries.

Table 6.4 Fish-friendly, round culverts for 30-ft-wide roads with 3:1 side slopes and 2 feet of road base above the culvert inlet.

		Bankfull stream width in feet					
		1 ½	2	3	4	5	6
		For warm water streams					
Corrugated Steel Pipe	Diameter ft.	1 ½	2	4	5	6	7
	Max. Channel Slope %	1.25	1.25	1	0.7	0.5	0.4
		For cold water streams					
Corrugated Steel Pipe	Diameter ft.	1 ½	2	4	5	5	6
	Max. Channel Slope %	2.5	2.5	1.9	1.9	1.9	1.9
		For either warm- or cold-water streams					
Arched Pipe	Ave. Diameter ft.		2	3	4	5	6
	Horizontal x Vert. inches		28x20	42x29	57x38	71x47	83x57

Culverts installed according to Table 6.4 will pass the 10- and 25-year storm with minimal flooding at the culvert inlet, and they will pass the 50-year storm with 2 feet of flooding above the culvert inlet. Thus, road fills must be at least 2 feet above the culvert top at the inlet. Always align culverts parallel with the floodplain to prevent excessive erosion of valley side slopes. Do not dig channel cut-offs across existing stream bends to ensure culvert alignment with the channel because this will induce stream head-cutting. Consider using box culverts, bottomless arches, or bridges for streams over 6 feet wide, or use double culverts for streams 8, 10, or 12 feet wide.

When road fills (and culverts) repeatedly wash out, consider placing a second "relief culvert" of the same size and at the floodplain elevation to relieve extreme pressures against tall fills during floods. Place the relief culvert on the floodplain midway between the channel and valley edge on the same slope as the floodplain (Figure 6.9).

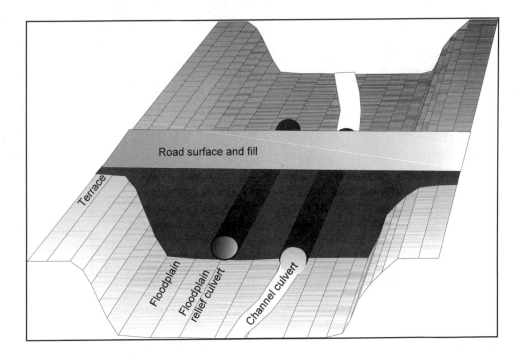

Figure 6.9 Where deep road fills block floodplain flow and culverts and roads wash out, consider placing a second "floodplain relief" culvert at the floodplain elevation midway between the channel and terrace slope.

Sandy Verry

Sandy Verry

Sizing of road culverts to match their diameter to the stream width will prevent stream channels below the culvert from becoming over-wide and shallow. Over-wide and shallow channels have reduced fish habitat during low flows in summer and winter, and often their pools fill with the sand eroded from the channel sides. Additonally, water velocities through undersized culverts at bankfull flow (typically during spawning runs) will exceed 5 feet per second and prevent fish migration. Instead of passing through to spawning areas upstream, the fish absorb their eggs while waiting. The 10 foot channel above the 3 foot culvert (top picture) becomes a 20 foot channel below the culvert (lower picture).

Chapter 7

Particulate Organic Contributions from Forests to Streams: Debris isn't So Bad

C. Andrew Dolloff and Jackson R. Webster

"Say you are in the country; in some high land of lakes. Take almost any path you please, and ten to one it carries you down in a dale, and leaves you there in a pool by a stream. There is magic in it. Let the most absent-minded of men be plunged in his deepest reveries — stand that man on his legs, set his feet a-going, and he will infallibly lead you to water... Yes, as everyone knows, meditation and water are wedded for ever."   Herman Melville, 1851— Moby Dick

As we meditate on the management of stream riparian areas, it is clear that the input of "debris" from terrestrial plants falling into streams is one of the most significant processes occurring at the interface of terrestrial and stream ecosystems. Organic matter — leaves, twigs, branches, and whole trees — provides energy, nutrients, and structure to streams flowing through forests. A host of vertebrate and invertebrate animals has adapted to life in flowing waters and depends on leaves and wood for food and habitat. Accumulations of leaves and wood also create refuges from the extremes of drought and flood and modify the downstream movement of sediment.

Despite all that we know about the importance of organic matter in streams, all too often wood and leaves in streams have been viewed as a liability at worst and a nuisance at best. Even the terms we use to describe it — debris, for example, — suggest something cast off or discarded. Although excessive amounts of organic matter have negative impacts in streams, such as lowering dissolved oxygen (Schneller 1955; Larimore et al. 1959; Hicks et al. 1991), buildup of toxic substances (Buchanan et al. 1976), and blocking fish migration (Baker 1979), most problems are local rather than symptomatic of a underlying pathology. All of these reasons aside, the main reason for our aversion to wood and leaves in streams is far more basic: it just plain looks bad (Dolloff 1994)! Peter Marshall's parable of the "Keeper of the

125

Spring" (Figure 7.1) illustrates the most common of many misconceptions about wood and leaves in streams. For swans a swimming, irrigation, hydropower, and pretty views, perhaps clean streams are desirable. But for diverse, productive invertebrates and fish, for preservation of natural sediment and water regimes, and for overall stream health terrestrial plant debris is not only desirable but essential.

THE KEEPER OF THE SPRING

This is the story of the keeper of the spring. He lived high in the Alps above an Austrian town and had been hired by the town council to clear debris from the mountain springs that fed the stream that flowed through the town. The man did his work well and the village prospered. Graceful swans floated in the stream. The surrounding countryside was irrigated. Several mills used the water for power. Restaurants flourished for townspeople and for a growing number of tourists.

Years went by. One evening at the town council meeting someone questioned the money being paid to the keeper of the spring. No one seemed to know who he was or even if he was still on the job high up in the mountains. Before the evening was over, the council decided to dispense with the old man's services.

Weeks went by and nothing seemed to change. Then autumn came. The trees began to shed their leaves. Branches broke and fell into the pools high up in the mountains. Down below the villagers began to notice the water becoming darker. A foul odor appeared. The swans disappeared. Also, the tourists. Soon disease spread through the town.

When the town council reassembled, they realized that they had made a costly error. They found the old keeper of the spring and hired him back again. Within a few weeks, the stream cleared up, and life returned to the village as they had known it before.

Modified from Peter Marshall's "Mr. Jones: Meet the Master."

Figure 7.1 Our notion of a healthy stream has been influenced by the popular media, including literature.

Our task in this chapter is to outline what we know about the functions and values of leaves and wood in streams. In doing so we hope not only to dispel the common misconception that wood debris in streams is undesirable, but also to instill the concept of organic matter as an asset to be husbanded.

Definitions

Organic material that falls into a stream from the surrounding land is known as allochthonous input. Combined with the instream accumulation of primary production by algae and vascular plants — the autochthonous input — allochthonous input provides the support system for all instream life. Allochthonous inputs span a broad range of sizes; from leaf fragments to branches and entire trees. Although the size range of these inputs is continuous, individual pieces typically are classified by size and function and grouped for convenience. Fine particulate organic material (FPOM) encompasses all particles that will pass through a .04-inch (1.0-mm) fine mesh sieve. The largest pieces of wood, known as large or coarse woody debris (CWD), typically are greater than three feet in length and at least 2 to 4 inches in diameter. In between is CPOM or coarse particulate organic material.

Input Mechanisms and Loads

Leaves and wood are transported into streams by various mechanisms, ranging from the predictable fall of leaves in the autumn to the catastrophic input of major storms. Factors such as species composition, forest health, size and type of stream channel, and land-use history influence the input rate and total loading of organic materials.

Inputs of allochthonous matter should decrease in importance as stream size increases (Vannote et al. 1980). Although there have been few measurements of allochthonous inputs to larger streams, several studies have confirmed this prediction (Cummins et al. 1983, Conners and Naiman 1984). In large streams with well-developed floodplains, allochthonous inputs initially fall to the floodplain but may be washed into the river during floods. This interaction between rivers and floodplain vegetation is not well understood, but Cuffney (1988) estimated that the flood plain was the major source of organic matter to the Ogeechee River in Georgia.

In the absence of a catastrophic event, the weight of leaf material that enters a stream each year frequently exceeds that of wood (Table 7.1). In a summary of other studies of streams in the Eastern United States, Webster et al. (1995) found that direct fall of leaves into upland streams ranges from 1800 to 4800 lbs/acre/y, averaging 3100 lbs/acre/y. The average is somewhat higher for floodplain streams and streams draining wetlands: 5200 lbs/acre/y. Leaf inputs account for about 58% of total inputs. When non-leaf materials are included, total inputs average 5300 lbs/acre/y to upland streams and 6100 lbs/acre/y to floodplain and wetland streams. These numbers become even higher if we include lateral inputs, that is, the allochthonous material that blows or rolls down the banks into streams. With lateral inputs included, leaf inputs to upland streams average 4700 lbs/acre/y and total inputs average 6000 lbs/acre/y. In deciduous forests, annual variation in these inputs is not great. While catastrophic events may cause large inputs at unusual times of the year, annual leaf input will not be greatly affected.

Wood inputs to streams in the Eastern United States have not been extensively measured. Webster et al. (1995) summarized 13 studies, showing an average of 1300 lbs/acre/y direct wood input to upland streams and 1100 lbs/acre/y to floodplains. Lateral inputs averaged about 19% of total wood inputs. Unlike leaf inputs, which are predictable in streams draining deciduous forests, wood inputs are highly variable in both space and time. One winter ice storm accounted for 80% of the annual wood input to the Sangamon River in Illinois, and wood input the year of the ice storm was three times that of the following year (Peterson and Rolfe 1982).

Table 7.1 Allochthonous inputs and standing crops of allochthonous material in two undisturbed streams at Coweeta Hydrologic Laboratory, Macon county, North Carolina. FBOM is fine (< .04 inch benthic organic matter. CBOM is coarse (> .04 inch) benthic organic matter. Small wood is 0.4 to 2 inches diameter and large wood is > 2 inches. Grady Branch is a first-order and Hugh White Creek is a second-order stream. Input data from Webster et al. (1990) and standing crops from Golladay et al. (1989).

	Grady Branch	Hugh White Creek
Litter fall (lbs/acre/y)		
Leaves	4303	3707
Wood	2317	942
Lateral input (lbs/acre/y)		
Leaves	1218	792
Wood	92	81
Standing crops (lbs/acre/y)		
FBOM	1311	1481
CBOM	2177	1900
Small wood	2677	2784
Large wood	40,854	45,816

Logging, especially when all trees including those adjacent to streambanks are removed, has a profound influence on the loading of large wood. Webster et al. (1992) presented a hypothetical model for wood loading following logging (Figure 7.2). Provided that slash and wood deposited by natural processes are not removed, wood loads should be highest immediately following logging.

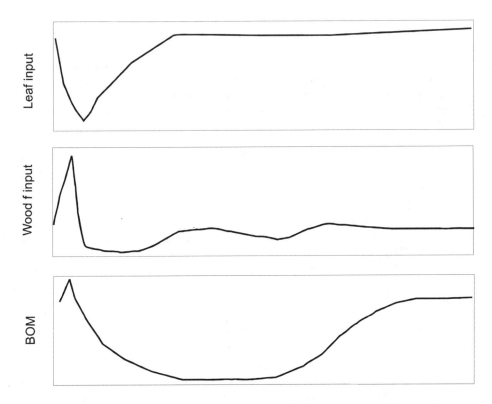

Figure 7.2 Hypothetical trends in allochthonous inputs and stream benthic organic matter (BOM) after riparian logging (modified from Hedin et al. 1988 and Webster et al. 1992).

These high loads should persist for 20 to 50 years before declining to lower levels. Loads should then gradually increase over many years as the riparian forest matures and provides a source of large wood. This last process may require centuries, depending on growth rates of riparian trees.

Wood inputs into wilderness streams provide the most reliable estimates of wood loading under undisturbed conditions as long as the wilderness encompasses true old-growth forest. Flebbe and Dolloff (1995) inventoried large wood in three North Carolina watersheds managed as wilderness. Right Fork of Raven Fork and Little Santeetlah drain true wilderness (never harvested or homesteaded), whereas about 80 years had passed since the second-growth forest surrounding Lost Cove had become established. Loadings of large wood in Right Fork (416 pieces/mi) and Little Santeetlah (291 pieces/mi) were at least three times greater than in Lost Cove (85/mi) which, along with logging, had experienced several major floods over the preceding 80 years.

Despite its obvious importance, however, logging history is but one of a host of factors that determine how much wood and what kind of wood a stream has. Insects, diseases, and storms, acting either singly or in combination, also influence wood loadings. Hedman et al. (1996) determined that the load of large wood was highly variable in 11 riparian forest-stream systems representing a 300-year range in the southern Appalachians. They attributed the lack of statistical difference among systems to particularly high variability among mid-successional systems. Wood loads in these systems were dominated by decay-resistant American chestnut and eastern hemlock derived from the pre-logging riparian forest. Although extirpated from the extant forest by the chestnut blight, American chestnut probably was a major component of the wood load in many eastern streams. Hedman et al. (1996) attributed the high variability in chestnut loads to the accelerated mortality and input of blight-killed trees which was offset in some systems by salvage logging.

Leaf inputs drop to nearly zero but recover fairly rapidly as trees regrow. Inputs of logging debris initially increase wood inputs and BOM, but inputs rapidly decline and remain at low levels. BOM declines more slowly as residual wood and debris dams decay and disintegrate. BOM remains low even after leaf input returns to near normal levels because of the lack of retention structures (wood and debris dams). In time, small wood is recruited as small early- and mid-successional trees die, but inputs of large wood, capable of forming debris dams and jams, don't occur for many years after logging. The time scale may range from about 50 to more than 200 years, depending on the rate of forest regrowth.

More dramatically, the high winds of hurricanes and tornadoes cause extremely high loadings. In 4 hours in 1989, Hurricane Hugo deposited more than a normal year's worth of leaves and woody debris on the forest floor and streams in Puerto Rico (Covich et al. 1991). Large accumulations of leaf litter were retained in debris dams in stream channels and remained in place for more than 8 months. Even after it was downgraded to a tropical depression, Hurricane Hugo more than doubled the load of large wood, from 76 to 186 pieces per kilometer, in streams of the Basin-Cove watershed, a tributary system of North Carolina's Yadkin River (Dolloff et al. 1994). Most of this input consisted of small trees and branches from the less than 60-year-old riparian forest. Because of its small size and susceptibility to rapid decay and transport out of the stream channel, this wood was unlikely to have long-term benefits for stream biota.

Floods, if severe enough, can produce large amounts of wood. A recent (1995) flood caused debris avalanches in several Shenandoah National Park watersheds. Torrents of water, rock, soil, and trees carved wide swaths from riparian areas many times the width of the stream channels. Beginning at the uppermost debris avalanche and continuing downstream beyond the Park boundary, the boulder-packed stream corridors were punctuated by huge piles of debris, composed of rocks, mud, and trees of all sizes, located at bends and other places where the flood's energy was dissipated.

Functions of Allochthonous Matter

Organic Matter for Trophic Processes

Although many benthic invertebrates feed on allochthonous organic matter (e.g., Hynes 1970), actually demonstrating the importance of allochthonous matter in streams is problematic. Fisher and Likens (1973) and many others demonstrated that allochthonous inputs can greatly exceed autochthonous production, but because of the low nutrient content of detritus compared to algae, the importance of allochthonous material was still unresolved. Recently, however, Wallace et al. (1997) eliminated allochthonous inputs to a small stream and documented the subsequent decline in invertebrate production. Increasing width in the downstream direction results in lower allochthonous inputs and allows more light to the stream, increasing autochthonous primary production (Vannote et al. 1980). However, even in larger streams the importance of allochthonous matter should not be underestimated. While leaves and sticks are usually retained and broken down near where they enter streams (e.g., Webster et al. 1994a and b), the initial consumption and use of the organic material is low. The unused material, now converted to fine particles, may be transported from small, headwater streams downstream into large rivers where it is ultimately consumed by invertebrates or assimilated by aquatic microbes and converted to CO_2.

All vegetative materials, including leaves, fruits and flowers, bark, roots, and boles ultimately contribute to the pool of FPOM and become available for further processing by microbes and macroinvertebrates. The proportion of fine organic matter derived from the decay and fragmentation of large wood is largely unknown in eastern streams but may be significant. In Oregon's western Cascades, Ward and Aumen (1986) estimated that instream wood processing could yield several times the fine material generated by needles and leaves. The amount of fine material from wood depends on the amount, type, and size of wood available for processing; small pieces of softwoods, with their relatively high surface-to-volume ratio, tend to disappear faster than large pieces of hardwood.

Large Wood

Before they are fragmented into FPOM or transported downstream, large pieces of wood play major roles in the habitats of invertebrates and fish (Bisson et al. 1987). Water flow around large wood forms pools and encourages scour from stream banks and bottoms (Shirvell 1990; Cherry and Beschta 1989; and many others). In addition to basic pool structure, wood provides complexity. Root wads, large branches, and multi-stem tree trunks partition the water column, providing cover and a measure of isolation for many species. Pools and other areas of slack flow created by wood provide refuges for obligate aquatic organisms during times of extreme high or low flows (Sedell et al. 1990).

Large pieces of wood influence flow velocity, channel shape, and sediment storage and routing (Harmon et al. 1986; Bisson et al. 1987). These individual pieces often accumulate to form the matrix of debris dams which are lined with leaves and finer particles of organic matter. The stairstep profile created by woody debris dams dissipates much of the energy in small, high-gradient streams (Heede 1972). For example, Bilby and Likens (1980) found that about 50% of the gradient drop in first- and second-order streams at Hubbard Brook occurred over debris dams (Table 7.2).

Table 7.2 Percent drop associated with sediment or rock and debris dams in Hubbard Brook (Bilby and Likens (1980). With permission.

Stream Order	% Drop Caused By		
	Organic Dams	Inorganic Dams	All Falls
First	52	16.3	68
Second	46	28.5	75
Third	10	28.0	38

Debris dams also increase water depth and decrease velocity, resulting in greater transit times for water, solutes, and suspended material (Trotter 1990; Gregory 1992; Wallace et al. 1995). A current study (Wallace et al. 1997) is also suggesting that leaves in streams also function in this way (J. R. Webster, unpublished). Woody debris dams have been shown to be extremely important in the retention of both particulate organic and inorganic matter (e.g., Bilby and Likens 1980; Mosley 1981; Speaker et al. 1984; Smock et al. 1989; Trotter 1990; Smith et al. 1993). These physical stream changes associated with large particulate matter indirectly affect stream communities. By adding logs to a stream at Coweeta Hydrologic Laboratory, Wallace et al. (1995) demonstrated major changes in the composition and production of the macroinvertebrate assemblage. Similarly, less than one year after Hilderbrand et al. (1997) added logs to two Virginia streams, they noted changes in macroinvertebrate assemblages associated with changes in habitat composition. In another Coweeta study, Tank et al. (Tank and Webster in press a; Tank et al. in press b) showed that elimination of leaf inputs to a stream decreased nutrient retention and indirectly affected microbial processes in the stream.

Impact of Land Use

When wood falls, blows, or rolls into a stream, particularly as a result of a storm or some other catastrophic event, the first reaction of most people and many governmental agencies is to pull it out. Our prejudice against wood in the water has caused us to anticipate the supposedly negative consequences of "allowing" wood into streams and rivers. The first reaction of most citizens and governments following a major influx of wood by floods or storms is to mobilize stream cleanup crews. Regulations and Best Management Practice in a number of states require that any wood entering streams as a result of timber harvest or other silvicultural activities be removed soon (Chamberlin et al. 1991; Hicks et al. 1991). These attitudes towards debris are not a product of recent ecological thinking but rather reflect the perspectives and practices of our forebears (Williams 1989; Maser and Sedell 1994; Whitney 1994; Verry and Dolloff Chapter 1), who viewed accumulations of debris as "unhealthy" (Figure 7.1.) or as obstacles to efficient movement of water and vessels. During the last 200 years, woody debris has been removed from nearly every major river in the continental United States (Sedell and Luchessa 1982). Sedell et al. (1982), Sedell and Froggatt (1984), and Triska (1984) documented the extensive geomorphological and habitat changes of rivers subject to woody debris removal. These and other studies conducted during the last half of the 20th century have caused us to revise our thinking about the effects of land use and the role of wood in water.

Aside from a complete loss of riparian forests, such as occurs when land is developed for highways, housing, or industry, the most obvious cause of woody input loss to forest streams is wood harvest. Following logging, organic inputs are greatly reduced (Webster and Waide 1982) although, with rapid regrowth of vegetation, leaf inputs may return to near pre-logging levels in 5 to 10 y (Webster et al. 1988, 1990). However, the composition of leaf inputs may remain altered for many years (Webster et al. 1983, 1990). Wood inputs are more drastically influenced (Likens and Bilby 1982; Webster et al. 1992). A one-time input of logging slash may be followed by many years of almost no wood input. After several years, self-thinning and competition may result in limbs and small trees dying and falling into the stream, but it may be hundreds of years before significant numbers of large trees begin to die and fall into the stream. Removing all trees down to the streambanks may result in a large one-time input of leaves (depending on the time of year) and large wood, which in time may be beneficial. But, in practice, much of the latter typically is removed either as part of the timber harvest or because of regulations requiring the removal of logging debris. Unfortunately, those who remove the debris may not be aware of the value and role of wood in the stream system and may accomplish far more than their assigned task when they remove all wood, regardless of how it got into the channel. Habitat damage from such removal or salvage may be long-lasting, resulting in changes in species distribution and fish production (Elliott 1986; Dolloff 1986). Particularly where riparian soils consist of unconsolidated or highly erodible sediments, pulling woody material out of streams will likely destabilize streambanks and channels and accelerate erosion.

Other disturbances to allochthonous inputs include defoliating insects such as gypsy moths. Severe and repeated defoliations kill trees and change the composition of riparian forests. Even though they may not kill trees, defoliations can drastically alter many instream processes. Most gypsy moth defoliations occur at the time of leaf-out in spring, resulting in large inputs or leaf fragments and frass, which may not be fully used by stream invertebrates adapted to autumn inputs. Less well described is the interaction of defoliation and water chemistry, particularly in watersheds that have been impacted by acid precipitation. In some areas, concentrations of nitrate in stream waters have increased dramatically following forest defoliation. Although nitrogen (N) compounds contribute to the acidity of precipitation, surface waters associated with forested upland watersheds typically have negligible nitrate concentrations due to plant and microbial use of nitrogen as a nutrient. However, defoliation disrupts the normally tight cycling of N and allows nitrate, an acid-anion, to "leak" from watersheds (Swank et al. 1981). Observed effects include increased frequency and severity.

Similar disruptions of normal inputs are caused by diseases such as chestnut blight, various forms of air pollution, and changes in tree species composition. In the Eastern United States, many forests have been converted from multiple-species deciduous forests to faster growing pines. The loss of the normal diversity of leaf species falling into the stream may severely disrupt the benthic invertebrate assemblage where various species are adapted to different types of leaf material (Cummins et al. 1989). For example, Woodall and Wallace (1972) found lower total weights of benthic invertebrates in a stream draining a white pine plantation than in a stream in a mixed deciduous forest.

Frequently Asked Questions

In this chapter we have only briefly reviewed the importance of organic materials in stream ecosystems. Because the intent of this book is to provide practical information for managers, we have attempted to answer some of the most frequently asked questions about the management of woody inputs to stream systems. Note: the views expressed reflect the biases of the authors and should be considered accordingly because each will change as new information becomes available.

OK, I'm convinced that organic debris — leaves, twigs, branches, and even whole trees — plays an important role in stream ecosystems. Now what? I need to know:

1.   How do I balance the seasonal needs for leaves and small woody materials with the long-term need for large wood? Do I need to worry about that?

> The answer to this and most questions begins with, "It depends." In this case, "it depends" on the goals of management (desired condition) for both the riparian forest and the stream system: the composition of the riparian forest and the specific activities planned for the streamside forest. Managers first need to ask if the area under consideration meets or exceeds desired conditions for such attributes as species composition and age-class distribution of the forest and pool: riffle ratio or number of pieces of large wood per mile for instream habitat. In forests that meet the goals and have tree and understory species that represent the area, little management may be necessary. In forests characterized by a few species that have low potential for providing large wood or where exotics have become established, some manipulation of the riparian vegetation may be desirable. For example, it may be desirable to remove (by mechanical means or fire) dense thickets of rhododendron which in monoculture provides neither high-quality detritus nor long-term potential for large wood recruitment.

2.   How much is enough? What sizes are appropriate?

> The simple answer is no one knows, or more accurately, no one has reported examples of streams that, at least from an ecological viewpoint, are so overloaded with wood that natural processes have been compromised. But, of course, where a catastrophic event has deposited large amounts of material directly upstream of a bridge, culvert, or some other structure, we could conclude that the system was indeed overloaded. All sizes of wood are desirable, but in general, the larger the better, both to promote stability (large pieces are more resistant to being moved) and persistence (large pieces have slower rates of fragmentation and decay).

3.   Can I (should I) control the amount of both small and large wood that enters my streams?

> To a limited extent it is possible, although costly, to control the amount of wood that enters a stream. In the context of ecosystem management, provision for recruitment of large wood can be incorporated as a component of other strategies such as gap management. Perhaps more importantly, because most major inputs of organic materials tend to be associated with storms and other catastrophic events, we need to be ready to "manage" the aftermath of disasters by carefully weighing the costs and benefits of stream cleanups; for example, removing debris jams that threaten buildings or roads but limiting other instream work to that necessary to maintain infrastructure such as water intakes.

4.   What tree species should I be managing for in my riparian areas?

      In general, management should be based on native species typical of riparian areas
      in the region. Much research is currently underway to address this question and to
      develop specific silvicultural prescriptions for "designer" riparian areas.

5.   How do I get organic materials into (and keep them in!) a stream?

      Simply putting leaves into a stream will not maintain or restore natural functions.
      Managers need to consider synergisms and feedback mechanisms. In most
      situations, the most appropriate method for managing organic inputs may be to
      manage the riparian forest to enhance natural inputs. On the other hand, there has
      been no shortage of creative solutions to the problem of how to get wood into
      streams, ranging from directional felling by chain saw or explosives to direct input
      using heavy equipment. Where recreational fishing or endangered species are a
      major consideration, we need to distinguish the passive riparian input strategy from
      the strategy of active habitat improvement. In streams where biologists have
      determined that populations are limited by habitat availability and quality, and where
      public interest is high, the placement of habitat structures such as k-dams, log weirs,
      and "lunker structures" can be effective (Hunter 1991). But habitat improvement
      is expensive and can be applied only to limited sections of a few streams.
      Additionally, recent studies have questioned the suitability and benefits of many
      habitat improvement projects (Frissel and Nawa 1992). While both active and
      passive strategies have their place, the passive strategy has the most potential for
      cost-effective, long-term, widespread benefits.

6.   Can I salvage dead and down trees in a riparian area?

      Meet objectives for riparian structure and function first! In general, it is probably
      better not to salvage trees because of the potential for damaging the integrity of
      streambanks. Depending on the characteristics of the floodplain, wood may be
      transported long distances from upland sites into stream channels during storms.
      Also, amphibians and other animals benefit from wood debris located on floodplains
      (see Chapter 10).

7.   If I build it, will they come? Will additions of large wood enhance the chances for species persistence or recovery?

> For some species, probably yes; particularly macroinvertebrates and salmonids. But for other fish and invertebrates such as mussels, it depends on their specific habitat requirements. Recovery of extirpated populations is more difficult if only habitat niches are provided, a necessary but not singular condition for success. In addition to questions of genetic integrity, presence of other species, and other considerations, there must be a source of recruitment, either by natural means from adjacent habitats or by artificial means provided by managers.

8.   Is control of exotic insect pests such as gypsy moth desirable from the standpoint of stream ecodynamics?

> This question usually arises when considering more comprehensive plans for pest management. Managers of aquatic systems need to consider such things as background water chemistry, the probability of changes in water quality, the effect of the pesticide on non-target macroinvertebrates, and the effect on stream water temperature if the riparian forest is allowed to defoliate.

9.   Should debris-removal BMPs be modified or eliminated?

> Turn it around — if a tree falls into a stream during logging, require that it not be removed without consulting a biologist or hydrologist. There would then be two new outcomes: one, fewer trees will "accidentally" be felled into streams; two, if they do fall in, good! This does not, however, suggest that streams can become dumping grounds for slash which tends to be unstable and easily transported in all but the smallest streams. And, of course, debris may be removed from areas immediately upstream of bridges, dams, and other structures.

10.   Why not simply stay out of riparian areas and allow natural succession to occur?

> To someone interested in preserving the status quo, this sounds like an attractive strategy; just say 'no' to any activity that could change the content or character of riparian areas. The problem is that change will occur and disrupt the status quo whether we do something or not. For some riparian areas, the prediction for the outcome of unmanaged change may be positive, resulting in the creation or maintenance of desirable attributes. The structure and function of many other riparian areas, however, in particular those that bear the legacy of land abuses, can more rapidly be restored by judicious application of the many tools at a manager's

disposal. But despite the recently accumulated wealth of scientific knowledge and practical experience with riparian areas, "to manage or not to manage" remains a social — rather than an ecological — question.

Bob Hollingsworth

Andy Dolloff

Large woody debris is a must for stream function in this bedrock system in northern Minnesota (above). In very large systems it is needed too, but some (in Virginia) would rather avoid it (bottom).

Chapter 8

Bird and Mammal Habitat
in Riparian Areas

Richard M. DeGraaf and Mariko Yamasaki

"Down a narrow pass they wandered, where a brooklet led them onward, where the trail of deer and bison marked the soft mud on the margin..."
Longfellow — Song of Hiawatha

Fifty-two species of birds and mammals are strongly associated with riparian areas in the Northeastern United States. There are profound regional differences in the current and potential conditions of riparian and adjacent upland habitats across this region. Three major factors that influence wildlife use of these areas are: (1) the degree of connectivity and extent of riparian forest, (2) water regime, and (3) key vegetation structures that are important to birds and mammals. We emphasize specific characteristics and how they are used by representative species and species-groups.

Although complete protection has been commonly used to "manage" riparian habitats, many riparian areas are amenable to active management. In fact, optimizing wildlife habitat quality over time requires active vegetation management; more species will occupy managed than unmanaged riparian forests. Table 8.1 lists several riparian management zone widths used by northern New England states and national forests.

Effects of timber harvesting in riparian areas are presented as effects on forest structure. Key considerations include the effects on softwood overstory composition, high perch/nest sites, cavity trees, understory shrub layers, herbaceous/wetland inclusions, and dead/down woody debris. Considering these factors and the applicable management zones, we present land management recommendations at landscape, stand, and within-stand scales to guide integrated forest and wildlife habitat management in riparian areas.

The continental Eastern United States is a diverse region comprised of various ecological conditions (McNab and Avers 1994; Albert 1995). Forest types, physical characteristics, and wildlife species of riparian habitats overlap throughout this area. The proportions of forest

139

Table 8.1 Guidance for riparian area widths and suggested management practices in northern New England

State or NF	Stream Order Classification	Associated Characteristics	Minimum Total Distance on Each Side of Stream	Management Practices
ME [a]	Intermittent, 1st, 2nd, & 3rd	Drains < 300 ac	--	Maintain stream shading; prevent soil erosion and sedimentation.
		Drains < 50 mi^2	100 ft	Leave first 25 ft uncut or apply light selection cut; in remaining width, use single tree or group selection, removing < 40% of the volume > 6 inches d.b.h. in 10 years.
	4th and greater	Drains > 50 mi^2	250 ft, 330 ft preferable	Leave first 100 ft uncut or apply light selection cut; in the remaining width, use single-tree or group selection, removing < 40% of the volume > 6 inches d.b.h. in 10 years.
	Lakes, ponds, coastal wetlands		250 – 330 ft	Leave first 75 ft uncut or apply light selection cut; in remaining width, use single-tree or group selection, removing < 40% of the volume > 6 inches d.b.h. in 10 years.
NH [b]	Intermittent	–	–	–
	1st, 2nd, & 3rd		50 ft	Maintain 50% of pre-harvest basal area within 50 ft.
	4th and greater		150 ft	Maintain 50% of pre-harvest basal area within 150 ft.
	Ponds < 10 ac		50 ft	Maintain 50% of pre-harvest basal area within 50 ft.
	Ponds > 10 ac		150 ft	Maintain 50% of pre-harvest basal area within 150 ft.
NH [c]	Intermittent		100 ft	–
	1st, 2nd, & 3rd		300 ft	Leave first 25 ft uncut.
	4th and greater		600 ft	Leave first 25 ft uncut.
	Ponds < 10 ac		100 ft	–
	Ponds > 10 ac		300 ft	Leave first 25 ft uncut.
NH [c]	Wetlands <10ac	nonforested	100 ft	–

Table 8.1 Continued

State or NF	Stream Order Classification	Associated Characteristics	Minimum Total Distance on Each Side of Stream	Management Practices
NH	Wetlands > 10 ac	nonforested	300 ft	Leave first 25 ft uncut.
VT[d]	All state waters		50 ft + 20 ft for each additional 10% slope	Maintain shade and stream bank stability.
VT[e]	Streams < 10 ft wide	seasonal or permanent	25 ft + 20 ft for each additional 10% slope	
	Streams > 10 ft wide	also rivers, lakes, ponds	50 ft + 20 ft for each additional 10% slope	
WMNF[f]	Upper perennial streams (steep)	in v-shaped valleys	50 ft + (4 x percent slope)	
	Transition streams	entrenched in flat-floored valleys	50 ft + (2 x percent slope)	Increase slope factor to 4 x for municipal watersheds.
	Lower perennial streams	bottomland streams	Floodplain to top of 1st terrace or 50 ft	
	Lakes, ponds, wetlands	no minimum size	Site-by-site determination	

[a]Land Use Regulation Commission regulations in the unorganized townships of Maine (in Elliott 1988); [b]New Hampshire Basal Area Law (in Cullen 1996); [c]New Hampshire Recomm. Voluntary Forest Mgt. Practices (in NH Div. Forests and Lands, DRED and SPNHF 1997); [d]Vermont Acceptable Management Practices for Maintaining Water Quality (in VT DFPR 1987); [e]Vermont Department of Fish and Wildlife Guidelines (in Regan and Anderson 1995); [f]White Mountain National Forest guidelines (in USDA 1986).

land and all other land (e.g., cropland, urban/industrial, and residential/suburban) (Table 8.2) vary greatly from east to west and north to south within the Northeastern United States (Powell et al. 1993). In the agricultural Midwest, riparian forests generally exist as narrow ecotones. In contrast, the riparian forests of the northern Lake States and mountainous regions of the East generally blend seamlessly into the adjacent upland forest.

Management options vary greatly with region and landscape character. In the fertile agricultural areas of the Midwest, bottomland hardwood forests, especially of large sawtimber, have been greatly reduced (P. Hamel, pers. comm.). Habitats for cerulean warbler (Hamel 1992a), Swainson's warbler, and prothonotary warbler are limited and fragmented (see scientific names in Table 8.5). Also, there are limited opportunities to provide habitat for large carnivores, such as black bears, in these landscapes (Hunter et al. [in prep.]).

Table 8.2 Forest coverage by region and state, 1992 (from Powell et al. 1993)

Region	State	Total Land	Total Forest	Other Land	Forest	Other
		-------------------- Thousand acres --------------------				Percent
NE	ME	19,753	17,533	2220	88.76	11.24
	NH	5740	4981	759	86.78	13.22
	VT	5920	4538	1381	76.66	23.34
	MA	5016	3203	1813	63.86	36.14
	CT	3101	1819	1283	58.66	41.34
	RI	669	401	268	59.94	40.06
	NY	30,223	18,713	11,510	61.92	38.08
	PA	28,685	16,969	11,715	59.16	40.84
	WV	15,415	12,128	3288	78.68	21.32
	MD	6295	2700	3595	42.89	57.11
	DE	1251	389	862	31.10	68.90
Total		126,816	85,380	41,437	67.33	32.67
NC	MI	36,358	18,253	18,105	50.20	49.80
	WI	34,761	15,513	19,248	44.63	55.37
	MN	50,955	16,718	34,237	32.81	67.19
	MO	44,095	14,007	30,088	31.77	68.23
	OH	26,210	7863	18,347	30.00	70.00
	IN	22,957	4439	18,518	19.34	80.66
	IL	35,580	4266	31,314	11.99	88.01
Total		286,674	83,108	203,566	28.99	71.01

In smaller, narrow Midwestern forests, cowbird parasitism greatly influences the nesting success of several Neotropical migratory species (Gates and Gysel 1978; Chasko and Gates 1982; Thompson et al. 1992). Cowbird parasitism is not a serious regional concern in extensive northeastern forests (Askins 1994; DeGraaf 1995; Hagan et al. 1997; Yamasaki et al. in press). Also, in extensive New England forests, red foxes establish home-ranges between those of coyotes (Harrison et al. 1989). Red fox may persist in this landscape because riparian areas delineate and limit coyote territories (Noble 1993). Bobcats, on the other hand, do not use riparian habitats differently than uplands (Noble 1993). Black bears use softwood cover types found in larger riparian corridors in Maine (Schooley 1990; Vander Haegen and DeGraaf 1996).

In addition to important differences between the Northeast and Midwest, riparian habitat values also differ across north-south gradients. For example, the importance of windfirm, coniferous, winter thermal cover for white-tailed deer in the Northern United States is well documented (Banasiak 1961; Verme 1965; Weber et al. 1983) but is less important south of New England and the Lake States. The amount and variety of hard mast species (e.g., oaks and hickories) found in riparian areas decreases farther north along the same latitudinal gradient (Powell et al. 1993).

Water regime has a major influence on riparian wildlife communities. Streams and rivers, ponds and lakes, emergent wetlands, shrub swamps, and wooded swamps often support different bird and mammal species in their respective riparian areas. While species such as green-backed heron occur in virtually all riparian situations, most wetland-dependent birds have more discrete habitat preferences. For example, common and red-breasted mergansers, wood ducks, bald eagles, and ospreys are primarily associated with lakes and rivers. Common loons are primarily associated with lakes and large ponds. Common snipe and screech-owls are primarily associated with the margins of emergent wetlands. Shrub swamp riparian areas are characteristically occupied by alder flycatchers, cedar waxwings, yellow warblers, prothonotary warblers, northern waterthrushes, and Wilson's warblers. Red-shouldered hawks, barred owls, and Acadian flycatchers commonly occupy riparian zones of forested wetlands (DeGraaf and Rudis 1986).

Among mammals, opossum, star-nosed mole, raccoon, fisher, weasels, mink, and moose occur in essentially all types of riparian habitats, but most riparian-associated species occur along the edges of emergent wetlands, shrub swamps, and forested wetlands (Seidensticker et al. 1987; Petersen and Yates 1980; Sanderson 1987; Thomasma 1996; Hall 1951; Eagle and Whitman 1987; Franzmann and Schwartz 1997). These species include masked shrew, smoky shrew, eastern cottontail, New England cottontail, deer mouse, white-footed mouse, red-backed vole, meadow vole, and southern bog lemming (DeGraaf and Rudis 1986; DeGraaf et al. 1992). The short-tailed shrew is characteristic of forested wetland riparian areas; otter and beaver inhabit riparian areas along lakes, streams, and rivers (Hill 1982; Melquist and Dronkert 1987).

Avian Communities

The vegetation structure of riparian areas largely determines the wildlife habitat values for the avian community (Table 8.3) and strongly influences mammal communities (Table 8.4) (DeGraaf et al. 1992). Greater differences in wildlife communities occur between riparian and adjacent upland habitats when the upland has a strikingly different land-use (e.g., agricultural) from the vegetation structure and condition of the riparian habitat. For example, in northern New England forests, some high-order streams have no adjacent, distinct, woody-vegetation community; in such cases there are few, if any, differences in breeding bird composition associated with the stream. Where a shrubby understory is present, alder flycatchers occur preferentially; where the channel includes slow-moving backwater swamps, northern waterthrushes occur; where large trees are left along stream courses, pileated woodpeckers commonly excavate cavities. These cavities are, in turn, used by common mergansers, wood ducks, screech-owls, and many other species; some of which do not depend upon riparian area vegetation *per se*, but most of their preferred cavities occur there. Pileated woodpeckers and wood ducks are commonly found along wooded stream courses in large cities for this reason.

Avian community studies in arid and semi-arid parts of the Western United States provide the most dramatic contrasts in species richness and abundance between riparian and adjacent upland habitats (e.g., Szaro 1980; Tubbs 1980). Similar contrasts in bird abundance occur between floodplain and upland woodlands and herbaceous habitats in the Central Plains (Stauffer and Best 1980). As adjacent upland conditions and riparian habitats become more similar (e.g., forest-forest edges as opposed to forest-nonforest edges) as in Maine (Hooper 1991) and the central Appalachians (Murray and Stauffer 1995), avian communities also become similar in adjacent habitats. Larue et al. (1995) found higher avian abundance and species richness in boreal riparian conifer stands than in stands farther from water, due to the presence of aquatic-dependent species and others associated with the shrub and grass wetland habitat in boreal riparian forests.

Table 8.3 Birds with a strong preference for riparian habitats in the Eastern United States.

Species	Status	Distribution		Habitat Use		Structural Habitat Features				
		Breeding	Winter-ing	Nest	For-age	Snag/ Cavity Trees	Stream-side Thicket	Back-water Swamp	Herbac./ Wetland Cover	Mature Forest
Green-backed Heron	M[a]	NE[b] NC SE	SE	X	X		X			
Common Merganser	M	NE NC	E	X		X				
Red-breasted Merganser	M	NE NC	SE	X			X		X	
Red-shouldered Hawk	M	E		X	X		X			X
Common Snipe		NE, NC, E	SE, SO	X	X		X	X	X	
Eastern Screech-Owl	R	E	E	X	X	X				X
Barred Owl	R	E	E	X	X	X				X
Bald Eagle	M	E	E	X		X				X
Osprey	M	NC, NE	SE, SO	X		X				X

Table 8.3 Continued

Species	Status	Distribution		Habitat Use		Structural Habitat Features					
		Breeding	Winter-ing	Nest	For-age	Snag/ Cavity Trees	Stream-side Thicket	Back-water Swamp	Herbac./ Wetland Cover	Mature Forest
Red-bellied Woodpecker	R	E	E	X	X	X				X
Pileated Woodpecker	R	E	E	X	X	X				X
Eastern Wood-Pewee	M	E		X	X					X
Acadian Flycatcher	M	E		X	X		X			X
Alder Flycatcher	M	NE, NC		X	X		X			
Veery	M	NC, NE		X	X		X			X
Cedar Waxwing	M	E	SE, SO	X	X		X			
Yellow Warbler	M	E (excpt.S		X	X		X			
Palm Warbler	M	NE, NC	SE	X	X		X			
Cerulean Warbler	M	Interior E		X	X					X
Prothonotary Warbler	M	E (excpt. NY)		X	X	X	X			X
Kentucky Warbler	M	Interior E		X	X		X			X
Swainson's Warbler	M	SE, SO		X	X		X			X
Northern Waterthrush	M	NC, NE		X	X			X		
Louisiana Waterthrush	M	Interior E		X	X				X	X

Table 8.3 Continued

Species	Status	Distribution		Habitat Use			Structural Habitat Features				
		Breeding	Winter-ing	Nest	For-age	Snag/Cavity Trees	Stream-side Thicket	Back-water Swamp	Herbac./Wetland Cover	Mature Forest
Yellow-breasted Chat	M	E (excpt. New Eng.		X	X		X			
Wilson's Warbler	M	Extreme NE		X	X		X			
Rusty Blackbird	M	NC, NE	SE, SO	X	X		X			

[a] M = migrant; R = resident [b] NE = Northeast; NC = North-central; SE = Southeast; SO = South; E = throughout Eastern United States

Small Mammal Communities

Riparian studies of eastern small mammal communities generally focus on alterations of riparian habitat (Possardt and Dodge 1978; Geier and Best 1980; Small 1986) (Table 8.4). The water shrew, star-nosed mole, and rock vole are known riparian habitat dwellers (Beneski and Stinson 1987; Petersen and Yates 1980, Kirkland and Jannett 1982). Stand type and age are probably not the best descriptors of small mammal habitats; most species in the East use a broad range of forest and nonforest types, stand conditions, and stand ages (Miller and Getz 1977; DeGraaf et al. 1991).

Microhabitats (especially for food and cover) may be key factors in small mammal distribution (DeGraaf et al. 1992); small mammal communities in extensive forest communities respond more dramatically to changes in annual food availability and weather than to silvicultural treatment (Healy and Brooks 1988). In urban and agricultural landscapes, small mammal communities are keenly sensitive to changes in vegetation structure. Increases in woody vegetation from cessation of grazing or increases in herbaceous vegetation from tree removal dramatically affect small mammal communities in midwestern riparian forests (Geier and Best 1980).

Within riparian forests, differences in mammal distribution are subtle and may vary regionally. For example, cumulative catch-per-unit-effort was similar in first-order streamside forest and adjacent upslope plots on the Bartlett Experimental Forest, New Hampshire (Yamasaki, unpublished data).

Table 8.4 Mammals with a strong preference for riparian habitats in the Eastern United States (from DeGraaf et al. 1992; Baker 1983; Kurta 1995)

Species	Status	Habitat Use		Structural Habitat Features				
		Forage	Shelter	Softwood Composition	Snag/ Cavity Trees	Understory Shrub Layers	Dead/Down Woody Debris	Herbaceous /Wetland Cover
Virginia Opossum	R[a]	X	X		X	X	X	
Arctic Shrew	R	X	X					X
Water Shrew	R	X	X					X
Southeastern Shrew	R	X	X					X
Northern Short-tailed	R	X	X			X		X
Star-nosed Mole	R	X	X					X
Little Brown Myotis	R	X	X		X			
Northern Long-eared	R	X	X		X			
Indiana Myotis	M	X	X					
Silver-haired Bat	M	X	X		X			
Eastern Pipistrelle	R	X			X			
Big Brown Bat	R	X			X			
Red Bat	M	X						
Hoary Bat	M	X						
Beaver	R	X	X			X		X
Muskrat	R	X	X					X
Black Bear	R	X		X	X	X	X	X
Raccoon	R	X	X		X	X	X	X
Fisher	R	X		X	X	X	X	
Ermine	R	X			X		X	X
Long-tailed Weasel	R	X	X			X	X	
Mink	R	X	X				X	X
River Otter	R	X	X				X	X
White-tailed Deer	R	X	X	X		X		X
Moose	R	X		X		X		X

[a] R = resident; M = migrant

In the Pacific Northwest, the woodland jumping mouse, short-tailed shrew, and masked shrew were caught at slightly higher rates at streamside plots than at upland plots; the southern red-backed vole was caught at a slightly higher rate on upslope plots than on riparian plots (e.g., McComb et al. 1993; Doyle 1990; Anthony et al. 1987). Neither streamside nor upslope habitats alone provided adequate habitat for all small mammal species sampled.

Disturbance Impact

In extensive agricultural areas, narrow forest riparian zones contain many species of birds and mammals that would not be present if the land were cleared to the banks. Some are riparian-dependent, others are forest-associated. It is useful to distinguish between the two. In forested riparian areas, many species are present because of the forest, and most forest-associated species are tolerant of a wide range of disturbances as long as the area remains essentially forested. In New York northern hardwood forests, for example, the breeding bird composition changed little after 25%, 50%, and even 100% of the trees over 14 inches d.b.h. were removed (Webb et al. 1977).

Species associated with the riparian habitat *per se* may not be as tolerant of disturbance as co-occurring species primarily associated with the forest vegetation in the riparian zone. Species that require moist soil, water for breeding, or that obtain prey primarily in aquatic habitats are likely to be severely impacted by riparian habitat changes and may be less widely distributed than primarily forest-associated species. Consider the needs of riparian-dependent wildlife by assessing six key habitats: softwood overstories, high perches and nest sites, cavity trees, forest vegetation layers, dead/down woody debris, and patches of shrub-grass-herbaceous wetland vegetation. By providing these, managers can maximize the number of wildlife species that potentially occupy forested riparian areas.

Effects of Timber Harvesting in Riparian Areas

Softwood composition
Ungulates in the Northern United States benefit greatly from coniferous habitats occurring in riparian areas. Eighty-five percent of 350 Maine winter deeryards occurred in dense conifer cover along streams or about pond and lake shores (Banasiak 1961). Conifer thermal cover consists of tall (>35 ft), closed canopy (>70%) conditions over at least 50% of the deeryard (Reay et al. 1990). Female moose use lowland softwood riparian areas (Leptich and Gilbert 1989) more than males in the summer, and both sexes prefer adjacent wetlands, open water, and bogs (Franzmann and Schwartz 1997).

Meeting the needs of wintering deer and wide-ranging moose requires managing stands in a landscape context (Verme 1965; Weber et al. 1983; Reay et al. 1990). Many more species than deer, such as marten and fisher, use these habitats (Thomasma 1996). Even bats are associated with softwood stands. More than 90% of insectivorous bat feeding occurred in softwood regeneration cuts (0.1 to 2-acres) using group selection methods (Krusic 1995). Increasing the softwood component throughout drainages for wildlife is a long-term effort requiring the cooperation of both foresters and biologists.

Nesting/perch sites

Riparian nest sites and high perches are important for several bird species. Many regulations have been imposed to protect the nest trees of ospreys and bald eagles. These birds commonly use the same nests year after year and are fairly sensitive to disturbance during the breeding season. Large, dead-topped white pines and hardwoods that clearly tower above the surrounding canopy (DeGraaf et al. 1992), especially within 0.5 mile of a river or shoreline, are often used as bald eagle or osprey nest sites. Screech-owls prefer to forage along forested riparian areas, especially those that include marsh habitats. Red-shouldered hawks and barred owls forage extensively in riparian woodlands and forested wetlands. Colonial waterbirds normally nest and roost in riparian woody vegetation, although some species, notably great blue herons, can nest far from water. For most of these species, human disturbance at the nest site is of primary concern. Management activities near nesting colonies should be curtailed at this time.

Cavity trees

Riparian areas present more opportunities to maintain higher densities of cavity trees for wildlife than do more intensively managed adjacent uplands. Higher densities of large-diameter cavity trees can often be found in riparian areas than in upland sites due to nutrient enrichment (Leak 1982; Naiman et al. 1993). Large diameter dead and live trees with cavity potential, especially those trees leaning over water, are of highest value to a variety of primary cavity-excavating birds and secondary cavity-dwelling birds and mammals (Gysel 1961; Conner et al. 1975; Evans and Conner 1979; DeGraaf and Shigo 1985; Tubbs et al. 1987).

Several woodpeckers are common in riparian habitats because dead wood is abundant. For example, the red-bellied woodpecker is often closely associated with riparian woodlands. Tree species composition in riparian woodlands, especially silver maple, red maple, and cottonwoods, are relatively short-lived and prone to crown breakage. Such trees provide abundant cavity substrate for downy and hairy woodpeckers.

The wood duck, common and hooded mergansers, and common goldeneye nest in tree cavities in riparian woodlands. Cavity trees used by wood ducks can be up to a half-mile or more from water, but cavity trees close to water are preferred by these species. Other secondary cavity nesters, such as eastern bluebirds and tree swallows, are most abundant in riparian habitats where dead trees with old woodpecker holes are common.

Recently-dead, large-diameter hardwood trees with old woodpecker holes, common near beaver ponds, provide roosting and maternity colony opportunities for northern long-eared bats in the White Mountain National Forest (Sasse and Pekins 1996). Indiana bat summer roosts are often found in riparian as well as upland habitats (Kurta et al. 1993; Gardner et al. 1991) in the upper Midwest.

Dead/down woody debris

Opportunities exist for coarse woody debris to accumulate along and within stream corridors with beneficial effects to the aquatic community (See Chapter 7). Rabon (1994) surveyed the timber resources along and in 60 streams on the Green and White Mountain National Forests in northern New England. He modeled several management strategies to increase the component of coarse (large) dead and down woody debris in stream channels and adjacent riparian areas. Such increases are important to the abundance and diversity of fish species and fish-eating mammals. Poorly stocked areas with few trees (less than B-line stocking), steep slopes, and some soil characteristics severely limit prescribing a series of silvicultural treatments (USDA 1986; Solomon et al. 1995). Accumulation of coarse woody debris in stream corridors was best done over time using no management. Riparian areas with few environmental constraints and high stocking levels may present opportunities to not only manipulate species composition but also maximize tree diameters through periodic harvests or thinning.

Shrub-grass-herbaceous wetland edges

Many bird species, such as grebes, bitterns, rails, and coots, are associated with waterside marsh vegetation in riparian areas. The American woodcock is an upland shorebird that requires moist soil containing earthworms; it breeds in moist riparian forests with brushy understories. The American woodcock can be found commonly or almost exclusively in wooded riparian habitats during dry summers and through the autumn migration. Other species, such as the cattle egret, glossy ibis, and double-crested cormorant, may depend on riparian vegetation only for nesting or roosting and not for food at all (Erwin 1996).

Various open-nesting birds are closely associated with riparian forest habitats: waterthrushes, several flycatchers including olive-sided, yellow-bellied, and Acadian flycatchers, veery, yellow-throated vireo, prothonotary warbler, palm warbler (on migration), Wilson's warbler, yellow-breasted chat, northern oriole, and rusty blackbird. Many other shrubland birds are common in streamside thickets or swales but are not riparian-dependent: song sparrow, chestnut-sided warbler, red-winged blackbird, and cardinal are examples. In boreal balsam fir riparian forests, streamside marshland vegetation and dense shrub thickets greatly increase the numbers of breeding birds (Larue et al. 1995).

Beavers alter habitats through their dam-building and foraging activities, benefitting a variety of aquatic, semi-aquatic, and terrestrial birds and mammals (Diefenbach et al. 1988). Small hardwood regeneration patches in flat areas along stream gradients <3% in wide valleys (>150 ft), rather than narrower or steeper ones, can increase the likelihood of beaver usage of an area (Hodgdon and Hunt 1953; Howard and Larson 1985). Young hardwoods, especially aspen and alder within 100 feet of water, are a critical forage requirement for continued activity at colony sites (Belovsky 1984), although beaver will forage several hundred feet from the water's edge (Howard and Larson 1985).

Riparian areas are important foraging sites for many mammals. Riparian areas adjacent to still or moving water were among the most favored feeding habitats for bats in the White Mountains of New Hampshire and Maine (Krusic et al. 1996). The big brown bat, red bat, eastern pipistrelle, and the myotids were particularly active in these habitats (Krusic and Neefus 1996). Otter, mink, and raccoon forage on fish, crayfish, mollusks, and invertebrates that flourish in the open water (Hill 1982). Long-tailed weasel and ermine, mink, raccoon, and black bear forage on the variety of small mammals and herbaceous plants that are regularly found in the open meadows of beaver flowages (Hill 1982; Elowe 1984). Muskrat can usually be found in riparian areas where cattails are present, even in roadside drainage ditches (Errington 1963).

Buffer zones

Buffer zones are commonly used to protect riparian areas. Criteria used vary greatly depending on agency or company needs, riparian type, topography, slope, and soils (Table 8.2). Common considerations in buffer-zone design generally include a no-cut or lightly cut area of variable width that minimizes soil erosion and maintains streambank stability (Small and Johnson 1985; Elliott 1988; NH Div. Forests and Lands, DRED and SPNHF 1997) and an adjacent zone where some of the overstory remains over time. Johnson and Brown (1990) found 59% of the overall avian community in both a partially cut lakeshore buffer strip and undisturbed lakeshore site in Maine. Avian density increases in buffer strips immediately after harvest and then declines until adjacent clearcuts regenerate for species dwelling in >50-year-old boreal forest stands (Darveau et al. 1995).

Nest predation potential

Predation at artificial nests suggests that nests in riparian buffer strips in Maine's industrial forests were preyed on twice as often as those in intact riparian forest. This pattern was similar for both ground and shrub nests; predation rates were similar in mainstream and tributary buffer strips. Predators were mostly forest species, not species directly associated with riparian habitats (Vander Haegen and DeGraaf 1996). Predation rates on artificial nests in even-aged stands in an extensive forest did not increase in early stages of growth and were not related to stand area (DeGraaf 1995).

Recommendations

There are no one-size-fits-all recommendations to guide habitat management in riparian areas. Variety in buffer widths, disturbance regimes, adjacent land uses, and vegetative structure is an important consideration.

Landscape-Level Considerations

At the landscape scale, several items need to be considered in developing habitat management plans:

- Consider variable riparian area management widths with some regard to stream order hierarchy (stream width is easier to identify on the ground and more consistent than map scale dependent stream order).

- Limit new roads in riparian areas and consider reducing the traffic on existing roads in riparian zones at certain times of the year (e.g., bear hunting seasons).

- Avoid long, linear clearcuts adjacent to riparian areas, especially if the other side of the drainage was cut recently (20 to 30 years ago).

- Consider tree species composition potential — are long-term changes in composition warranted, possible, or necessary?

- Consider using riparian management zones wider than those normally prescribed to protect streambank stability, provide brook shading, and limit sedimentation where agricultural or urban landscapes predominate.

- Consider: (1) limiting grazing activities at the water's edge (use fencing when necessary) and (2) limiting borrow pit development and reclaiming existing borrow pits with native species.

Stand-Level Considerations

Birds and mammals will benefit from a variety and diversity of vegetative conditions, forest cover types, sizes, and size classes (DeGraaf et al. 1992). Again, there are no one-size-fits-all solutions. Site, slope, aspect, soil types, and seasonal limitations (e.g., raptor nesting concerns) all bear on potential stand-level prescriptions. Opportunities are normally present with both even-age and uneven-age management systems to meet wildlife habitat landscape goals; consider how you might apply landscape goals at the individual stand level.

Within-Stand or Structure Considerations

The vegetation structures to be maintained or developed need to be based on the specific site potential. For example, in seasonally flooded drainages, it might be difficult to establish and maintain a dense shrub zone or dense herbaceous ground cover; yet in other less frequently disturbed drainages, the likelihood of success is much greater. To provide an array of structural components:

- Consider high densities of cavity trees and snags, especially large-diameter trees; think hard before immediately prescribing salvage harvests.

- Consider the opportunity to increase the dead and down coarse woody debris component in drainages not only for stream channel modifications but also for terrestrial wildlife.

- Consider a variety of canopy closures: raptor nesting and perching tree potential, softwood-to-hardwood or mast-to-nonmast basal area ratios.

- Encourage the development or maintenance of distinct shrub layers, thickets, and grass/sedge and herbaceous ground cover. These add important habitat elements to any riparian area.

- Consult DeGraaf et al. (1992), Green (1995), Hamel (1992b) or numerous state publications (Elliott 1988; Regan and Anderson 1995; and others) for additional information and suggestions on the variety of structural conditions that produce a habitat-friendly riparian area.

Table 8.5 A comparison of common and scientific names for birds and mammals.

Common Name	Scientific Name
Common Loon	*Gavia immer*
Double-crested Cormorant	*Phalacrocorax auritus*
Great Blue Heron	*Ardea herodias*
Green Heron	*Butorides virescens*
Cattle Egret	*Bubulcus ibis*
Glossy Ibis	*Plegadis falcinellus*
Wood Duck	*Aix sponsa*
Common Goldeneye	*Bucephala clangula*
Common Merganser	*Mergus merganser*
Red-breasted Merganser	*Mergus serrator*
Osprey	*Pandion haliaetus*
Bald Eagle	*Haliaeetus leucocephalus*
Red-shouldered Hawk	*Buteo lineatus*
Common Snipe	*Gallinago gallinago*
American Woodcock	*Scolopax minor*
Eastern Screech-Owl	*Otus asio*
Barred Owl	*Strix varia*
Red-bellied Woodpecker	*Melanerpes carolinus*
Downy Woodpecker	*Picoides pubescens*
Hairy Woodpecker	*Picoides villosus*
Pileated Woodpecker	*Dryocopus pileatus*
Olive-sided Flycatcher	*Contopus cooperi*
Yellow-bellied Flycatcher	*Empidonax flaviventris*
Acadian Flycatcher	*Empidonax virescens*
Alder Flycatcher	*Empidonax alnorum*
Yellow-throated Vireo	*Vireo flavifrons*
Tree Swallow	*Tachycineta bicolor*
Eastern Bluebird	*Sialia sialia*
Veery	*Catharus fuscescens*
Cedar Waxwing	*Bombycilla cedrorum*
Yellow Warbler	*Dendroica petechia*

Table 8.5 Continued

Common Name	Scientific Name
Chestnut-sided Warbler	*Dendroica pensylvanica*
Palm Warbler	*Dendroica palmarum*
Prothonotary Warbler	*Protonotaria citrea*
Kentucky Warbler	*Oporornis formosus*
Swainson's Warbler	*Limnothlypis swainsonii*
Northern Waterthrush	*Seiurus noveboracensis*
Wilson's Warbler	*Wilsonia pusilla*
Yellow-breasted Chat	*Icteria virens*
Song Sparrow	*Melospiza melodia*
Northern Cardinal	*Cardinalis cardinalis*
Red-winged Blackbird	*Agelaius phoeniceus*
Rusty Blackbird	*Euphagus carolinus*
Virginia Opossum	*Didelphis virginiana*
Masked Shrew	*Sorex cinereus*
Arctic Shrew	*Sorex arcticus*
Water Shrew	*Sorex palustris*
Smoky Shrew	*Sorex fumeus*
Southeastern Shrew	*Sorex longirostris*
Short-tailed Shrew	*Blarina brevicauda*
Star-nosed Mole	*Condylura cristata*
Little Brown Myotis	*Myotis lucifugus*
Northern Long-eared Myotis	*Myotis septentrionalis*
Indiana Myotis	*Myotis sodalis*
Silver-haired Bat	*Lasionycteris noctivagans*
Eastern Pipistrelle	*Pipistrellus subflavus*
Big Brown Bat	*Eptesicus fuscus*
Red Bat	*Lasiurus borealis*
Hoary Bat	*Lasiurus cinereus*
Eastern Cottontail	*Sylvilagus floridanus*
New England Cottontail	*Sylvilagus transitionalis*
Beaver	*Castor canadensis*

Table 8.5 Continued

Common Name	Scientific Name
Deer Mouse	*Peromyscus maniculatus*
White-footed Mouse	*Peromyscus leucopus*
Southern Red-backed Vole	*Clethrionomys gapperi*
Meadow Vole	*Microtus pennsylvanicus*
Rock Vole	*Microtus chrotorrhinus*
Muskrat	*Ondatra zibethicus*
Southern Bog Lemming	*Synaptomys cooperi*
Meadow Jumping Mouse	*Zapus hudsonius*
Woodland Jumping Mouse	*Napaeozapus insignis*
Coyote	*Canis latrans*
Red Fox	*Vulpes vulpes*
Black Bear	*Ursus americanus*
Raccoon	*Procyon lotor*
Marten	*Martes americana*
Fisher	*Martes pennanti*
Ermine	*Mustela erminea*
Long-tailed Weasel	*Mustela frenata*
Mink	*Mustela vison*
River Otter	*Lontra canadensis*
Bobcat	*Felis rufus*
White-tailed Deer	*Odocoileus virginianus*
Moose	*Alces alces*

Richard Buech

A riparian obligate!

Chapter 9

Managing Riparian Areas for Fish

Carl Richards and Bob Hollingsworth

"From the days when I angled for minnows with a pin, the delights of the running brooks have held me with a gentle firmness from which I have not escaped and never shall. One kind of fishing may be better than another, yet all are good. For me there is no answer to the question: 'What would you rather do than go a-fishing?'"
Gifford Pinchot — Fishing Talk, 1936

The fisheries resources of the continental Eastern United States are among the most varied in the world. Ranging from walleyes and northern pike in the northwoods, to flathead catfish in the Ohio and Mississippi valleys, to trout in secluded New England brooks or Appalachian "forks," the region has something for everyone. Ozark streams offer smallmouth bass fishing, while the Great Lakes tributaries feature world-class salmon and steelhead runs. Non-game fish include a rich array of cold- and warm-water forms as well as several threatened, endangered and sensitive species, most of which are found south of the Ohio River.

Although good fishing is found in the region, it used to be better. Aquatic habitats are broadly degraded from pre-settlement conditions. Human activity altered the structure and function of aquatic and riparian systems, simplifying habitats and reducing the energy available to aquatic communities. Waterways, including sensitive headwaters, were cleaned and straightened on a massive scale (Verry 1992). Riparian areas and adjacent uplands were cleared, interrupting the accrual of forest litter and woody debris in lakes and streams. Accelerated sedimentation filled pools and covered food-producing and spawning substrates. Increased peakflows reduced streambank stability. These and other factors combined to significantly diminish aquatic biological diversity and productivity. Streams are now wider, shallower, and in many cases, warmer than they used to be.

Managing the biological potential of streams, lakes, and riparian areas to recover threatened or endangered species; to maintain the viability of native and desired nonnative species, and to provide quality recreational fishing opportunities is a significant challenge in forest land in the Eastern United States. Many of these lands, especially those in the public domain, are recovering from previous land use impacts. While agencies such as the USDA Forest Service have worked for several decades at restoring the terrestrial component of "the

157

lands nobody wanted"; the structure and function of the aquatic and riparian systems embedded within those lands were more recently understood. Restoration of aquatic and riparian systems has lagged behind. Some protection needs, especially in headwater streams, remain unmet. Paradoxically, the forests drained by those headwaters appear "healthy" from a wood-production perspective, if not ecologically. These finite aquatic resources are reduced in biological diversity and productivity from historic levels and are at less than ecological potential. Coincidentally, there is high demand on specific aquatic resources; primarily, the so-called gamefish and panfish species.

   The health of aquatic systems reflects the management of adjacent riparian areas and uplands. Put another way, simply because water runs downhill, the quality of fish-habitat reflects the quality of land management — or the lack of it. Aquatic systems depend on riparian and terrestrial systems for food, energy, temperature regime, regulation of sediment, habitat diversity, and water flows. Land management, especially management of forest land, affects all these factors, frequently far downstream.

A Brief Review of Fish Ecology

Before examining the influence riparian management has on fish, it is useful to review the components of physical habitats and processes in streams that foster healthy, diverse fish populations and communities. Although habitat requirements vary with individual species, all streams have common elements that fish require during various life stages. In addition, some basic processes are fundamental to maintaining fish populations. The basic processes and physical features are generally the first things noticed and assessed by fisheries biologists in the field.

Spawning Habitat

All fish require some sort of spawning habitat to successfully propagate. In streams, the most critical areas for spawning are riffles and flooded backwaters, and the fishes that require them generally can't spawn anywhere else. Riffles are stream sections in which relatively fast water flows through and over large substrate particles such as gravel and cobble. Eggs deposited in the substrate of riffles have a constant source of fresh water for embryo development. Many game and nongame species require these habitats for spawning. When spawning gravels become clogged with fine materials such as sand and clay, water is no longer able to flow through the substrate, and eggs are unable to survive. Developing eggs require a constant flow of water to supply dissolved oxygen supply and to wash away metabolic wastes. The extent and quality of riffles are important for understanding the status of a stream reach.

Since time variation in flow elevation is a natural component of stream ecosystems, several species have evolved to utilize flooded areas in spring and summer for spawning substrates. These areas may not be part of the stream in low-flow periods but are still vital for successful reproduction of some species. This is but one important way that streams interact with their floodplain. It's important to understand that the response of streams to seasonal rainfall, temperature and other climatic conditions is essential to the maintenance of healthy fish populations.

Rearing Habitat

Juvenile and adult fishes require a variety of microhabitats within streams. These are the habitats where the fish spend most of their lives and live and grow to maturity. Some fishes may prefer the eddies behind boulders, some may prefer the fast water of a riffle, and some may require a deep pool. As with spawning habitats, the rearing habitat of stream fishes is species specific, and different age classes of the same species often may require different rearing habitats. However, some generalizations can be drawn about channel features that enhance the availability of rearing habitat for stable, diverse fish communities. First, due to the variety of microhabitats that are chosen by different species, stream channels that contain many different habitats support more species than channels with little variation. The presence of a series of riffles, pools, and runs within a stream reach provides this type of complexity. In addition, variation along the water/land interface provides habitat complexity. Features such as undercut banks, shallow shorelines, and other shoreline features provide additional microhabitats for fish.

Coarse woody debris (CWD) is one of the most important features of stream channels contributing to habitat complexity. Although CWD performs several important functions in stream ecosystems, the provision of microhabitats afforded by woody debris is one of the most important. Many fishes can use wood as a place of refuge or a place to ambush prey. In addition, the hard substrates and associated invertebrate production provided by woody debris is an important source of food for fishes in many streams. Accumulations of CWD are associated with increased abundance of many stream fishes. The presence of woody debris in stream channels is vital to fish community health.

Food

The abundance and quality of food resources are important for growth and reproduction of all fish species. In streams, the dominant food for fishes is invertebrates. Almost all species seek invertebrates from stream substrates or other surfaces in the stream or actively forage on invertebrates suspended in the water column. Consequently, maintaining stream attributes that enhance invertebrate production is important in stream management. Two general concepts should be understood.

First, there are strong links between invertebrate production and energy supply in streams. Detrital material that enters the stream from trees and other vegetative material near the stream provides the basic nutrients and energy to feed much of the invertebrate food web. The maintenance of the timing, type, and quantity of detrital inputs to streams is extremely important. The biological processes that condition detritus as a source of food for invertebrates is seasonally dependant and follows natural patterns of leaf fall. Invertebrates have adaptations that allow them to exploit this resource only during certain times of the year. Furthermore, leaves from different types of trees or other herbaceous vegetation have different decomposition rates and suitability as food to invertebrates. Consequently, altering riparian vegetation can change invertebrate community composition. In addition to detrital inputs to streams, instream algal production is crucial for invertebrate production. The availability of light and nutrients often controls algal production. Removing riparian shading and altering nutrient supply to a stream will change invertebrate community composition and production.

Second, the input of fine sediments to streams will decrease the production of invertebrates significantly. Greatest invertebrate production comes from large, hard substrates, such as gravel and cobble, as well as other hard surfaces, such as submerged limbs. When these surfaces are covered with fine sediments, their suitability for invertebrate production is greatly diminished. With sedimentation, interstitial spaces needed by invertebrates within the substrate are lost, and detrital decomposition and algal production are decreased. Different regions of the country have different natural levels of fine sediments in streams due to local geologic conditions. Addition of fine sediments above natural levels can have persistent detrimental effects that last for decades or even centuries. Fine sediment is the most widespread pollutant to streams.

Water Quality

Water quality encompasses many different types of physical and chemical processes. The most widespread alteration of water quality comes from the addition of nutrients from land use practices. Nitrogen and phosphorus are important to primary production in streams and therefore influence many aspects of stream ecosystems that affect fish production.

Temperature is one of the most fundamental physical characteristics of water. All biological processes are regulated by temperature. Increasing or decreasing temperature alters primary and secondary production in a stream as well as decomposition. In addition, all species of organisms in streams, including fish, have specific temperature tolerances. When temperatures deviate from a species' preferred range, production or reproductive success of that species will decline. Land-use conversions, riparian alterations, impoundments, and other activities can significantly alter temperatures and therefore the suitability of a stream for certain fishes.

Water Quantity

The quantity and timing of flow are probably the most crucial aspects of habitat to fishes. The volume of water flowing through a stream reach determines many aspects of the availability of microhabitats within the stream channel. All fish have velocity preferences within the stream channel that vary with the life stage of the fish. Seasonal variation in streamflow is a natural component of the streams that fish communities have adapted to and require. Streamflow patterns typically reflect local climate conditions and geology.

Riparian and Instream Linkages

Viewing streams within watersheds and watersheds within landscapes provides a context for understanding the effects of riparian management on stream ecosystems. Biotic communities in streams are influenced by many physical and chemical conditions that are in turn modified by local and watershed-scale features. Management actions in riparian areas may have different effects depending on landscape setting. Quantifying the relative roles of regional (landscape, watershed) and local (riparian) regulation for stream communities and biotic processes is important for developing both a theoretical and a quantitative understanding of the interactions of stream ecosystems with their adjacent terrestrial landscapes (Gregory et al. 1991; Richards et al. 1996).

Geology and land cover are the fundamental components of watersheds. In a given climatic setting, land use influences delivery of water and material to streams. Land use also determines much of the variation in the physical character of stream channels (Maxwell et al. 1995; Richards et al. 1996). Infiltration and erosion are controlled by geology and vegetative cover (Bloom 1991). Export of organic and inorganic materials from watersheds to streams is ultimately limited by the soil's physical and chemical characteristics and by precipitation. Geologic control of water and resource delivery to streams is greatly modified by vegetative cover.

Land use is superimposed on the geologic template. It mediates stream-water chemistry (Omernik et al. 1981; Peterjohn and Correll 1984; Osborne and Wiley 1988), fish community structure (Schlosser and Karr 1981a, b; Roth et al. 1996), macroinvertebrates (Richards et al. 1996), and periphyton (Leland 1995; Kutka and Richards 1996). For example, urban development increases the amount of impervious surface, resulting in more variable stream discharge, increases in suspended sediment and dissolved solids, and higher chloride concentrations, erosion, and nutrient inputs (e.g., Burton et al. 1977a, b). Timber harvest and associated activities can affect stream water quality, increasing suspended sediments and turbidity (Packer 1967; Bormann et al. 1969), water temperature (Beschta and Taylor 1988), export of phosphorus and nitrogen (Hall et al. 1980; Golladay et al. 1992), and loss of soil nutrients (Brown et al. 1973; Triska 1984; Verry et al. 1983; Golladay et al. 1992). Stream organic matter dynamics are altered through short-term reductions in leaf and litter fall and a long-term decrease in coarse woody debris (Webster and Waide 1982; Gregory et al. 1987; Golladay et al. 1989). These changes negatively affect flow rates, riparian structure, and instream habitat. These, in turn, influence primary production, organic matter processing and retention, and community structure.

The riparian corridor moderates larger scale features through its influence on the form and quantity of organic matter input to adjacent streams (Conners and Naiman 1984; Lamberti et al. 1991; Merritt and Lawson 1992; Gregory et al. 1991). In general, riparian vegetation modifies the effects of land-use activities, such as urban development, crop and animal cultivation, and timber harvest, by trapping sediments and nutrients and by providing shade and organic matter to streams. Riparian areas contribute CWD to the stream channel, influencing stream geomorphology and many ecosystem processes (Bilby 1981; Speaker et al. 1984; Triska 1984; Gregory et al. 1991; Webster et al. 1994; Wallace et al. 1995). CWD causes channel migration, which in turn forms backwaters, pools, and riffles. It also reduces flow velocity (Leopold et al. 1964; Keller and Swanson 1979; Triska, 1984; Harmon et al. 1986). Increased channel complexity is linked to habitat and biotic diversity (e.g., Angermeier and Karr 1984; Benke et al. 1984; Smock et al. 1989; Fausch and Northcote 1992; Richards and Host 1994) and the retention of organic matter (Bilby and Likens 1980; Webster et al. 1994; Raikow et al. 1995; Lamberti and Gregory 1996). The riparian area also provides organic matter, the dominant energy source for many streams (e.g., Cummins and Klug 1979; Webster and Benfield 1986).

In agricultural and urbanized environments, riparian forests are often fragmented or eliminated, and woody debris is removed from streams as a flood-control measure. These disturbances directly affect organic matter processing by reducing the potential sources, pools, and fluxes of debris within the channel and jeopardizing biotic communities. Land use changes affecting the riparian area alter detrital inputs and, thereby, alter stream community structure and function (Tuchman and King 1993; Johnson et al. 1994; Richards et al. 1996).

Instream primary production is strongly influenced by land use and land cover patterns, as well as by geologic factors, and it is strongly modified by riparian management. In agricultural catchments, a surplus of nutrients results in eutrophied systems where primary production may be light rather than nutrient-limited (Wiley et al. 1990). In forested catchments of the Upper Midwest and elsewhere, streams are limited both by nutrient availability (e.g., Elwood et al. 1981; Pringle 1987) and light (see Lamberti et al. 1989; 1991; Rosemond et al. 1993). Underlying geology dictates which nutrient is limiting (Grimm and Fisher 1986), but some streams vary between N and P limitation on a seasonal basis (Tate 1990; Allen and Hershey 1996).

Fish and macroinvertebrates are strongly related to all the above instream processes and conditions. Macroinvertebrate community structure is sensitive to predominant organic matter forms (e.g., Cummins 1973; Vannote et al. 1980), algal abundance and patchiness (e.g., Lamberti and Resh 1983; Richards and Minshall 1988; Hart and Robinson 1990), stream habitat (e.g., Gurtz and Wallace 1986; Richards et al. 1993, 1996, 1997; Richards and Host 1994), riparian vegetation (Woodall and Wallace 1992; Molles 1982), and water quality factors (e.g., Lenat 1984; Richards et al. 1996; Lamberti and Berg 1995). Fish respond to these same variables but are particularly sensitive to reduced habitat complexity and increased temporal variability in flows (Berkman and Rabeni 1987; Schlosser 1982; Bilby and Bisson 1987).

Instream processes and biotic community structure are, therefore, influenced by local (e.g., riparian management) or regional factors (land use and land cover patterns, surface geology, topography, soils). The influence of riparian and regional-scale environmental factors may vary with the degree of human disturbance within the watershed and the type of watershed. However, different physical and chemical features of stream environments, and consequent biological characteristics may be controlled by watershed features at entirely different scales. For example, the amounts of CWD and light available for primary production may be largely controlled by local riparian management. On the other hand, nitrogen availability and overall stream morphometry are often more a function of watershed hydrology and geology and may be less influenced by local riparian conditions (Richards et. al 1996, 1997; Johnson et al. 1997).

General Harvest Management Principles

Although the interactions between watersheds and stream ecosystems are complex, a relatively simple set of guidelines can protect many of the essential connections between terrestrial and aquatic systems.

Proportion of the Watershed Harvested

As noted earlier, several instream physical features are influenced more by watershed hydrology than by local conditions, and hydrology drives many other instream biological and physical processes. Changes in vegetative cover can profoundly impact stream hydrology and fisheries. In forested regions of the northern lake states, Verry (1992), noted that the snowmelt rate in aspen stands less than 15 years old was much higher than the rate in older stands. Streams in watersheds with a high proportion of young stands have flow rates that increase in-channel erosion and sedimentation. He suggested that the effect was realized when the 60% of the watershed was covered by younger stands.

In watersheds with mixed land uses, where agricultural and other open land uses exceed 30%, the critical threshold is reached sooner (Verry 1992) and the proportion of forested land that can be maintained in young stands is less.

Headwater Streams: The Maximum Interface Between Aquatic and Terrestrial Systems

Within any watershed, there are many more miles of small than large streams. These headwaters (stream orders 1 through 3) represent the maximum interface between the terrestrial and aquatic systems. Many headwater streams also harbor relict populations of indigenous species that may be vulnerable to land-use changes. Timber harvest and other vegetative manipulations may be significant on land drained by these small streams (Karr and Schlosser 1978). Although intermittent or seasonally dry channels are often ignored or regarded as insignificant by managers, most sediment enters stream systems through these small streams during storms. Riparian management must include guidelines and best management practices (BMPs) designed to protect the integrity and functions of headwater streams.

Fine Sediment — The Oldest Insult

One of the most serious influences of human activity in riparian areas is increased yield of fine sediment — silt and sand — to stream channels. Fine sediment reduces aquatic habitat diversity and biological productivity. It fills pools and covers or infiltrates spawning and food-producing substrates (Waters 1995). Extensive deposits can cause a channel to aggrade and widen, exposing it to more sunlight with concomitant increased temperature (Alexander and Hansen 1983; OR/WA Wildlife Comm. 1979). Even small amounts of sediment can be damaging. Although turbidity rarely exceeds levels high enough to kill fishes (Bennett 1971), the indirect effects on ecosystem processes are apparent at much lower concentrations (Waters 1995). Since low-gradient headwater streams have limited flushing capacities, cumulative effects of sedimentation can persist for decades even when contemporary sediment production has returned to natural levels.

All earth-disturbing activities can yield sediment to waterbodies, and a lot of research has focused on techniques and practices designed to prevent sediment from reaching stream channels. Most sediment generated by forest practices is associated with transportation systems, and there are many examples of BMP manuals and guidelines prepared by state and federal government agencies. In general, water and sediment should be diverted from road surfaces and ditches to the forest floor and away from stream channels. Roads and trails should be managed as extensions of the landscape drainage network, and ditch lines often function as ephemeral or even intermittent channels. Stream crossings should be designed so that bridges or culverts are not low points in the road grade. Culverts and other structures in the stream channel should be designed to assure fish passage and eliminate the potential for failure under highly variable flows.

Coarse Woody Debris and Habitat Diversity in Streams

As previously discussed, woody debris is a fundamental component of eastern stream ecosystems. Prior to 1970, fisheries managers had little appreciation for the role of coarse woody debris as fish habitat. Streamside blow-downs with upturned root wads were viewed as no more than potential sediment sources. The functions of large limbs and boles as stable overhead cover and as pool-forming agents went largely ignored, and debris jams were viewed as barriers to fish movement. Aside from the fact that very few jams are complete barriers, they have beneficial features that were overlooked. Our understanding of the role of CWD in fish habitat is incomplete, but fish clearly need trees — dead and down in stream channels and lake basins — for habitat diversity.

How much woody debris is enough? Researchers feel that they have not seen an upper limit. Since many rivers in their pristine condition had debris jams up to 5 miles long (Sedell et al. 1982), it seems likely that the upper limit for large woody debris will be determined by social factors, not by fish habitat objectives. Power boating, rafting, water skiing, canoeing, and flood control are but a few of the activities that can be adversely affected by high densities of CWD.

Managing for big trees in riparian areas can speed the accrual of woody debris to streams, including intermittent and ephemeral channels, and to lake basins. Big trees also contribute CWD to the non-aquatic portions of riparian areas and are consistent with the needs of many riparian-dependent wildlife species. Promoting big trees will succeed best where local site conditions and ecological potential are appropriate. Bisson et al. (1987) discuss ways to favor big trees in riparian areas:

- Leave an undisturbed buffer strip of old-growth timber in the riparian area. While assuring a supply of long, large-diameter logs, this may be the most costly approach in terms of timber value forgone.

- Leave the estimated amount of timber required to meet habitat diversity needs and allow it to accrue to the stream channel or lake basin naturally.

- Remove timber from the riparian area on a long rotation-age basis, where trees are harvested every 100 to 150 years rather than on a 50 to 80 year-cycle. This approach allows trees to reach larger size and to have a much greater chance of accruing to a channel, a lake basin, or to the non-aquatic riparian area.

- Manipulate riparian vegetation stand structure and composition using silvicultural techniques designed to (a) maintain a relatively even delivery of large woody debris to the channel, basin, or riparian area, and (b) provide a mix of riparian tree species. This approach could include deliberate introductions of unmerchantable trees and debris (cull logs, stumps with root wads, or large branches and tops) during harvest or other treatment.

Adequate Riparian Buffers are Essential Fish Habitat

Riparian buffers along streams act not only as sediment filters, areas for material processing, and sources of organic material but also as sources of CWD. Riparian buffers are commonly prescribed in forest harvest management, but the suggested width of buffers varies greatly. Large and Petts (1994) reviewed more than 27 documents and research reports that recommended widths for naturally vegetated zones that would protect ecosystem function for stream margins. All but three studies suggested minimum widths of 10 meters. Fourteen of the reports recommended riparian buffers of 30 meters or more. These studies support the suggestions of Verry (1992) in the Lake States region for establishing zones of approximately two mature tree lengths in which harvests and other land-use activities are intensively managed.

Identification of Fisheries Goals

Clear goals are essential for effective management of fisheries. Problem identification, evaluation of specific management activities, and monitoring of fisheries resources all require that the status of aquatic communities be inventoried and described and the desired condition clearly articulated. Since streams are complex ecosystems, multiple measurements are required to determine the status of an individual stream. These involve ecosystem process rates (e.g., primary production, organic matter storage), physical and chemical condition (e.g., channel morphology, substrate composition, nutrient supply), and biota (e.g., trout biomass, macroinvertebrate species composition). Obviously, such complex sets of measurements are usually not within the realm of everyday management and certainly could not be considered for all or even many of the streams in forests of the Eastern United States. Fish or invertebrates are frequently used to gauge the impacts of human activities in aquatic systems and can sometimes provide answers to management questions. But temporal fluctuations in species abundance typically require data on long-term trends to facilitate sound conclusions. Long-term information is essential when managing for high-value fishery stocks, threatened and endangered species, and when the effects of stressors must be evaluated. However, more often these types of intensive studies are neither warranted nor practical.

Community-based information indices offer an alternative for examining stream systems that have many advantages over other assessment techniques. Indices using aspects of community structure in streams and lakes are not new, examples have been around for at least 100 years (Davis 1995). However, a renewed interest in biotic indices has occurred in the United States over the last decade. In particular, the application of multimetric indices, such as the Index of Biotic Integrity (IBI) (Karr and Chu 1999), has shown promise for monitoring stream ecosystems. These indices are composed of a series of metrics (biological attributes) that assess the condition of an ecosystem. Such indices are valuable for:

- Identifying streams with degraded ecosystems that warrant restoration
- Designating high-quality streams for additional protection
- Classifying streams on the basis of biotic integrity
- Monitoring streams for changes in ecosystem quality over time
- Assessing responses of streams to management and other human activities

Metrics are chosen that reflect ecosystem characteristics and are sensitive to human-caused changes from natural variation. The critical aspects of stream condition are water quality, habitat structure, flow regime, energy sources, and biotic interactions (Karr 1991). The most common IBIs for streams have been developed and tested using fish (Karr 1981; Miller et al. 1988; Lyons 1992; and Lyons et al. 1995, 1996). Most multimetric biological indexes for fish in streams consist of 8 to 12 metrics, each selected because it reflects an aspect of the condition of the stream biological system. Usually these have involved species in rich warm-water fish communities, but cool water communities are also amenable to this approach typically with a smaller number of total metrics (Lyons et al. 1996).

Invertebrates can also be used to develop IBIs (Fore et al. 1994; ICI reference). Their use may be just as effective, particularly when fish sampling is impractical or where only a few fish species are present. IBIs may also be developed for algal communities (Bahls 1993) although the identification of metrics is still being developed (Kutka and Richards 1996). Rather than being fixed, the selection and development of metrics can be fine tuned to a particular region or purpose. Sampling protocols for fish and invertebrates are well described (e.g., Ohio EPA 1988; Klemm et al. 1993; Lyons et al. 1996).

A critical feature of the use and application of IBIs is the development of a regional reference (Barbour 1996). The IBI requires quantitative predictions of what a biological community should look like under reference or least impacted conditions. By identifying the range of variation in an IBI or in individual metrics of an IBI, the status of any stream or stream reach can be readily identified by comparison to the regional reference. Reference streams should include the range of conditions typical of the area, including natural geologic features and land-management activities. The development of a regional reference requires much work in that many streams need to be surveyed; however, the consequent advantages of having these reference conditions for interpretation of present and future management actions outweigh the initial investments in resources. IBIs have proved effective for examining a variety of stream characteristics related to riparian functions in the United States. Response signatures of metrics can be developed that discriminate between stressors of various types (Yoder and Rankin 1995) so that, for example, the effects of sedimentation in a stream reach might be differentiated from the effects reduced shading.

Caught and released. Secret Smallmouth bass waters. It's the fishing!

Chapter 10

Ecology and Management of Riparian Habitats for Amphibians and Reptiles

Thomas K. Pauley, Joseph C. Mitchell,
Richard R. Buech and John J. Moriarty

"If it is granted that biodiversity is at high risk, what is to be done? The solution will require cooperation among professions long separated by academic and practical tradition."   E.O. Wilson — The Diversity of Life

Native vertebrate faunas in North America use riparian areas more than any other habitat (Thomas et al. 1979; Brinson et al. 1981; Ohmart and Anderson 1986). Studies on the ecology of amphibians and reptiles in riparian habitats have lagged behind those on birds and mammals (see references in Brinson et al. 1981; Wigley and Melchiors 1994) and indeed were almost nonexistent until the 1980s. Since then, most research on this group has been conducted in the West and Northwest (e.g., Vitt and Ohmart 1978; Bury 1988; Bury and Corn 1988; Jones 1988; Corn and Bury 1989; Gomez and Anthony 1996). With few exceptions (Pais et al. 1988; Rudolph and Dickson 1990; Foley 1994), knowledge of amphibians and reptiles in eastern riparian habitats has been limited largely to species richness (e.g., Tinkle 1959; Wharton et al. 1981; DeGraaf and Rudis 1986; Mitchell et al. 1994).

Amphibians and reptiles are important components of riparian food webs, perhaps more so than birds or mammals (Ohmart and Anderson 1986), because many use both aquatic and terrestrial habitats and spend a portion of their lives in riparian areas. Amphibians, especially, are considered important indicators of environmental change for the following reasons: (1) most have complex life cycles with aquatic and terrestrial stages that expose them to perturbations in both environments, (2) they have permeable skin, gills, and eggs that are susceptible to environmental alterations, (3) their dependence on ectothermy for temperature control makes them vulnerable to environmental fluctuations, (4) many hibernate

169

or aestivate in soils that may expose them to toxic conditions, and (5) they are both predators and prey in terrestrial and aquatic food webs (Dunson et al. 1992).

Most reptiles are covered by epidermal scales and are less sensitive to moisture loss than amphibians. They also are ectotherms and, thus, their activities and behaviors are tied closely to environmental temperatures. The production of shelled eggs and retention of young in females until birth enable reptiles to occupy many habitats and, unlike most amphibians, release them from the reproductive connection to water. However, many species in Eastern North America use aquatic and semiaquatic habitats and often occupy riparian areas.

Increasing numbers of amphibians and reptiles that occur in eastern riparian habitats are listed as endangered, threatened, or sensitive species under federal, state, or agency mandates (Levell 1997). Species-level mandates and requirements under such U.S. congressional actions as the National Forest Management Act and Section 404 of the Federal Water Pollution Control Act require that wetlands and, by association, some riparian habitats and species be protected. However, it is not always clear to land managers how these requirements can be met. This is especially true of amphibians and reptiles which only recently have been included in some agency management objectives (Gibbons 1988).

In this chapter, we provide an overview of amphibian and reptile associations in riparian habitats of the warm continental divisions of the Humid Temperate Domain (Bailey 1995) in eastern North America, review briefly the range of threats to these animals; and discuss conservation and management options that may be applied by managers. A primary objective is to demonstrate that amphibians and reptiles use aquatic, riparian, and terrestrial habitats and require ecologically functional habitat complexes to maintain population viability. Riparian management zone width and landscape-level management are thus important parts of the discussion. A secondary objective is to provide a window into the herpetological and ecological literature so that field biologists and natural resource managers will be able to obtain information on these two groups of animals. We take a broad view of riparian habitats in this chapter because many amphibians and reptiles use these areas for only a portion of their life cycles. In some cases, this ecotone simply lies between the aquatic and terrestrial habitats that play important roles in their lives.

Herpetofaunal Diversity in Riparian Habitats

Riparian habitats have unique properties not found in other ecosystems and, because they have characteristics of both aquatic and terrestrial ecosystems, may be considered ecotones between them (Johnson et al. 1977, 1979). The microclimates of riparian zones differ from those in adjacent terrestrial habitats because of increased humidity, moisture, and shade (Thomas et al. 1979). Temperature of water entering riparian habitats from streams and seepages is more similar to subsoil temperature than terrestrial temperature (Beschta et al. 1987). These environmental conditions fall within the temperature and moisture tolerance limits of many amphibians and reptiles. Many of these animals spend all or parts of their lives in riparian habitats. Indeed, many species usually considered characteristic of upland habitats frequent riparian systems (Buhlmann et al. 1994). Of the 143 amphibians and reptiles in the continental forests of eastern North America (based on Conant and Collins 1991), 71 amphibians and 50 reptiles are known to use riparian habitats sometime in their life cycles.

The variety of wetland and riparian habitats that occur in eastern North America can be grouped in many ways (Mitsch and Gosselink 1993). For this review, we grouped possibilities into seven primary riparian habitat types that, in our opinion, capture most types of aquatic-terrestrial interfaces used by amphibians and reptiles of the region. We realize that the list of variations of habitats within several of the seven types we have selected is long. Our list of habitats is limited by space and serves primarily to focus manager attention on some of the major differences among riparian habitats. We emphasize freshwater wetlands in this review and, thus, not riparian habitats associated with estuaries or marine ecosystems. For each type, we provide several examples of taxa that occur in these habitats and include pertinent ecological notes. Table 10.1 lists the affinity of these 121 taxa for seven habitat types: seeps, streams, rivers, lakes and ponds, beaver ponds, bogs and fens, and ephemeral ponds in riparian areas (see Table 10.4 for a common name to latin name list). Sources of information on species and their natural history in riparian habitats in eastern North America include Conant and Collins (1991); Vogt (1981); Degraaf and Rudis (1986); Green and Pauley (1987); Ernst and Barbour (1989); Klemens (1993); Ernst et al. (1994); Mitchell (1994); Oldfield and Moriarty (1994); Palmer and Braswell (1995); Harding (1997); Green (1997); and Petranka (1998). A full bibliography of regional checklists and identification guides is in Moriarty and Bauer (in press).

Table 10.1   Examples of amphibians and reptiles known to use riparian zones for reproduction, foraging, or shelter in the eastern continental forest. Abbreviations: S = Seeps, St = Stream, R = River, L/P = Lakes and Ponds, BP = Beaver Ponds, B/F = Bogs and Fens, EP = Ephemeral Pools. Common names follow Crother (in press).

Species	S	St	R	L/P	BP	B/F	EP
Salamanders							
Lesser siren		X	X	X	X		
Hellbender			X				
Mudpuppy		X	X	X			
Eastern newt		X	X	X	X	X	X
Ringed salamander				X	X		X
Streamside salamander		X					X

Table 10.1 Continued

Species	S	St	R	L/P	BP	B/F	EP
Jefferson salamander					X	X	X
Blue-spotted salamander					X	X	X
Spotted salamander					X	X	X
Marbled salamander				X	X		X
Mole salamander				X	X		
Eastern tiger salamander		X		X	X		X
Small-mouthed salamander				X	X		X
Allegheny mountain dusky salamander	X	X					
Blue Ridge dusky salamander	X	X					
Carolina mountain dusky salamander	X	X					
Ocoee salamander	X	X					
Imitator salamander	X	X					
Santeetlah dusky salamander	X	X			X		
Northern dusky salamander	X	X			X		
Pygmy salamander	X	X					
Black-bellied salamander	X	X			X		
Seal salamander	X	X					
Black Mountain salamander	X	X					
Shovel-nosed salamander	X	X					
Spring salamander	X	X			X		
Red salamander	X	X				X	
Mud salamander		X	X			X	
Northern slimy salamander		X	X	X			
White spotted slimy salamander		X	X	X			
Western slimy salamander		X	X	X			
Ravine salamander	X	X					
Valley and Ridge salamander	X						
Wehrle's salamander	X	X					
Cow Knob salamander	X						
Red-backed salamander	X	X	X	X	X		
Cumberland Plateau salamander	X	X	X	X			
Yonahlossee salamander		X	X				

Table 10.1 Continued

Species	S	St	R	L/P	BP	B/F	EP
Four-toed salamander		X			X	X	X
Junaluska salamander	X	X					
Many-ribbed salamander	X	X					
Northern two-lined salamander	X	X			X		
Southern two-lined salamander	X	X			X		
Blue Ridge two-lined salamander	X	X					
Long-tailed salamander	X	X					
Three-lined salamander	X	X			X		
Cave salamander	X	X					
Frogs and Toads							
Eastern spadefoot					X		X
American toad	X	X	X	X	X	X	X
Fowler's toad		X	X	X	X		X
Northern cricket frog	X	X	X	X	X		X
Bird-voiced treefrog				X	X		X
Green treefrog		X	X	X	X		X
Barking treefrog				X	X		X
Gray treefrog	X	X	X	X	X	X	X
Cope's gray treefrog	X	X	X	X	X	X	X
Mountain chorus frog	X	X			X		X
Southeastern chorus frog	X	X	X	X	X	X	X
Spring peeper	X	X	X	X	X	X	X
Eastern narrow-mouthed toad							X
Mink frog				X	X		
American bullfrog		X	X	X	X	X	X
Green frog		X	X	X	X	X	X
Northern leopard frog		X	X	X	X	X	X
Southern leopard frog		X	X	X	X	X	X
Plains leopard frog		X		X	X		
Carpenter frog		X			X	X	X
Crawfish frog				X	X		X
Pickerel frog		X	X	X	X		

Table 10.1. Continued.

Species	S	St	R	L/P	BP	B/F	EP
Wood frog				X	X	X	X
Turtles							
Alligator snapping turtle			X				
Snapping turtle		X	X	X	X	X	X
Eastern mud turtle		X	X	X	X		
Loggerhead musk turtle		X	X	X			
Eastern musk turtle		X	X	X	X		
Eastern box turtle	X	X	X	X			X
Spotted turtle	X			X	X	X	X
Wood turtle		X	X		X		
Bog turtle		X				X	
Northern map turtle		X	X				
False map turtle		X	X				
Ouachita map turtle		X	X				
Painted turtle		X	X	X	X	X	X
Blanding's turtle		X	X	X	X	X	
River cooter			X	X			
Northern red-bellied cooter			X	X	X		
Slider			X	X	X		
Spiny softshell		X	X				
Smooth softshell		X	X				
Lizards							
Five-lined skink	X	X	X	X	X		
Southeastern five-lined skink	X	X	X	X	X		
Broad-headed skink		X	X	X			
Snakes							
Northern watersnake		X	X	X	X	X	X
Plain-bellied watersnake		X	X	X			
Diamond-backed watersnake		X	X	X			
Queen snake		X	X	X			
Kirtland's snake	X	X	X				

Table 10.1 Continued

Species	S	St	R	L/P	BP	B/F	EP
Eastern Ribbonsnake	X	X	X	X	X		X
Common gartersnake		X	X	X	X	X	X
Short-headed gartersnake		X	X	X	X		
Butler's gartersnake		X	X	X	X		
Plains gartersnake		X	X	X	X		
Dekay's brownsnake		X	X				
Red-bellied snake		X	X			X	
Rough earthsnake		X	X				
Smooth earthsnake		X	X				
Ring-necked snake		X	X	X			
Western wormsnake		X	X	X	X		
Eastern wormsnake		X	X	X	X		
Rough greensnake		X	X	X	X		
Eastern ratsnake		X	X	X	X		
Western fox snake		X	X	X			
Milksnake			X	X			
Prairie kingsnake			X	X			
Common kingsnake		X	X	X	X		
Cottonmouth			X	X	X		X
Copperhead		X	X	X	X		
Massasauga	X	X	X		X		
Pygmy rattlesnake		X	X				
Timber rattlesnake		X	X				

Headwater Wetlands and Seepages

Riparian zones associated with headwater (first order) streams and seepages are narrow transitional areas between shallow water and terrestrial habitat. The emerging groundwater and the canopy cover create a cool, moist environment. Flooding occurs typically in spring when many amphibians breed. The flushing effect renders these wetlands less acidic than other wetlands (Welsch et al. 1995). This habitat supports a diversity of salamanders, including widespread species such as northern dusky (*Desmognathus fuscus*), four-toed (*Hemidactylium scutatum*), spring (*Gyrinophilus porphyriticus*), and red (*Pseudotriton ruber*) salamanders. Allegheny Mountain dusky (*Desmognathus ochrophaeus*) (Marcum 1994) and pygmy salamanders (*Desmognathus wrighti*), for example, deposit eggs in seeps and headwater streams (Organ 1961). After a short larval period, juveniles seek refuge beneath leaves and woody debris on the forest floor. Although some adults stay near the water, more are terrestrial and move into the adjacent forest (J.C. Mitchell, personal observations). Pickerel frogs (*Rana palustris*) may breed here, as well as wood frogs (*Rana sylvatica*), if pools of water are present in late winter and spring. No reptiles are permanent residents, but several use these zones for foraging (e.g., common gartersnakes, *Thamnophis sirtalis*) and escape from summer heat (eastern box turtles, *Terrapene carolina*). Approximately 30 species of salamanders, 7 species of frogs, 3 species of turtles, 2 species of lizards, and 3 species of snakes occur in this habitat.

Streamside Riparian Zones

Riparian habitats associated with second- and third-order streams usually occur in narrow floodplains and may or may not have standing water in pools. Adjacent seepages may provide another habitat component. These habitats and associated streams provide environments for nesting, larval development, foraging, and refuge for several genera of salamanders, including dusky salamanders (*Desmognathus*), brook salamanders (*Eurycea*), the spring salamander (*Gyrinophilus*), and red and mud salamanders (*Pseudotriton*). Brook salamanders (e.g., northern two-lined [*Eurycea bislineata*] and long-tailed [*Eurycea longicauda*]) deposit eggs beneath rocks in streams or seeps, and adults move into adjacent moist, terrestrial habitats outside the breeding season (T. K. Pauley and J. C. Mitchell, personal observations). Pickerel frogs commonly occur along first-order streams. Other species of frogs use this habitat, including northern cricket frogs (*Acris crepitans*) and green frogs (*Rana clamitans*). Wood turtles (*Clemmys insculpta*) use first- to fourth-order streams for overwintering and the riparian areas of adjacent uplands for foraging and nesting (Kaufmann 1992). Box turtles are commonly encountered in riparian areas along streams. Northern watersnakes (*Nerodia sipedon*) and gartersnakes (*Thamnophis* spp.) forage for fish and amphibians along the edge. Approximately 38 species of salamanders, 16 species of frogs, 13 species of turtles, 3 species of lizards, and 25 species of snakes use this habitat.

Riverine Floodplains

Wide floodplains are characteristic of large riverine systems. These riparian areas often have wooded swamps, marshes, oxbows, pools of standing water, meandering tributaries, and seepages depending on geographic location and local land use (Mitsch and Gosselink 1993). These habitats are inundated periodically by floods, and the amphibian and reptile populations here almost certainly fluctuate dramatically. These riparian systems offer a diversity of habitats to amphibians and reptiles, including American bullfrogs (*Rana catesbeiana*), mole salamanders (*Ambystoma* spp.), two-lined salamanders (*Eurycea* spp.), northern watersnakes, plainbellied water snakes (*Nerodia erythrogaster*), river cooters (*Pseudemys concinna*), and softshell turtles (*Apalone* spp.). Species richness in river floodplains is high (e.g., Wharton et al. 1981). Approximately 10 species of salamanders, 13 species of frogs, 17 species of turtles, 3 species of lizards, and 28 species of snakes use this habitat.

Large Ponds and Lakes

Riparian areas around large ponds and lakes vary in width and character, are commonly terrestrial-like, and can include other aquatic habitats such as bogs. Soils range from hydric to xeric and thus support a diversity of plant communities comprised of herbaceous (grasses, forbs) and woody (shrub, tree) species. These plant communities, in turn, support many species of amphibians and reptiles. Margins of large reservoirs may be relatively barren and offer little vegetative cover; rocks become cover objects. Fish may or may not be present in the ponds. Frogs are conspicuous features of this zone throughout eastern North America. Widespread examples include bullfrogs, green frogs, northern leopard frogs (*Rana pipiens*), cricket frogs (*Acris* spp.), and American toads (*Bufo americanus*). These amphibians forage in and call from the grassy margins, lay eggs in shallow water where larval development occurs (2 years or more for the bullfrog), and overwinter beneath grass mats or under water. Some of these ponds and lakes have an annual anoxic period and are without fish which results in a higher diversity and density of amphibians (Vogt 1981). Snakes, such as northern watersnakes and garter snakes, prey on amphibians in all life history stages. Painted turtles (*Chrysemys picta*) and snapping turtles (*Chelydra serpentina*) use the riparian area for foraging and in some cases for nesting. Approximately 13 species of salamanders, 19 species of frogs, 11 species of turtles, 3 species of lizards, and 20 species of snakes utilize this habitat.

Beaver Ponds and Small Impoundments

Impoundments constructed in small streams by beavers (*Castor canadensis*) and humans create wetland habitats with riparian areas of various widths and with differing vegetation communities, depending on the nature of the surrounding watershed. Riparian habitats range from narrow transition zones with thick shrub vegetation to wide sedge meadows between the open water and surrounding forests (Welsch et al. 1995). These areas support numerous species of frogs, salamanders, turtles, and snakes. Widespread examples are gray treefrogs (*Hylachrysoscelis, H. versicolor*), spring peepers (*Pseudacris crucifer*), cricket frogs, green frogs, red-spotted newts (*Notophthalmus v. viridescens*), northern water snakes, rough green snakes (*Opheodrys aestivus*), painted turtles, and snapping turtles. Amphibians lay their eggs in shallow water where larvae develop and otherwise spend their lives in the surrounding vegetation. Snakes forage along the margins and turtles dig nests in and beyond riparian areas. At least 21 species of salamanders, 22 species of frogs, 9 species of turtles, 2 species of lizards, and 14 species of snakes use this habitat.

Bogs and Fens

Bogs are northern and high-elevation peatlands characterized by cool temperatures, anaerobic soil water, and low acidity (pH < 5) that retards decomposition (Richardson and Gibbons 1993; Welsch et al. 1995). These features result in an accumulation of organic matter and the formation of deep peat substrates. The primary water source is precipitation. The acid species of *Sphagnum* moss is the characteristic plant in this community. Fens (also known as mires, glades, and forested swamps) are peatland habitats but differ in that they receive water from streams and groundwater containing bicarbonates, and they span a water pH range from 5 to 9 (Welsch et al. 1995). Organic substrates (peats) also form in these wetlands for the same reasons, and a few alkaline species of *Sphagnum* may also be present. The riparian areas of bogs and fens extend from the open water of pools and streams to the edge of the upland habitat and contain shallow, standing water seasonally. The bog turtle (*Clemmys muhlenbergii*), now protected under the U.S. Endangered Species Act (USDI Fish and Wildlife Service 1997), is the premier reptile in these habitats. Gray treefrogs, spring peepers, chorus frogs, wood frogs, blue-spotted salamanders (*Ambystoma laterale*), dusky salamanders, brook salamanders, snapping turtles, and spotted turtles (*Clemmys guttata*) are examples of habitat associates of the bog turtle. Approximately 7 species of salamanders, 11 species of frogs, 5 species of turtles, and 3 species of snakes occur in this habitat.

Ephemeral Pools

These temporary pools are small bodies of water that may support a surprisingly large number of amphibians and reptiles (Roble 1989; Mitchell 1997a, b). A few amphibian species, such as marbled salamander (*Ambystoma opacum*) and eastern spadefoot (*Scaphiopus holbrookii*) must breed in ephemeral pools. Some mole salamanders in the genus *Ambystoma* and fairy shrimp (Crustacea, Amphipods) are common species that avoid breeding sites with fish (Hopey and Petranka 1994). These fishless ephemeral pools occur throughout Eastern North America and come in a wide variety of configurations, including natural sinkhole pools, woodland depressions, and shallow depressions in fields and in dirt roads (Adams and Lacki 1993). Ephemeral pools fill from fall and winter rains and usually dry up by summer through evapotranspiration and lowering of the water table (Welsch et al. 1995). Variable hydrologies have dramatic effects on the amphibian fauna (Rowe and Dunson 1995), and duration of the hydroperiod is a key determinant of the fauna of ephemeral ponds (Schneider and Frost 1996). The riparian zone associated with this wetland is narrow and can be as simple as the margin of the depression. Most ephemeral pools occur, or did so in the past, in forested areas of the Continental Eastern United States. The exception is the prairie pothole region. Amphibians that breed in these wetlands and reptiles frequenting these sites for foraging use the surrounding terrestrial habitat extensively. Over 10 species of salamanders, 19 species of frogs, 4 species of turtles, and 4 species of snakes have been found in association with ephemeral pools. Table 10.2 summarizes the number of species found in all of the habitats.

Table 10.2 The number of amphibian and reptile species occurring in riparian habitats in the Continental Eastern United States

Amphibian or Reptile Group	Habitat Location							
	Seep	Stream	River	Lakes	Beaver Pond	Bog or Fen	Ephem. Pool	All
Amphibians								
Salamanders	30	38	10	13	21	7	10	48
Frogs & Toads	7	16	13	19	22	11	19	23
Total	37	54	23	32	43	18	29	71
Reptiles								
Turtles	3	13	17	11	9	5	4	19
Lizards	2	3	3	3	2	0	0	3
Snakes	3	25	28	20	14	3	4	28
Total	8	41	48	34	25	8	8	50
Amphibians and Reptiles								
Total	45	95	71	66	68	26	37	121

The Terrestrial Connection

Amphibians and reptiles are not sedentary animals. Even the smallest of them may travel long distances during its life (Dodd 1996). Although many species are conspicuous at wetland breeding sites when male frogs call to attract females and salamanders mate in water at night during the breeding season, adults and juveniles disperse to upland terrestrial habitats during nonreproductive times of the year. For many species, time spent at breeding sites is a small portion of their adult life. The width of the riparian area is an important component of this landscape-dynamic behavior. Narrow areas will be traversed quickly and play small roles in the ecology of amphibians, whereas wide riparian areas, such as those associated with riverine floodplains, may be occupied for longer periods for reproduction, foraging, and shelter. Distances covered by amphibians as they disperse away from breeding sites vary from only a few yards for small salamanders to a kilometer or more for large salamanders and frogs (Table 10.3). Because all but the completely aquatic amphibians (e.g., mudpuppies [*Necturus maculosus*], sirens [*Siren* spp.]) use upland habitats extensively, the type and quality of those habitats become extremely important for maintenance of population viability.

Table 10.3 Examples of distances moved by selected eastern North American salamanders, frogs, freshwater turtles, and snakes from aquatic habitats overland into terrestrial habitats. All distances are in feet. Single numbers are maximum distances reported, and ranges are provided where available. Distances for amphibians and snakes represent terrestrial locations irrespective of behavior and distances for turtles are from nest locations, except for *Clemmys guttata*, which are upland aestivation sites. Several references for this table were derived from Dodd (1996).

Species	Location	Movement	Reference
Salamanders			
Jefferson salamander	Ohio	10-810	Downs (1989)
Jefferson salamander	New York	5282	Bishop (1941)
Spotted salamander	Michigan	515-817	Kleeberger & Werner (1983)
Spotted salamander	Kentucky	20-722	Douglas & Monroe (1981)
Spotted salamander	New York	50-690	Madison (1997)
Marbled salamander	Indiana	0-1477	Semlitsch (1998)
Mole salamander	S Carolina	265-856	Semlitsch (1981)
Eastern tiger salamander	S Carolina	532	Semlitsch (1983)
Northern dusky	Kentucky	10	Barbour et al. (1969)

Table 10.3 Continued

Species	Location	Movement	Reference
Ocoee salamander	N Carolina	200	Hairston (1987)
Allegheny Mt. dusky	W Virginia	0-72	Ordiway (1994)
Northern two-lined salamander	Quebec	328	MacCulloch & Bider (1975)
Red-spotted newt	Mass.	2625	Healy (1975)
Frogs			
Northern cricket frog	Texas	550	Pyburn (1958)
American toad	Minnesota	32800	Ewert (1969)
Eastern narrow-mouthed	Florida	138-3000	Dodd (1996)
Southeastern chorus frog	Indiana	328	Kramer (1974)
American bullfrog	New York	350	Raney (1940)
Pickerel frog	Minnesota	1640	Oldfield & Moriarty (1994)
Northern leopard frog	Minnesota	4921	Oldfield & Moriarty (1994)
Eastern spadefoot	Florida	1319	Pearson (1955)
Turtles			
Snapping turtle	Michigan	3-600	Congdon & Gatten (1989)
Painted turtle	Michigan	3-538	Congdon & Gatten (1989)
Spotted turtle	Connecticut	10-541	Perillo (1997)
Wood turtle	Illinois	2100-3000	Rowe & Moll (1991)
Wood turtle	Minnesota	ave=656	R.R. Buech (unpub. data)
Chicken turtle	S Carolina	ave=230	K.A. Buhlmann (p. comm.)
Blanding's turtle	Michigan	7-3658	Congdon & Gatten (1989)
False map turtle	Wisconsin	15-500	Vogt (1981)
Eastern mud turtle	S Carolina	ave=160	V.J. Burke (p. comm.)
Northern red-bellied cooter	Virginia	820	Mitchell (1994)
Snakes			
Northern water snake	W Virginia	2297	T.K. Pauley (unpub. data)
Common gatersnake	Kansas	1139-2300	Fitch (1958, 1965)
Queen snake	Ohio	8	Wood (1944)

In contrast to many amphibians, freshwater turtles spend their lives more closely associated with water. Although wood turtles may become terrestrial in July and August (Buech 1995), other freshwater turtles leave aquatic sites only to disperse or find suitable nesting sites. All female turtles nest on land and most dig cavities in the ground to deposit their eggs. Nest location distances from aquatic sites vary greatly. Some turtles nest within the riparian area (Jackson and Walker 1997), but others will travel long distances to find a suitable location (Table 10.3). Loss or degradation of such upland nesting sites may cause turtles to lay eggs in less suitable places. Consequences include lower egg survival, reduced juvenile recruitment, and the production of all male or all female offspring, as sex in many species of freshwater turtles is determined by nest temperature during egg incubation. Nest temperature is influenced by nest location (Bull and Vogt 1979; Vogt and Bull 1984).

Distances moved by aquatic and semi-aquatic snakes are based on recapture data obtained from individuals caught in wetland habitats and upland hibernacula. The variation in the few available records for viviparous species (Table 10.3) is at least partly due to the location of suitable overwintering sites. Data on distances of nest sites from wetlands for oviparous species are, as far as we can ascertain, nonexistent.

Threats to Amphibians and Reptiles

Riparian habitats are vulnerable to a wide array of perturbations. Major disturbances include surface water withdrawal, road construction, bridge crossings, dam construction, flooding, logging, mining, agricultural and urban development, recreation, grazing, exotic species introduction, human-associated debris, and pollution, including acid rain, pesticides, and trace metals (Mitsch and Gosselink 1993; Naiman and DeCamps 1997; Dodd 1997; Buhlmann and Gibbons 1997). Loss, degradation, and fragmentation of riparian habitats from these human sources reduce amphibian and reptile populations. Other types of threats are more taxa-specific.

Populations of some amphibians have declined precipitously and some species have even become extinct in parts of western North America and other parts of the world from habitat loss and degradation, pollution, introduced species, pathogens, and increased ultraviolet radiation due to ozone thinning (Blaustein and Wake 1990; Blaustein et al. 1994a, b; Ovaska 1997; Berger et al. 1998). In eastern North America, the northern cricket frog has disappeared from parts of its range in the upper Midwest (Lannoo et al. 1994), striped newts (*Notophthalmus perstriatus*) have been impacted from wetland loss in the Southeast (Dodd and LeClaire 1995), and the tiger salamander (*Ambystoma tigrinum*) is rare in the northeastern portion of its range (McCoy 1985; Mitchell 1991). Habitat alteration is apparently the most prevalent cause of amphibian population decline or loss in the East. Acid precipitation and pollution causes amphibian decline in some areas, such as the Appalachian Mountains and the Northeast (Freda et al.1991; Rowe et al. 1992). Populations of some vertebrate predators of amphibians and reptiles, such as raccoons (*Procyon lotor*), opossum (*Didelphis viriginiana*) and ravens (*Corvus* sp.), have reached unnaturally high levels because of food and shelter inadvertently provided by humans. Survival and reproductive success of predators are enhanced by these subsidies when natural resources are low (Boarman 1993). Raccoons are well known subsidized predators of frogs and turtles and their eggs (Chapman and Feldhamer 1982; Ernst et al. 1994). Removal of raccoons in an Iowa study was the apparent cause of increased survival in freshwater turtle populations (Christiansen and Gallaway 1984). Corvids (crows and ravens) eat freshly laid turtle eggs (Mitchell 1994), and the managed wild turkey (*Meleagris gallopavo*) is known to eat salamanders and turtle eggs (Dickson 1992; Mitchell and Klemens *in press*). In the Midwest, striped skunks (*Mephitis mephitis*) are major predators of turtle eggs (Vogt 1981). Uncontrolled subsidized predator populations in some riparian areas could decrease the viability of amphibian and reptile populations.

Collection of amphibians and reptiles for the pet trade is an ever-increasing threat to many populations. Snakes and turtles are prime targets, but frogs and some salamanders are being collected as well (Buck 1997). Turtles in the genus *Clemmys* (bog, spotted, wood), all residents of riparian habitats, bring high prices in the wildlife trade. For example, 4692 *Clemmys* (all four species combined) were exported from the United States during 1989-1994, despite the fact that several were protected by state law and international treaty, and sold for a total market value of $102,658 (Salzberg 1995). During that same period, 317,156 painted turtles (*Chrysemys picta*) were exported for a total of $759,685. Such pressures on wild populations are known to cause declines in these vertebrates (Klemens 1993).

Conservation and Management

The ecological health of riparian habitats is reflected in high herpetofaunal diversity and numbers of individuals. A shift in species richness, decline of a species, or drastic change in population or community structure may indicate habitat alteration. Without amphibians and reptiles, energy flow would be reduced and species interactions would be lessened, resulting in ecological imbalances. For example, salamanders in a northeastern forest have been shown to be 60% efficient at converting ingested energy into new tissue (Burton and Likens 1975), predatory salamander larvae are important in determining types and amounts of zooplankton and insects (Dodson 1970; Dodson and Dodson 1971), and tadpoles are important in determining types and amounts of phytoplankton, magnitude of nutrient cycling, and levels of primary production (Seale 1980). Fluctuations of amphibian populations are natural, but they make detection of long-term trends difficult (Pechmann et al. 1991; Pechmann and Wilbur 1994). Thus, the importance of these taxa to riparian systems, and the need for long-term monitoring, suggest that integrated management guidelines should be developed by all appropriate agencies and organizations.

Perturbations that affect riparian habitat integrity also affect herpetofaunal integrity. Thus, conservation efforts to maintain riparian health will benefit amphibians and reptiles. Several directly benefit these vertebrates. Hardwood forest canopies should remain intact and be restored where forests once occurred naturally. Canopy removal alters the microclimate (especially temperature and moisture), reduces leaf litter and coarse woody debris (CWD), and compacts the soil, which in turn, results in restricted movement of terrestrial and some streamside salamanders (deMaynadier and Hunter 1995). High sediment loads in water smother amphibian eggs and larvae, and low levels of oxygen due to high biological oxygen demand (BOD) will suffocate them (T. K. Pauley, unpublished data). Siltation in streams from logging and from road and bridge crossings should be avoided. Like native species, cattle use riparian areas for water, shade, and food (Kauffman and Krueger 1984), but their use of these areas often severely degrades habitats. Pollution from upstream and upland sources can have devastating consequences for amphibians and can result in malformations of larvae and metamorphs (Bonin et al. 1997; Quellet et al. 1997) and, by extension, can cause population declines.

Because many riparian areas are used only temporarily by amphibians and reptiles, managers must seriously consider the composition of the landscape beyond riparian areas to ensure that they remain ecologically functional. The use of buffer zones has been recommended by numerous ecologists concerned with water quality and fisheries resources. One model designed to ameliorate agricultural impact proposes a three-tiered buffer strip totaling 70 ft wide (Naiman and Decamps 1997). Forested buffer widths of 30 to 200 ft have been recommended to protect water quality (Brazier and Brown 1973; Aubertin and Patric 1974; Welsch 1991) and widths of 10 to 350 ft for fish populations (Castelle et al. 1994). Buffer widths $\geq$100 ft have been recommended for maintaining macroinvertebrate species diversity in streams within logged watersheds (Erman et al. 1977; Newbold et al. 1980). Castelle et al. (1994) concluded from a literature survey that buffers of 50 to 100 ft will protect streams and wetlands under most circumstances.

Less is known about appropriate sizes and effectiveness of buffer zones as a means of protecting the herpetofauna within riparian habitats. Available results from the limited studies conducted in eastern North America are insufficient to suggest general management recommendations. In Kentucky, similar numbers of amphibians and reptiles were found in a clearcut bisected by a stream with a 50-ft forested buffer zone and one without a buffer (Pais et al. 1988). Numbers of species were highest in a mature forest. In eastern Texas, Rudolph and Dickson (1990) found fewer numbers and species of amphibians and reptiles in narrow (<80 ft) buffer strips than in wider strips (100 to 300 ft). In southeastern Texas, Foley (1994) found no clear differences in numbers of amphibians and reptiles inside and outside of 65 ft wide streamside management zones among control, select-cut, and clearcut silvicultural treatments. Distances moved by amphibians and reptiles into riparian and upland habitats from aquatic habitats (Table10.2) suggest that the narrower buffer widths do not completely encompass the areas used by many of these animals. Burke and Gibbons (1995) suggest a buffer width of 900 ft but do not address habitat management within buffer zones.

Dodd and Cade (1998) determined that amphibian movements away from a temporary breeding pond were nonrandom and influenced by differences in terrestrial habitat preferences. Obviously, the type of wetland with which the riparian area is associated and the composition and needs of the herpetofauna influence directly the effective size of the buffer zone for these animals.

The use of buffer zones as management tools to ensure population viability in riparian areas is a complex issue. We cannot always expect that buffers will be wide enough despite the best efforts of managers. Thus, the nature of the habitat beyond the buffer zone, and whether it meets the needs of the amphibians and reptiles in the watershed, becomes an important consideration. Some species in the local assemblage will be generalists and able to use a variety of natural and managed habitats (e.g., American toads in wildlife openings). Others will be more specialized and less able to tolerate managed or altered systems (e.g., two-lined salamanders in clearcuts). The geographic location, nature of the local ecosystem, and the range of microhabitats in and out of the riparian zone will dictate the composition of the amphibian and reptile assemblages in a particular area. Adequate buffer zone width depends on the type of wetland and riparian habitats present and on the home ranges, movement, and habitat selection of local species. In addition, between-habitat variation in survival and life-history traits in amphibians that occupy a variety of habitat types determines local population dynamics (Holomuzki 1997).

Many amphibians and reptiles need both aquatic and terrestrial habitats for reproduction, foraging, or shelter. In some landscapes, the riparian area may be too narrow to accommodate their spatial needs, and forested buffer zones added to them may be too wide and provide inappropriate habitat for some species. Thus, the temporal and spatial needs of species during their nonbreeding periods present a complex problem when those needs cannot be met by simply adding more forest buffer. Wide buffer zones to protect riparian habitats may jeopardize species that require nonforest terrestrial habitats (e.g., Jackson and Walker 1997). Simply adding buffers to the landward side of riparian areas, as suggested by some authors (e.g., Schaefer and Brown 1992), may not achieve management goals for these animals. Perhaps the most reasonable management activity would be to mimic historic (prior to European settlement) disturbances that provided evolutionary basis for natural community structure. Thus, understanding the natural history and dynamics of the local herpetofauna at the landscape scale becomes the first requirement for developing management actions on their behalf.

The ecological health and management of the aquatic side of riparian systems should not be ignored. Amphibians and reptiles require clean water and specific microhabitats for key life functions. For example, wood turtles require open areas, often provided by eroding streambanks, for nesting (Buech 1995). Stabilization of all eroding streambanks in a watershed in the name of trout management can negatively affect this declining species by eliminating these nesting areas and creating populations of non-reproducing adults.

We recommend that a holistic view of habitat protection and management be adopted by all managers faced with conflicting demands. The needs of all taxa must be considered when developing effective management plans and conducting restoration projects. Such plans and options for local herpetofaunas will not be effective if information is derived solely from generalized field guides and regional management guides. While such publications provide useful background information and general recommendations, they do not substitute for a thorough understanding of the natural history, population ecology, and landscape movement dynamics of the amphibian and reptile species in an area. We cannot overemphasize how important such knowledge is to the development of management plans. Recent advances in assessing habitat suitability for wildlife that incorporate horizontal and vertical habitat strata into GIS add predictive power to such models (Short et al. 1996). However, these models do not substitute for accurate, quantitative knowledge of local population dynamics.

A number of specific recommendations can be made to managers charged with protection and management of eastern herpetofaunas that, in our experience, enhance the production of effective management plans.

Obtain a thorough baseline data set on the life history of the species in your area to answer questions related to who, what, where, and when. Understand the population and seasonal dynamics of the target taxa and the range of threats to them, including the problems of overcollecting and subsidized predators.

Develop monitoring studies to determine population trends and detect changes in species composition among sites (see Heyer et al. 1994; Fellers and Freel 1995; Olson et al. 1997). Train field personnel extensively to conduct such studies and allocate sufficient time to meet these obligations.

Engaging experienced field herpetologists to assist in all phases of the program ensures accuracy and provides the best return for limited funds. Many amphibians and reptiles are secretive, active for only limited periods, in some cases difficult to identify (especially larvae and immatures), and hard to monitor effectively. However, because these animals are so closely tied to aquatic, riparian, and terrestrial habitat complexes, understanding populations in the target area will ensure that management actions will be effective.

Summary and Conclusions

Riparian habitats in eastern North America are used extensively by amphibians and reptiles. Such ecotones associated with freshwater wetlands provide shelter, food, breeding and egg-laying sites, and places for larval and juvenile growth and development. Where riparian zones are wide, individuals may spend their entire lives there. Where the zones are narrow, these animals may simply pass through on their way to water or, conversely, to terrestrial habitats. Daily or seasonal movement distances may far exceed the width of the riparian zone and the buffer strip designed to protect wetlands and riparian habitats. Knowledge of how local populations of amphibians and reptiles function in particular riparian habitats and settings is crucial to the development of effective management options to protect them from decline or local extirpation.

Studies of amphibians and reptiles in the East lag behind such studies in western North America. Clearly, there is a wealth of opportunity for research on species movements from a landscape perspective, how amphibians and reptiles use different types and sizes of riparian habitats, and how the ecological structure of riparian zones (e.g., CWD) enhances or inhibits herpetofaunal populations. Riparian habitats are parts of larger landscapes and fully functional ecological complexes (Gregory et al. 1991; Malanson 1993). Management of amphibians and reptiles from such a holistic perspective will be more effective than one that focuses only on single species or habitats.

Sandy Verry

Mike Becker

A permanent pool in West Virginia (top), or a vernal pool in Minnesota (bottom) are home to amphibians and reptiles.

Table 10.4 A comparison of common and scientific names for amphibians and reptiles

Common Name	Scientific Name
Salamanders	
Lesser siren	*Siren intermedia*
Hellbender	*Cryptobranchus alleganiensis*
Mudpuppy	*Necturus maculosus*
Eastern newt	*Notophthalmus viridescens*
Ringed salamander	*Ambystoma annulatum*
Streamside salamander	*Ambystoma barbouri*
Jefferson salamander	*Ambystoma jeffersonianum*
Blue-spotted salamander	*Ambystoma laterale*
Spotted salamander	*Ambystoma maculatum*
Marbled salamander	*Ambystoma opacum*
Mole salamander	*Ambystoma talpoideum*
Easter tiger salamander	*Ambystoma tigrinum*
Small-mouthed salamander	*Ambystoma texanum*
Allegheny mountain dusky	*Desmognathus ochrophaeus*
Blue Ridge dusky salamander	*Desmognathus orestes*
Carolina mountain dusky salamander	*Desmognathus carolinensis*
Ocoee salamander	*Desmognathus ocoee*
Imitator salamander	*Desmognathus imitator*
Santeetlah dusky salamander	*Desmognathus santeetlah*
Northern dusky salamander	*Desmognathus fuscus*
Pygmy salamander	*Desmognathus wrighti*
Black-bellied salamander	*Desmognathus quadramaculatus*
Seal salamander	*Desmognathus monticola*
Black Mountain salamander	*Desmognathus welteri*
Shovel-nosed salamander	*Desmognathus marmoratus*
Spring salamander	*Gyrinophilus porphyriticus*
Red salamander	*Pseudotriton ruber*
Mud salamander	*Pseudotriton montanus*
Northern slimy salamander	*Plethodon glutinosus*
White spotted slimy salamander	*Plethodon cylindraceus*
Western slimy salamander	*Plethodon albagula*
Ravine salamander	*Plethodon richmondi*

Table 10.4 Continued

Common Name	Scientific Name
Valley and Ridge salamander	*Plethodon hoffmani*
Wehrle's salamander	*Plethodon wehrlei*
Cow Knob salamander	*Plethodon punctatus*
Red-backed salamander	*Plethodon cinereus*
Cumberland Plateau salamander	*Plethodon kentucki*
Yonahlossee salamander	*Plethodon yonahlossee*
Four-toed salamander	*Hemidactylium scutatum*
Junaluska salamander	*Eurycea junaluska*
Many-ribbed salamander	*Eurycea multiplicata*
Northern two-lined salamander	*Eurycea bislineata*
Southern two-lined salamander	*Eurycea cirrigera*
Blue Ridge two-lined salamander	*Eurycea wilderae*
Long-tailed salamander	*Eurycea longicauda*
Three-lined salamander	*Eurycea guttolineata*
Cave salamander	*Eurycea lucifuga*
Frogs and Toads	
Eastern spadefoot	*Scaphiopus holbrookii*
American toad	*Bufo americanus*
Fowler's toad	*Bufo fowleri*
Northern cricket frog	*Acris crepitans*
Bird-voiced treefrog	*Hyla avivoca*
Green treefrog	*Hyla cinerea*
Barking treefrog	*Hyla gratiosa*
Gray treefrog	*Hyla versicolor*
Cope's gray treefrog	*Hyla chrysoscelis*
Mountain chorus frog	*Pseudacris brachyphona*
Southeastern chorus frog	*Pseudacris feriarum*
Spring peeper	*Pseudacris crucifer*
Eastern narrow-mouthed toad	*Gastrophryne carolinensis*
Mink frog	*Rana septentrionalis*
American bullfrog	*Rana catesbeiana*
Green frog	*Rana clamitans*
Northern leopard frog	*Rana pipiens*

Table 10.4 Continued

Common Name	Scientific Name
Southern leopard frog	*Rana sphenocephala*
Plains leopard frog	*Rana blairi*
Carpenter frog	*Rana virgatipes*
Crawfish frog	*Rana areolata*
Pickerel frog	*Rana palustris*
Wood frog	*Rana sylvatica*
Turtles	
Alligator snapping turtle	*Macroclemys temmincki*
Snapping turtle	*Chelydra serpentina*
Eastern mud turtle	*Kinosternon subrubrum*
Loggerhead musk turtle	*Sternotherus minor*
Eastern musk turtle	*Sternotherus odoratus*
Eastern box turtle	*Terrapene carolina*
Spotted turtle	*Clemmys guttata*
Wood turtle	*Clemmys insculpta*
Bog turtle	*Clemmys muhlenbergii*
Chicken turtle	*Deirochelys reticularia*
Northern map turtle	*Graptemys geographica*
False map turtle	*Graptemys pseudogeographica*
Ouachita map turtle	*Graptemys ouachitensis*
Painted turtle	*Chrysemys picta*
Blanding's turtle	*Emydoidea blandingii*
River cooter	*Pseudemys concinna*
Northern red-bellied cooter	*Pseudemys rubriventris*
Slider	*Trachemys scripta*
Spiny softshell	*Apalone spinifera*
Smooth softshell	*Apalone mutica*
Lizards	
Five-lined skink	*Eumeces fasciatus*
Southeastern five-lined skink	*Eumeces inexpectatus*
Broad-headed skink	*Eumeces laticeps*

Table 10.4 Continued

Common Name	Scientific Name
Snakes	
Northern watersnake	*Nerodia sipedon*
Plain-bellied watersnake	*Nerodia erythrogaster*
Diamond-backed watersnake	*Nerodia rhombifera*
Queen snake	*Regina septemvittata*
Kirtland's snake	*Clonophis kirtlandii*
Eastern Ribbonsnake	*Thamnophis sauritus*
Common gartersnake	*Thamnophis sirtalis*
Short-headed gartersnake	*Thamnophis brachystoma*
Butler's gartersnake	*Thamnophis butleri*
Plains gartersnake	*Thamnophis radix*
Dekay's brownsnake	*Storeria dekayi*
Red-bellied snake	*Storeria occipitomaculata*
Rough earthsnake	*Virginia striatula*
Smooth earthsnake	*Virginia valeriae*
Ring-necked snake	*Diadophis punctatus*
Western wormsnake	*Carphophis vermis*
Eastern wormsnake	*Carphophis amoenus*
Rough greensnake	*Opheodrys aestivus*
Eastern ratsnake	*Elaphe obsoleta*
Western foxsnake	*Elaphe vulpina*
Milksnake	*Lampropeltis triangulum*
Prairie kingsnake	*Lampropeltis calligaster*
Common kingsnake	*Lampropeltis getula*
Cottonmouth	*Agkistrodon piscivorus*
Copperhead	*Agkistrodon contortrix*
Massasauga	*Sistrurus catenatus*
Pygmy rattlesnake	*Sistrurus miliarius*
Timber rattlesnake	*Crotalus horridus*

Examples are the nearly formed Mink Frog (top) and the Central Newt (middle). The Painted Turtle (bottom) is living on the edge as she buries her eggs on the side of a gravel road.

Chapter 11

The Human Dimensions of Riparian Areas: Implications for Management and Planning

John F. Dwyer, Pamela J. Jakes and Susan C. Barro

I was born upon thy bank, river,
My blood flows in thy stream,
And thou meanderest forever,
At the bottom of my dream.
Henry David Thoreau, Journals *(1906)* 1842 entry

This chapter introduces an important dimension in building our understanding of how riparian systems function — people. The human dimensions of natural resource management concerns how people value and interact with these ecosystems, their processes and functions. People as users, managers, owners, or involved citizens are integral components of riparian ecosystems and are interconnected with the physical and biological dimensions in many ways.

Managers are finding themselves spending more time working with the public. A better understanding of people's values, attitudes, beliefs, knowledge, and expectations about riparian areas can facilitate efforts to involve a wider segment of the public, making management more effective. We hope the information in this chapter is helpful in guiding this important endeavor and that it will also encourage researchers from the physical, biological, and social sciences to work more closely together to support the management of riparian systems.

We begin the chapter by defining two fundamental components for understanding the human dimensions of riparian ecosystems: people-resource interactions and the uniqueness of riparian areas. We then present four case studies that illustrate applied human dimensions research carried out at the request of managers or planners. In the last section we discuss how to effectively use human dimensions research in guiding planning and management efforts.

People-Resource Interactions

Most people interact with riparian ecosystems in some way through work, play, or day-to-day living. These interactions are almost universal because many cities, towns, businesses, homes, recreational areas, and other important parts of people's lives are located in riparian areas. Many people's livelihoods are tied directly or indirectly to riparian areas, especially as fishing and hunting guides, farmers, loggers, resort owners, miners, marina operators, municipal water supervisors, and others. People's leisure time is often linked with riparian resources. Those living near riparian areas tend to have more direct contact with them on a day-to-day basis, while those farther away usually experience riparian areas less intimately or less frequently. Even people far from lakes, rivers, streams, and other riparian areas may still relate to them regularly through memories, photographs, and other means. Riparian resources are often highly significant to these individuals.

Relationships between people and riparian areas change as the population and its distribution over the landscape changes. Changes in land use, economic development, modifications in transportation corridors, and shifts in people's values, attitudes, and behaviors may significantly affect the character of human-riparian interactions.

Uniqueness of Riparian Areas

Two distinguishing characteristics of riparian areas have important implications for the human dimensions of these ecosystems. First, riparian areas offer opportunities for *unique experiences* that depend on the presence of a land/water interface. People seeking different experiences often converge on the same, sometimes limited, geographic area. Outdoor settings with both land and water have a high esthetic appeal and are preferred for a range of outdoor activities (Dwyer et al. 1989; Kaplan 1977; Schroeder et al. 1990; Schroeder 1996; Stynes 1997). As a result, riparian areas are the focus of much conflict. For example, kayakers and jet skiers, managers protecting endangered species and trappers, individuals seeking solitude, and resort developers all compete for access and control.

Water plays a key role in linking geographically dispersed areas, carrying and sometimes magnifying the impacts of actions in one area to another. The *presence of linkages* is the second major characteristic of riparian areas that makes them unique from a human dimensions standpoint. Linkages increase the significance, complexity, and scale of riparian resource management. While many interactions among people and resources are concentrated at the land/water interface, the resulting impacts extend beyond adjacent geographic areas. Social linkages with riparian resources are likely to extend the scale of analysis for planning because many of those who influence or are influenced by riparian areas reside outside of the physical boundaries of the riparian area or watershed. Management of riparian areas is complex and collaborative, and it occurs at several scales. Some issues are largely local in nature; others must be considered at a regional or perhaps national scale.

Given the diverse and dynamic interactions people have with riparian areas, one of the fundamental questions for management is "How do people value particular areas?" Bengston and Xu (1995) provide evidence of recent shifts in the way that people value forest ecosystems. They found that people's values related to forests are shifting from economic or utilitarian views to an appreciation of the life support (e.g., environmental), esthetic, and moral or spiritual qualities of the forest. Understanding the diverse ways in which people value riparian areas and incorporating this information into resource planning facilitates sound decision making.

Human Dimensions in Riparian Management: Four Case Studies

Each of the following case studies illustrates human dimensions research undertaken at the request of land managers or planners to aid in riparian area management and decision making. These studies show how human dimensions research can improve efforts to consider people's concerns in the management of forests and other ecosystems. The social settings for these efforts range from rural to urban, and various ecosystems are represented.

Enhancing Public Involvement: Black River, Michigan

The Black River flows through the Ottawa National Forest in northern Michigan. Forest personnel were aware of the special character of the Black River area and wanted to include people's values and feelings about the area plan for the future. Schroeder (1996) worked with forest staff to find out more about people's attachment to the Black River area. Fliers about the study were sent to members of the public involved in planning for the Black River Opportunity Area and were also posted in local businesses and at recreation sites. People responded to the fliers by sending the researcher their names and addresses. Study participants were then asked to write about the area and the features and experiences that have led them to special feelings about the Black River area.

A qualitative analysis of these written descriptions indicated that esthetic, cultural, and natural values are highly important (Schroeder 1996). Visitors and residents reported the rustic and peaceful character of the Black River area central to their experience of a special place. The culture and history of the area as a commercial fishing village is also important to their sense of the Black River as a special place. They greatly appreciated the well-maintained recreational facilities and easy access to waterfalls and other scenic resources.

The study identified specific places and features in the Black River area that are especially important to people, such as waterfalls, large trees, clean water, and wildlife. By identifying what particular features of the area are considered special and why, this study helped clarify what visitors and residents desire. Results enabled managers to corroborate and extend information obtained in other public involvement activities.

National Forest Planning: Functional Communities of Northern Wisconsin

Riparian areas and their resources are important to communities. Many settlements tied their initial identity to riparian resources as a "river town" or "port city," for example. Strong ties still exist between communities and riparian resources.

As part of an effort to revise the forest management plans on the Chequamegon-Nicolet National Forest in Wisconsin, Jakes et al. (1998a; 1998b) conducted a series of face-to-face interviews with long-time residents of northern Wisconsin. The purpose of these interviews was, first, to identify functional communities — geographic areas in which people share perceptions of, and relationships to, forests and natural resources — in and around the national forest. Researchers then analyzed and described communities' ties to the landscape.

Residents of northern Wisconsin indicated that activities in, and issues related to, riparian areas are very important in defining or characterizing local communities. For example, residents along the south shore of Lake Superior focused on that Great Lake and its role in defining their community (especially in terms of maintaining a quality of life for area residents). One resident in northeastern Wisconsin labeled her community the "Silent Sport Capitol of Northern Wisconsin," reflecting the importance of canoeing, kayaking, fly fishing, and other recreational activities on wild and scenic rivers in establishing an identity for that community. In identifying land management issues, those interviewed in northern Wisconsin emphasized their ties to riparian areas. For example, residents of one community expressed concern about the effects of a proposed mining development on the area's lakes and rivers. Many residents stressed the importance of publicly owned land in maintaining access to lakes and rivers that were becoming off limits to local users due to development. Findings from this study helped guide, and sometimes refocus, the revision of the forest plan for the Chequamegon-Nicolet Forest.

Policy Development and Regional Planning: Seasonal Homeowners in Michigan

On any Friday evening in Michigan, Interstate-75 traffic is bumper-to-bumper as residents travel "up north to the cabin." Seasonal homes are a significant presence in counties across northern Minnesota, Wisconsin, and Michigan, where they account for more than half of the houses. The demands of seasonal residents for new or improved services, and their values (sometimes different from local residents) are the impetus for many of the changes observed in northern Lake States rural communities.

Stynes et al. (1997) surveyed 1,300 seasonal homeowners from six counties of Michigan's northern Lower Peninsula by mail to identify the characteristics of seasonal homes and their owners, measure patterns of use and associated recreation activity; and estimate the local economic impacts of seasonal homes. Results indicated 80% of the properties are on lakes, which suggests significant implications for riparian resource management.

Almost half of the homeowners in the survey cited outdoor recreation as an "extremely important" reason for owning their seasonal home. Many of their recreational activities occur in riparian areas including fishing from shore, cross-country skiing, hiking, nature study, and the use of snowmobiles and other off-road vehicles. The conversion of seasonal homes to permanent residences also has important implications for local and regional planning. Stynes et al. found that 40% of the seasonal homeowners listed the potential use as a retirement home as an important reason for owning their seasonal home. Twenty percent said they are likely to convert their seasonal home to a permanent residence within the next 5 years. If half of those likely to convert do so, the resident population of the region would increase by 10% in 5 years and by 20% in 10 years. Much of this increase could take place in riparian areas, particularly sites along lakes.

As a result of this research, Michigan and neighboring states have increased their efforts to more fully account for and include the needs of seasonal homeowners in local and regional planning and to consider the potential impacts of seasonal home development on rural communities and natural resources.

Urban Restoration: The Chicago River

The Chicago Rivers Demonstration Project was begun in the early 1990s as a national model for the enhancement of urban waterways. The project established an extensive partnership coordinated by the USDI National Park Service Rivers, Trails, and Conservation Assistance Program and by the Friends of the Chicago River. Physical, biological, and social components of the riparian ecosystem were analyzed along the 156-mile river corridor to obtain baseline data on ecosystem health and to identify the diversity of uses and perceptions of the corridor (Gobster and Westphal 1998a).

Analysis of the social component of the project included several methods of gathering information from people who had different types of associations with the river. People recreating along selected reaches of the river were interviewed on-site to gather information on the range of activities that take place there as well as the different opportunities provided. Focus groups were conducted with residents who lived near the river to gain a deeper understanding of how they perceived and used the river and how they thought it could be improved. Riparian corridor residents were sampled by telephone about their awareness and perceptions of the river to obtain a statistical representation for all corridor residents. Canoeists, kayakers, and rowers were surveyed by mail about their use of the river. Finally, in-depth personal interviews were conducted with resource experts in positions potentially influential in determining recreational use of the corridor.

Results of the various studies, which reinforced each other, indicated that many outdoor activities are associated with the Chicago River and its corridor (Westphal 1998). Some of these activities involve active use of the water or land/water interface, but a great many do not. Respondents reported that "as long as the river doesn't smell too bad" it was a great recreation resource for many activities from biking to relaxing to boating. For many, the river and associated environments are an important part of the broader setting for their experiences, creating a richer neighborhood or workplace. For others, the river is the place to go for a break from hectic city life. Visual access was as important as physical access to the river (Gobster and Westphal 1998b). But results of the Chicago River study also highlighted the fact that many urban residents, even those living near the river, do not feel connected to it and, thus, are unaware of its condition. Information gained about how people use and perceive the river as well as how the river corridor could be enhanced is already being used in long-range plans for river management, and efforts to increase people's awareness of the river and knowledge of related issues are underway.

Human Dimensions Research

Results of the case studies discussed above gave managers new insights to help them more fully integrate the human dimensions into riparian resource management. As more people interact with riparian resources and feel compelled to be more involved in the management of these areas, the need to better understand the human dimensions becomes increasingly important for managers. Consequently, research on the human dimensions of riparian ecosystem management merits at least as much attention from researchers as the physical and biological dimensions do (Jakes and Harms 1995). In many instances, management questions will require research that links the biological, physical, and social sciences. When research is integrated across disciplines, the high standards of scientific quality established for each scientific discipline must be maintained.

Building an understanding of the human dimensions of riparian areas requires systematic and scientifically valid approaches. According to one of the standard social science research texts, *The Practice of Social Research* (Babbie 1998), we must engage in the three major elements of scientific enterprise — theory, data collection, and data analysis. In the four case studies described above, researchers used these elements to make sense of what they observed about people living, working, and recreating in natural environments.

As managers increasingly work with researchers, it is important to understand the process by which a managerial question is answered through research. First, the managerial question is translated into a research question. Once the research questions are clearly understood, researchers can select the methods of data collection and analysis that will best answer them. In essence, then, the specific research question(s) determine the research approaches and methods used (Bickman and Rog 1998). The answers to some questions require in-depth information about people's emotions, beliefs, relationships, and values (e.g., esthetic, moral, or spiritual) that is best gathered and analyzed with qualitative approaches. Alternatively, questions related to such things as behaviors, activities, and preferences are often best answered by collecting and analyzing data using quantitative approaches.

In the case studies we have presented in this chapter, researchers drew on theory from a variety of social sciences including environmental psychology, rural sociology, political science, geography, and economics. Researchers used one or more data collection techniques that produced both qualitative and quantitative data to help answer research questions (Table 11.1). The methods of data analysis ranged from content analysis to standard statistical analysis. If you'd like more detailed information about human dimensions research, consult the literature listed in Table 11.2.

Human Dimensions Information in Management and Planning

The goal of human dimensions research in riparian ecosystems is to help managers and planners make decisions that will maintain or improve the health and sustainability of these systems and increase their contributions to the quality of life. In the following discussion, we highlight a few examples of how human dimensions information can be useful in planning for and managing riparian areas.

Table 11.1 Advantages and limitations of various social science methods

Methods Applied in Case Studies	Advantages of Methods	Limitations of Methods
Focus groups - Chicago River	• Relatively inexpensive ($3,000 - $5,000 per group). • Interaction among group members enriches information. • Has flexibility to pursue issues discovered during discussion. • Can be a source of detailed information.	• Requires trained facilitator to conduct groups. • Strong/vocal individuals may dominate flow of discussion. • Results cannot be generalized to a larger population. • Special training needed to interpret results.
In-depth interviews - Wisconsin functional communities - Chicago River	• Can get in-depth and rich information. • Is flexible to allow pursuit of interesting topics. • Allows interviewees concerns and expectations to emerge.	• Requires trained interviewers. • Is time consuming (1-2 hours each). • Usually only a small number of people can be interviewed. • Results cannot be generalized to a larger population. • Generates large amounts of detailed data that may be difficult to interpret.
Content analysis Text Interpretation - Black River	• Can be applied to a variety of written materials such as letters, newsletters, and newspapers. • Method is not hampered with details of recruitment, sampling, or scheduling interviews.	• Time consuming to repeatedly read through text, identify themes or keywords, and code responses. • Requires special training to select materials for analysis and to conduct and report on analysis.
Mail survey - Chicago River - Seasonal homeowners	• Findings can be generalized to a larger population. • Relatively low level of intrusion (People can fill out survey at their convenience).	• Tends to be expensive due to costs of acquiring a random sample, designing and printing surveys, and a need for repeated mailings. • Knowledge of statistics is needed for analyzing and interpreting results. • Surveys of non-respondents may be necessary.
Telephone survey - Chicago River	• Results can be generalized to a larger population. • Can get quick responses. • In comparison to a mail survey, it is easier to ensure that people answer all the survey questions.	• Tends to be expensive because it often involves contracting with a phone survey lab. Costs are associated with designing the survey, paying trained telephone interviewers, and analyzing data, etc. • Can inconvenience or intrude on people. • Some of the people you want to talk with may be less likely to agree to an interview over the telephone.

Developing Management and Planning Options

There is a tendency to regard human dimensions information as a tool for supporting or "selling" projects, programs, or policies that have already been agreed upon by managers and planners. However, such information can play a key role in the development of management and planning alternatives. The traditional "after the fact" approach for using human dimensions information is not nearly as effective as one where human dimensions information is used to *develop* projects, programs, and policies. In many instances, the time and effort spent early on in incorporating human dimensions information into development of management and planning options reaps huge dividends later in reduced effort needed to "sell," "revise," "re-plan," or "reconsider."

The following questions provide a guide to help managers and planners evaluate riparian resource management options from a perspective that connects the physical, biological, and social dimensions of riparian systems within the context of resource management. If these questions cannot be answered in a management situation, additional information and research may be needed.

- What benefits associated with riparian areas are important to people?

- What characteristics (structure and function) of the riparian area are critical in providing these benefits?

- How will changes in the structure and function of the riparian area influence the benefits made available to people?

- To what extent are individuals and groups (including communities) willing to accept or support changes in the riparian area necessary to accomplish management goals, given that these changes may affect the array of benefits available to people?

Establishing Dialogue with the Public

Involving the public in planning and management is critical for effective management and use of riparian areas. For the public to provide meaningful guidance, we need to develop more effective techniques to (1) inform citizens about riparian resources and their management, (2) involve the public in developing and assessing management options, and (3) draw a wider range of individuals and groups into the planning process and into monitoring the results once plans are implemented.

Information about ecosystems (How do they change over time?; How do they respond to management programs and practices?) is particularly relevant to public involvement. Jakes

(1998) outlined how public involvement (and research) can help build a better understanding between the public and planners/managers and bring their expectations and goals for resource management decisions more into sync.

For effective dialogue between managers, planners, and the public to occur, some common understanding is necessary. Managers cannot assume that their knowledge and beliefs are shared by a large segment of the public. For example, riparian area managers may feel there is ample justification for conducting ecological restoration in riparian areas to achieve biodiversity (as well as other) goals. However, the public may not support this type of effort because many people do not understand the ecological restoration process or the reasons behind it. Barro and Bopp (1996) report that the college students they surveyed in the Chicago area were not aware of the subtleties of ecological restoration, such as the concept of increased biodiversity and enhanced ecosystem function. Two factors that have been identified as barriers to communicating about and gaining support for ecological restoration and increasing biodiversity are (1) the complexity of the issue, and (2) people's lack of experience in natural areas (Bidwell and Barro 1997). Other significant barriers to overcome relate to the perceptions that managers and planners have about what the public wants or expects from riparian ecosystems. Promising approaches to reducing these barriers include more dialogue between managers, planners, and the public, as well as increased research on what the public wants and expects from the management of riparian resources.

Deciding Whether to Apply Human Dimensions Information

In analyses of resource planning and management options, there are questions about whether information from other situations or places can be applied in the current analysis. This is of particular concern when considering information on the human dimensions of resource management.

People's interactions with riparian systems may vary significantly with place and time, depending on the riparian resources, management context, and people's values, beliefs, and expectations about riparian areas and their management. For example, Chicago residents are likely to have quite different standards of quality for the Chicago River than for a mountain brook. And a visitor from Minnesota (Land of 10,000 Lakes) may perceive the water quality in the Chicago River differently from a Chicago resident. This is not to say that information from other areas and experiences cannot sometimes supply helpful guidance for management decisions, but planners and managers need to clearly understand and consider the context in which that information was generated.

Given the complexity of riparian resource management and use, managers will seldom, if ever, have complete information on which to base their decisions. This holds true for physical and biological as well as human dimensions information. Thus, riparian management decisions will continue to be made in the face of a significant degree of uncertainty, but this uncertainty can be reduced by gathering information from a variety of sources and by maintaining a strong dialogue with the public.

Using Secondary Data to Describe the Human Dimension

Resource managers and planners frequently ask about using the information related to people and their activity that is readily available from the U.S. Bureau of the Census as well as from state, regional, and local agencies. This includes information on population, employment, housing, manufacturing, and other factors. Although this information (which is readily available at http://www.census.gov) provides useful insight into the human resources in particular areas, it is not linked to specific resources or their management. Consequently, while we might be able to gain insight into the trends in population, employment, and housing in a particular area, we would not know how that information was tied to the management of riparian resources. For example, (1) how might changes in the local population influence riparian resource management, and (2) how might riparian resource management influence the future population in the area. The overlay of physical, biological, and social resources in a GIS format shows the association among variables and helps to outline the context for management, but it does not tell us anything about cause and effect relationships that are critical in evaluating planning and management options. Analysis of management and planning options needs to focus on these cause and effect relationships between the key attributes of people and management options.

Displaying Human Dimensions Information

Human dimensions information should be presented in a manner that facilitates its use in planning and management. To reach a broad audience among resource managers and the public, researchers must look for improved ways to display results of human dimension research. At public meetings on resource management, people often cluster around maps, discussing the information being displayed. This reflects how graphs, charts, and maps enhance the communication process (Bickman and Rog 1998, 527). At the same time, some information does not lend itself to graphic displays, such as information about people's values, feelings, and links to the land. These may be best presented in narrative form or by other methods.

Summary

"People are part of [riparian] systems; they derive material and nonmaterial goods and services from them; they live, work, and play in [riparian areas], and their attitudes, behavior, and knowledge of the [riparian] ecosystem affect it in both direct and indirect ways. Thus, [riparian] management systems that alter the structure and processes of the biological component will alter the human system that interacts with it. Conversely, the way in which people are organized and the processes through which they make decisions will lead to alterations in the [riparian] ecosystem." (FEMAT 1993, VII-110)

The human dimension is a critical component of the planning and management of riparian ecosystems that promises to increase in significance over time. We have presented case studies that illustrate the depth, complexity, and significance of people's interactions with riparian ecosystems across a rural to urban continuum. These illustrations include a wide range of social science research techniques that can facilitate and improve riparian resource planning and management. We have provided some general guidance for using these techniques to help steer the management and use of riparian areas. In the development and analysis of planning and management options, the human dimensions should receive attention comparable to that given to the physical and biological dimensions; and high-quality science and analysis need to be applied. In many instances, it is crucial to identify physical, biological, and social linkages. The effectiveness with which human dimensions are integrated into resource planning and management is likely to determine the effectiveness of riparian area management and planning in the years ahead.

Sandy Verry

Boating the Great Lakes and Great Rivers of the East is a major recreation pursuit.

Table 11.2 References for human dimensions research methods

Bailey, K. O. 1994. Methods of social science research. New York: Free Press. 588p.

Creswell, J. W. 1994. Research design: qualitative and quantitative approaches. Thousand Oaks, CA: Sage Publications. 228p.

Delbecq, A. L., A. H. Vandeven, and D. H. Gustafson. 1975. Group techniques for program planning: a guide to nominal group and delphi processes. Glenview, IL: Scott, Foresman. 174p.

Denzin, N. K. and Y. S. Lincoln, eds. 1994. Handbook of qualitative research. Thousand Oaks, CA: Sage Publications Inc. 643p.

Dey, I. 1993. Qualitative data analysis: a user-friendly guide for social scientists. London: Routledge. 285p.

Dillman, D. A. 1978. Mail and telephone surveys: the total design method. New York: Wiley. 325p.

Fink, A. 1998. How to conduct surveys. 2nd ed. Thousand Oaks, CA: Sage Publications. 160p.

Kruger, R. A. 1994. Focus groups: a practical guide for applied research. Thousand Oaks, CA: Sage Publications. 255p.

Lavrakas, P. J. 1993. Telephone survey methods: sampling, selection, and supervision. 2nd ed. Newbury Park, CA: Sage Publications. 181p.

Morgan, D. L. and R. A. Kruger. 1997. The focus group kit. Thousand Oaks, CA: Sage Publications. 5 - volume set.

Morgan, D. L. 1997. Focus groups as qualitative research. 2nd ed. Thousand Oaks, CA: Sage Publications. 255p.

Patton, M. Q. 1982. Practical evaluation. Beverly Hills, CA: Sage Publications, Inc. 313p.

Rivard, R. T. 1997. A sense of place. The Bee. 113:1-3.

Vacationers, members of watershed districts, soil and water conservation districts, lake associations, county citizens, township citizens, want and need to be part of planning for riparian area management. Besides the sense of ownership, common ground alternatives gain acceptance and implementation through peer networking.

Clockwise from upper left: fencing and rock ford for cattle, discussing rootwad installation for bank stabilization, restoring the banks of the AuSable in Michigan, public use of national forest lake access, and public fishing on private lands with easements.

Chapter 12

Lake Riparian Areas

David W. Bolgrien and Timothy K. Kratz

*"A lake is the landscape's most beautiful and expressive feature. It is earth's eye;
looking into which the beholder measures the depth of his own nature. The fluviatile
trees next to the shore are the slender eyelashes which fringe it, and the wooded hills
and cliffs around are its overhanging brows. (...) The forest has never so good a
setting, nor is so distinctly beautiful, as when seen from the middle of a small lake."*
Henry David Thoreau — Walden

Lake riparian is not a well-defined concept among ecologists and managers. Even Thoreau
considered the forest surrounding lakes to be "fluviatile" (related to flowing waters). In
common usage, streams have riparian areas (from the Latin *riparius*, "belonging to the river
bank"), but lakes have shores. (See Chapter 2 for our definition of riparian areas that
includes streams, lakes, wetlands, and their surrounding uplands). Limnology textbooks
typically discuss littoral zones (shallow, nearshore areas) but not riparian areas (Ruttner
1953; Wetzel 1983). Malanson (1993) explicitly excluded lakes from his discussion of
riparian landscapes. Compared to streams, surprisingly little ecological literature deals
specifically with lake riparian areas. But the fundamental function of the riparian area to
modulate important hydrologic and material fluxes between terrestrial and aquatic systems
certainly apply to lakes. To effectively manage lake riparian areas, we must understand their
complexity and their interconnectedness with adjacent and distant wetlands and uplands.
Like riverine riparian areas, lake riparian areas simultaneously act as filters, barriers, sources,
and sinks for numerous biological, physical, and chemical processes at diverse space and time
scales. Frey (1990) reminds us that lakes are essential parts of their surrounding
environments that respond perceptibly to any changes in that environment.

To complete the definition of lake riparian area, certain legal and social issues also must
be considered. A lake and its shorelands are objects to be owned and enjoyed as private
property. In fact, people who own shorelands are referred to as "riparians" by social
scientists (including lawyers). The legal definition of riparian land varies among states. In
general it is a parcel that adjoins a watercourse and is in the same watershed as that
watercourse (WIDNR 1971). To develop and implement effective management plans,

ownership and property rights, as well as the attitudes and behaviors of riparian stake holders, must be considered. Managers need to identify institutions and organizations that might assist (or hinder) management plans.

Thus, lake riparian areas are frequently the intersection between environmental processes and social activities. The diverse lake riparian area definitions necessitate a management perspective that is spatially variable (e.g., regional, watershed, and local) and encompasses the range of values associated with the land (e.g., ecologic, economic, and aesthetic). We first will discuss the legal and social instruments that may be incorporated into the management of lake riparian areas. These include riparian rights, institutions for community-based management, and prescribed "Best Management Practices" affecting riparian conditions. Second, we will discuss ecological processes occurring in lake riparian areas that affect water quality and shoreline habitat. These include the distribution of its coarse woody debris, dissolved organic carbon, and littoral habitat. Here we consider riparian areas to include uplands and wetlands adjacent to a lake, including the lake's littoral zone.

Components of the riparian area are hydrologically connected. Specific dimensions are intentionally lacking in this functional definition. Our discussion will focus on natural lakes in forested regions. Because this research is part of the North Temperate Lakes Long Term Ecological Research project (Magnuson and Bowser 1990), many examples will be from northern Wisconsin. Human-made lakes, reservoirs, flowages, and lakes for municipal water supplies are beyond the scope of this discussion because many have manipulated hydrologic regimes resulting in highly disturbed riparian/littoral areas and special rules governing access, use, and development. However, certain management strategies presented here may be applicable.

Social Institutions and Lake Riparian Ecology

North temperate lake riparian areas are places of complex ecological and social interactions. Of profound importance is the fact that they may be privately owned. The same is true for streams, but there are fundamental differences. Historically, streams have powered commerce and transportation to a much greater degree than lakes (excluding the Great Lakes and reservoirs). The societal value of inland lakes is largely based on recreation. For example, in the Northern Highlands Lake District of Wisconsin, most of the old-growth forest was clearcut by the end of the 19th century (Davis 1996). This resulted in a land-use conflict between area boosters promoting farming and forest conservationists. By the 1930s, chronically poor farming conditions, limited market access, and increased mobility by the growing middle class shifted the economic base to tourism.

That human activities are the driving forces impacting lakes and their riparian areas is evident in current demographic and economic trends in northern Wisconsin. A recent survey by the Wisconsin Department of Natural Resources (WIDNR 1996) found that tourism in northern Wisconsin is a major industry and in some areas is the sole basis for economic viability. All lakes attract development, and few large lakes in Wisconsin remain undeveloped. Private lands account for > 80% of the shoreline on the vast majority of lakes (>10 ha) in northern Wisconsin (Table 12.1). Over the past 30 years, the number of shoreline dwellings has increased on lakes regardless of size (Figure 12.1).

Table 12.1 Distribution of lake riparian ownership in northern Wisconsin by lake size. Data are from lakes north of Hwy. 29 (WIDNR 1996). A "private" lake was defined as having < 20% of its shoreline owned by county, state, federal, or tribal governments.

| Lake size | Total | Public | Private | Public | Private |
Acres	No.	No.	No.	%	%
10 - 50	2500	435	2065	17	83
51 - 100	629	60	569	10	91
101 - 200	402	35	367	9	91
>200	488	20	468	4	96
Total	4019	550	3469	14	86

The legal definition of a "navigable" water body in Wisconsin emphasizes the evolution of lakes into important recreational resources. In 1868, a water body was navigable if it could float a saw-log on a recurring basis (WIDNR 1971, 1982). A practical and symbolic change occurred in 1952 when the sawlog was replaced by the canoe as the float standard. In addition, the enjoyment of navigable water bodies for recreation, including the enjoyment of natural beauty in the riparian area, was declared a legal right that was entitled to protection (WIDNR 1971).

This social history created and continues to influence environmental conditions in lake riparian areas and, therefore, guides potential management strategies. Forested lake riparian areas of the Eastern U.S. are dominated by logging and urbanization consisting largely of residential construction (i.e., cottages and retirement homes) and tourist services. Lake managers are under growing pressure not only to accommodate development but also to retain the environmental and scenic values that attracted the development in the first place. The environmental issues important to lake riparian management must be understood within the context of the human ecology of a region.

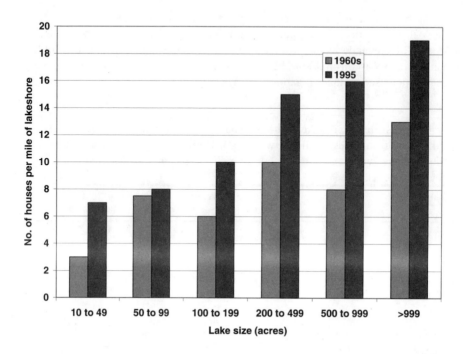

Figure 12.1   Change in the number of dwellings per mile of shoreline of 235 "private" northern Wisconsin lakes from the 1960s to 1995. A private lake was defined as having < 20% of its shoreline owned by county, state, federal, or tribal governments (data from WIDNR 1996).

Riparian Rights

Lake riparian property rights present an interesting example of conflict between the fundamental tenet of American culture to own property and the inherent need for natural resource management. In most eastern states, water rights evolved directly from the Northwest Ordinance of 1787 in which navigable waters were declared common highways held in trust by the state (WIDNR 1971). This Trust Doctrine is a contrast to the "prior appropriation" doctrine held in many western states (dominated interestingly by streams, not lakes). Different rules apply to reservoirs. States retain ownership of the water and allocate its use according to the relative reasonableness of the competing uses (Dresen and Kozak 1995).

Lakebed ownership and other specific riparian rights vary greatly among states. Without riparian rights, cottage owners in Wisconsin would be unable to build docks on public lakebeds (O'Connor et al. 1995). In Michigan, lakebeds belong to the riparian landowner. States are increasingly exercising regulatory power to maintain waterflow and ecological integrity. The riparian rights of private owners may be constitutionally protected by "due process" and "just compensation" clauses from government actions that result in "the taking of private rights for the public good." Further, they are not subject to forfeiture, cannot be lost by nonuse, and cannot be conveyed separately. The latter point is being legally stretched by developers to increase the value (and price) of back-lot properties that are technically nonriparian (Bartke and Patton 1979).

Community-Based Management Strategies

The legal status of riparian lands must be balanced with the need to manage and conserve lake riparian areas. Most states have community-based institutions to deal with lake environmental issues. For example, shoreland owners around Lake George (NY) formed an association in 1885 to protect water quality. Overabundant aquatic vegetation on Lake Geneva (WI) led to the formation of a lake association in 1898 (Schrameyer 1997). Many lake associations (and associations of lake associations) have web sites on the Internet (e.g., North American Lake Management Society (www.nalms.org). State-specific legal and institutional mechanisms to empower associations to achieve conservation objectives should be used by riparian managers. Volunteer monitoring networks may also be important components of community-based management strategies (Rumery and Vennie 1988; Maas et al. 1991; Korth and Klessig 1990; Heiskary 1989; Heiskary et al. 1994). These networks are typically composed of shoreland owners who are trained to collect basic limnological data. Program goals are to develop databases from which water quality trends can be inferred at variable spatial and temporal scales and to instill a sense of stewardship in lake users. The Great American Secchi Dip-In (US EPA Office of Water Clean Lakes Program; http://humbolt.kent.edu/~dipin/) is an excellent example.

In contrast to lake associations, which are largely voluntary, special purpose units of government may be created to protect, rehabilitate, and improve lakes for recreation. Since 1974, Wisconsin has allowed the formation of "lake districts" with taxation and some regulatory authority (Klessig et al. 1989; O'Connor et al. 1995). District boundaries are flexible and typically include all riparian properties and possibly the entire watershed. They may also include areas directly affecting the lake such as its "service" area — commercial and transportation corridors that extend outside of the watershed. Unlike in other units of government, both resident and nonresident (seasonal) property owners may vote in lake districts.

Ecological stress resulting from the combination of the high value of lake riparian areas and riparian property rights is intensified by the possessiveness shoreland owners feel for "their" lake and their proprietary desire to determine management objectives. Effective leadership (possibly provided by state managers or lake associations) is necessary to ensure adequate participation from nonriparian lake users. Public lakes used by everyone, but managed by no one, are classic candidates for the "tragedy of the commons" (Korth and Klessig 1990). Community-based management strategies emphasizing the relationship between riparian condition and water quality are the key to successful resource management.

Regulations for Lake Riparian Protection

Minimizing disturbances in riparian areas is an important means of preserving scenic beauty, controlling soil erosion and compaction, and reducing the flow of nutrients to lakes (Osborne and Kovacic 1993; Hillbricht-Ilkowska 1995; WIDNR 1995). Vegetated buffer strips significantly reduce phosphorus runoff from lakeside residential construction (Woodard and Rock 1995). Forested riparian buffer zones, where logging methods are modified for sensitive terrain and harvests are limited, provide for long-term sequestering of nutrients and carbon. Best Management Practices (BMPs) for forestry, agriculture, and zoning regulations dictate buffer strip widths, construction set-backs, and allowable riparian timber harvests. While BMPs have evolved as a compromise between environmental protection and commercial natural resource use, they generally have regulatory status only on public lands. On private property, they may serve only as nonregulatory guidelines. Local governments may modify BMPs and zoning codes but usually must adopt more, not less, restrictive rules. In practice, conditional use permits and regulatory variances issued by local authorities can systematically undermine the effectiveness of broad-scale controls.

Buffer zone configurations should be integrated with lake classification schemes to optimize their effectiveness. Lakes classified by their depth, shape, and flushing times have an expected (normal) level of algae production (trophic status). Those with a greater algae production, and those that have shown a particular sensitivity to increased phosphorous and nitrogen, are candidates for more stringent riparian protection. In British Columbia, the size and function of riparian management zones vary according to lake size, the rarity of the lake type in a particular biogeoclimate unit, and rankings of ecologically sensitive variables (endangered fish species, economically important fishery, recreational value, and visual absorption capability). Small lakes (2 to 12 acres) that are uncommon in a particular biogeoclimate unit or are ecologically sensitive, have the widest buffers. As lakes become more numerous and diverse within biogeoclimate regions, restrictions are decreased (Canadian Forest Practice Codes 1995).

Environmental Processes in Lake Riparian Areas

Human activities in lake riparian areas clearly are major driving forces that must be considered for effective management. Lake response to development historically has focused on nutrient loading, especially phosphorus (Dillon and Rigler 1975; Hutchinson et al. 1991; Dillon et al. 1994). Nationally, siltation and excess nutrients are the major sources of pollution in lakes and streams (USEPA 1998). Maintaining nutrient concentrations within ranges appropriate for specific laketypes is a more practical management objective than simply reducing the sum of inputs. Nutrient loading is intimately associated with riparian condition and remains a high management priority. In an effort to broaden the discussion of the multidimensional relationships between lake riparian and littoral areas, and lake riparian areas and watersheds, we have elected to highlight coarse wood debris, dissolved organic carbon, and littoral habitats. Substantially more research has been done on these topics in streams than in lakes.

Coarse Woody Debris (CWD) and Littoral Fish Habitats

We have long understood the value of coarse woody debris (CWD; wood >2 inches in diameter-see Chapter 7) for stream restoration and the maintenance of forest productivity (Harmon et al. 1986). However, the value of CWD to lakes, especially shallow water areas, is underappreciated. CWD promotes habitat complexity (on the shore and in the water), deflects waves, and traps sediment and litter. Its value varies with biomass, size, spatial arrangement, and degree of decay. Through time, riparian activities alter the quantity and distribution of CWD along the entire shoreline by directly modifying the landscape (Christensen et al. 1996) and altering forest disturbance patterns (Harmon et al. 1986). Net lake CWD biomass is the balance of input processes (e.g., windthrow, logging, erosion, fire) and output processes (e.g., decomposition, removal by humans, sinking/transport to deep waters). These processes contain a mix of environmental (e.g., climate, soils, water level) and social factors. Understanding how this mix results in the presence or absence of CWD in a lake is a prerequisite for effective riparian management planning in forested regions.

In lakes of northern Wisconsin and Michigan, one study found a significant and direct correlation between CWD and riparian tree densities and an indirect correlation between CWD and lake shore cottage densities on a whole-lake basis (Christensen et al. 1996). Assuming shoreline development reduced CWD supplies, these results suggest that input processes dominate CWD dynamics. However, we found that windthrow probability in riparian forests was unrelated to littoral CWD distributions. Although windthrow is the major source of CWD in terrestrial forests (Harmon et al. 1986), it did not appear to be important for delivering CWD to lakes. It is likely that the removal processes determine net CWD concentrations in lakes. CWD densities were significantly lower than expected around boat docks where property owners removed downed trees and logs. People are generally intolerant of debris cluttering up their swimming and boating areas. Similar results were reported by Brown and Collins (1997). The rapidly increasing amount of shoreline on northern lakes vulnerable to such actions may create a cumulative effect where intentional removal becomes more important than *in situ* decomposition (but see Guyette and Cole 1997). In general, CWD supply processes operate at longer time scales and over larger areas than removal processes. Supply variables include forest stand properties (managed mono-specific, same-age stand with lengthened disturbance regimes) and economic trends that permit expanded lakeshore development. Output processes, such as the actions of an individual cottage owner, operate much more quickly and more locally. This means that once CWD stocks are diminished or widely fragmented, their replenishment will require long-term management investment to secure adequate supplies and limit removal.

Riparian stakeholders must be motivated to conserve coarse woody debris and include it as part of overall lake management plans. This is not a simple proposal. In Wisconsin, three sets of regulations affecting coarse woody debris in lakes are confounded. First, the removal of material from lakebeds is illegal (S. 30.20 Stats.). However, CWD traditionally has not been considered bed material (Christensen et al. 1996) and, therefore, its removal is largely unregulated. Second, Wisconsin's Shoreline Management Program (S. NR 115.05(3)(c) Wis Adm Code) limits shoreline clearcutting to <30% of lake frontage on a per parcel basis. However, dead trees and snags (main sources of CWD) are exempt from this restriction and may be summarily removed. Third, by law (Protection of Forest Lands statutes (S. 26.12(6) Stats), logging slash must be immediately removed from and kept away from lakes and streams. Often, indiscriminately placed slash is detrimental to aquatic systems and its removal is required in the "public interest." But the strategic placement (or replacement) of CWD is commonly used in bank stabilization and habitat rehabilitation projects. The situation in Wisconsin may not be typical of all eastern states, but it is likely that few regulatory bodies recognize the ecological value of CWD in lakes.

Riparian processes (such as the supply of CWD) create complex habitats in shallow near-shore waters which is important from a management, conservation, and biodiversity perspective. This relationship is well illustrated by fish communities in lakes. Lake riparian areas influence the structure and integrity of fish populations and communities by controlling the distribution of food and cover (Chick and McIvor 1994; Irwin et al. 1997; Hunt 1997). Bryan and Scarnecchai (1992) found that fish abundance decreased with decreased aquatic vegetation density resulting from shoreline urbanization. Fish species richness is directly related to habitat complexity (Tonn and Magnuson 1982; Rahel 1984; Benson and Magnuson 1992) and inversely related to shoreline development (Jennings et al. 1996, 1997; Bozek et al. 1997). Jennings et al. (1996) analyzed communities and individual species to demonstrate that simplifying the habitat had a negative impact on fish.

Measuring how riparian disturbances degrade fish distribution and diversity is difficult because the effects may be cumulative through space and time. Benson and Magnuson (1992) reported that measures of fish community structure at different spatial scales were potentially more sensitive to changes in habitat structure (and therefore riparian condition) than changes in species over time. Since monitoring the presence or absence of individual species over time is the basis of most monitoring programs, monitoring alone may be unable to detect the impact of changing riparian conditions. Few multi-scale studies that quantify specific habitat characteristics and fish communities have been conducted. Even if community structure could be a more informative management tool than focusing on single species, the selection of "desired" fish species will remain an important, often politically-driven, exercise. The challenge is to manage ecosystems, not specific resources.

Watershed Delivery of Dissolved Organic Carbon to Lakes

The importance of dissolved organic carbon (DOC) in north temperate lakes, and hence the need to consider it for riparian management, stems from its impact on light penetration, biotic productivity, and metal chemistry. The net amount of DOC in a lake is a balance between watershed, in-lake supply and removal processes (Engstrom 1987). The former is related to landcover, wetlands, soil properties, hydrologic connectivity, and slope in the watershed (Gorham et al. 1983; Rasmussen et al. 1989; Wetzel 1992; Houle et al. 1995; Watras et al. 1996; Gergel 1996; Dillon and Molot 1997; Kratz et al. 1997; Molot and Dillon 1997). Innovative management schemes should consider watershed variables, such as wetland extent and forest fragmentation. In-lake DOC pools are largely determined by morphometric (shape) properties of a lake, such as depth, perimeter, deep/shallow area ratios (Wetzel 1990), and surface area (Fee et al. 1996). For example, the ratios of either sediment area or littoral area to water volume are higher in small lakes than in large lakes, leading to a proportionally greater release of DOC from organic matter in sediments (Rasmussen et al. 1989).

DOC is the primary cause of water color that controls light penetration in lakes and, thus, the depth of the thermocline (the zone of transition between warm surface waters and cold bottom waters). The depth of the thermocline is a key factor in determining fish thermal habitat and algal productivity (Kratz et al. 1997). In lakes with low DOC concentrations (therefore deep light penetration), fish spawn deeper to avoid exposing their eggs to UV radiation (Williamson et al. 1996, 1997). Watershed processes controlling the amount of DOC entering a lake can also influence the concentrations of dissolved mercury (Kortelain 1993). Watras et al (1996) estimated that water export of methyl mercury (co-transported with DOC) to northern Wisconsin lakes was similar to net internal methyl mercury production and much greater than that entering the lake in precipitation. In general, methyl mercury exports to lakes were highest from wetland-dominated watersheds and lowest from predominantly forested watersheds (St. Louis et al. 1994).

Thermocline depth in the relatively calm waters of a small lake also can be affected by riparian forest densities. In larger lakes, thermocline depth is determined by wind mixing. Schindler et al. (1996); Rask et al. (1993); Fee et al. (1996) and France (1997a, 1997b) found that lakes with deforested riparian areas had deeper thermoclines than lakes with intact, mature riparian forests. Presumably, deforestation resulted in more wind exposure and less shading (a minor component) which allowed more heat to be mixed to greater depths. Slow tree regeneration can result in significant long-term loss of habitat for fish intolerant of warm water. The mechanism of riparian deforestation is also important. DOC and sediment loading to a lake increases more after a fire than after logging or blowdowns. Forested riparian buffer zones can be used to protect lakes from wind stress. To be effective, buffers should be oriented to intersect prevailing winds and wide enough to protect trees from windthrow. Typically, riparian management strategies focus on the protection of shorelines and littoral species. However, it is clear that conditions in riparian areas also can affect offshore, deepwater processes.

Watershed characteristics are also important regulators of DOC in lakes. High-relief watersheds draining coarse, sandy soils with thin acid humus layers are associated with lakes with low DOC contents. Low-relief watersheds containing saturated, peaty, noncalcareous deposits are associated with high DOC contents (Gorham et al. 1983; Rasmussen et al. 1989; Kortelainen 1993; Haule et al. 1995). Regardless of the amount of DOC available in soils, it is the flow of water through the watershed that can ultimately determine the amount of DOC delivered to a lake (Schindler et al. 1996). Therefore, drought or water diversions may significantly alter lake clarity, fish communities, and mercury inputs.

It is generally accepted that the extent of wetlands in the watershed is directly related to the amount of DOC delivered to a lake (Dillon and Molot 1997; Molot and Dillon 1997 and references therein). However, the arrangement of wetlands is also important. For example, in 119 lakes in northern Wisconsin, the proportion of wetland area in a watershed and those within 300 ft of a lake explained a similar amount variation in DOC concentrations (Gergel 1996). Differences in wetland arrangement were apparently based on lake type. Small seepage lakes with no outlet stream were more influenced by riparian wetlands. Drainage lakes (with an outlet stream) were more influenced by wetlands throughout the watershed.

The diversity of important environmental processes in lake riparian areas precludes a one-size-fits-all management approach. The flow of nutrients, pollutants, and material between the watershed and lake can be interrupted or facilitated by riparian forests. Subtle changes in riparian conditions can have synergistic impacts in all areas of a lake and on all lake biota. Management strategies must be ecologically sound but must also accommodate competing uses and values of riparian resources.

Summary

Ecological processes in lake riparian areas are not as well studied as those in streams. There are many similarities between the systems, but lake-specific research is needed to develop effective management strategies. Ecological processes, social and legal institutions, and historical societal values of lake riparian areas must be considered in planning.

Lake riparian management plans that adopt a community-based perspective take advantage of the possessive feeling of riparian owners to protect "their" lake and a proprietary desire to participate in the management process. Lake associations are particularly useful. Riparian buffer strips, in conjunction with other shoreline protection regulations, can protect lake riparian areas from activities that are detrimental to environmental health.

Coarse woody debris is an example of the competing value systems operating in lake riparian areas. Long-term and large-scale variables (such as forest management) control the supply of coarse woody debris to lakes. Removal is largely based on quick, local decisions of riparian property owners. Even if people appreciate the high ecological value of coarse woody debris, the replenishment of stocks will take a long time.

Modifying riparian vegetation and shorelines can simplify the structure of littoral habitats which decreases their value to fish. Management objectives should consider the effects of erosion control structures, docks, coarse woody debris, and riparian vegetation on the use of the littoral zone by a diverse fish community.

Dissolved organic carbon in a lake directly and indirectly affects important physical and biological processes. The delivery of dissolved organic carbon is determined by the flow regime and the extent and distribution of wetlands in the watershed.

Wetlands adjacent to lakes provide a measured input of dissolved organic carbon to the lake's food chain.

Chippewa National Forest

James O. Sneddon

Cutfoot-Sioux resort on Little, Little Cutfoot Lake. Boat access ramp (top center) is a Chippewa National Forest facility that sees over 1000 boaters a day on the "walleye opener" weekend in Minnesota.

Roger Bay

Enjoying a wilderness lake experience in the Boundary Waters Canoe Area Wilderness, Superior National Forest, Minnesota.

Chapter 13

Integrated Management
of Riparian Areas

Steven T. Eubanks, Samuel Emmons
and Heather A. Pert

"Great discoveries and improvements invariably involve the cooperation of many minds."   Alexander Graham Bell

In years past, natural resource management was simpler than it is today. With less public interest, less demand for resources, and a more limited view of ecological systems, there were fewer conflicts. It was common for a single person to plan projects and make resource decisions in riparian areas where multiple resources are affected.

Today, our approach to natural resource management is changing. Previous chapters discussed the complexity, high values, and critical functions represented by riparian areas. In addition to our knowledge of natural resources, we now understand more about the complexity of social values tied to those resources. This all leads to one conclusion: resource management, in general, and certainly management of riparian areas, increasingly involves tough choices — choices that make single-person decisionmaking a questionable approach. Brown and Harris (1991) and MacKenzie (1997) point-out that with diverse interests having stakes in biological, social, economic, and political resources, integrated approaches to management have become increasingly important.

What is integrated management? Slocombe (1993) identified several key characteristics of integrated management: most importantly, (1) an interdisciplinary and collaborative approach that seeks individual and institutional cooperation and integration; (2) actions that define and move toward long-term goals; (3) dissemination and use of information; and (4) adaptive management that monitors and evaluates information and modifies practices where needed to achieve long-term goals.

In this chapter, we will discuss techniques for developing integrated resource management plans. Key steps include: (1) setting management direction; (2) inventorying values and resources; (3) developing alternatives; (4) selecting a preferred alternative; and (5) implementing the plan through an adaptive management approach.

Integrated Resource Management Planning

Developing an integrated plan for an area or project begins by clearly defining management direction. Management direction is the combination of long-term goals and the planned short-term actions needed to achieve these goals.

Set Management Direction

The concept of Desired Future Condition (DFC) is one real-world approach for long-term goal setting. DFC is used to describe what an area should look like in the future as well as to give an "integrated portrayal of land allocations, ecosystems functions, and human interactions (USDA FS 1993)." Since forest landscapes can be quite variable, DFCs need to be flexible enough to fit the variety of conditions present. It may take decades or longer to actually meet long-term goals. In the interim, short-term actions or projects are used to move an area or resource toward the desired long-term condition.

In this chapter, we will use a hypothetical example to illustrate the points raised. In our example, a private landowner recently purchased the 300-acre Greenwoods Forest (Figure 13.1). Greenwoods Forest is bordered on all sides by public lands, and a farm is located farther upstream. Although resource information is spotty, the owner knows that Shady Creek runs through the middle of the forest which consists primarily of 60 to 80-year-old jack pine (*Pinus banksiana*) with a mixture of hardwoods. Many of the older jack pine trees are dying. In addition, Shady Creek is a state-designated Blue Ribbon Trout Stream, is popular for canoeing, and a domestic water source. The landowner bought the property as an investment that would appreciate over time and would provide periodic income in the longer term. Due to the increasing mortality, there is interest in beginning to harvest the jack pine immediately. Since the property is newly acquired, no overall DFC or long-term goals currently exist, but the landowner is interested in sustainable resource management.

Develop Goals

The process of developing long-term goals for an area can be simple or complex depending largely on the complexity of the resources or resource issues involved, applicable laws, policy or regulation and, in some cases, on whether the land is publicly or privately owned.

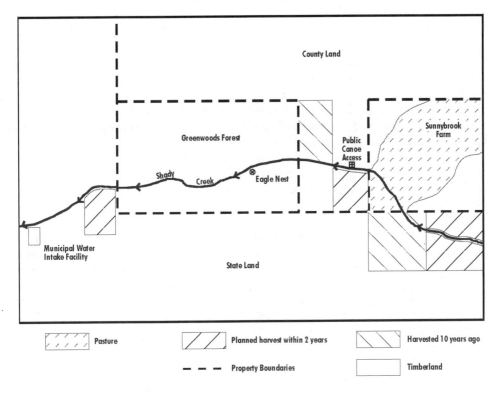

Figure 13.1 A GIS map showing a landscape view of Shady Creek area.

Consider the Missions of People and Organizations

The most effective integrated management involves all agencies, adjacent landowners, or individuals who have a specific interest in the area. But it is important to recognize that development of management goals may be limited by the mission or values of the organization or entity involved in making resource management decisions (MacKenzie 1997). For instance, a given landowner or agency may have an interest or mandate to focus on certain resources such as timber, water, or wildlife; or they may instead need to show equal attention to all resources. Up front acknowledgment that the other people and organizations involved will be guided by their own mission or values will avoid difficulties later in the process.

Consider Existing Legal or Regulatory Restrictions

Laws, regulations, or higher level plans and policy often direct the scope of desired future conditions. These restrictions or broader goals must be considered early in the planning process to avoid wasting time seeking an outcome that may be illegal or incompatible with higher goals for the area. Certainly, laws and higher level plans can be changed, but this can be a lengthy and cumbersome process. The starting point for planning and decisionmaking is to work within existing sideboards.

In our Shady Creek example, the landowner is considering timber harvest in a forest bisected by a stream. Before setting long-term goals for the area or proposing projects that could affect the stream, the owner checks for applicable laws, policies, or regulations. For example:

1. Formal designations: e.g., Wild and Scenic River, Blue Ribbon Trout Stream, or domestic water-supply watershed. In this case, Shady Creek's designation as a Blue Ribbon Trout Stream would need to be reflected in the long term goals for the area.

2. Management guidelines: Many states have mandatory or voluntary Best Management Practices that guide project implementation. The presence of a state or federally listed threatened or endangered species will also require special consideration.

3. Previous commitments: These include lease arrangements or other stipulations on the use of the land such as scenic or recreational easements, rights-of-way, and agricultural or mineral leases.

Based on what is known so far, the Greenwoods landowner drafts initial long-term goals for the property. Initial goals include:

1. Healthy forest ecosystems that provide sustainable timber production.

2. Water quality in Shady Creek that meets temperature and sediment guidelines for domestic water use.

3. Shady Creek meets Blue Ribbon Trout Stream standards.

These goals provide the management that guides further planning.

Inventory and Assess Values and Resources

Resource managers are challenged to understand the often conflicting values associated with riparian areas, acquire relevant and sound social and ecological information about the area, and use this information in developing management plans. Recreation, food and fiber production, wildlife, fisheries, water quality, floodplain hydrology, and aesthetics are just some of the values associated with riparian areas (Brinson et al. 1981). If not carefully managed, inventorying these resources can be overwhelming, expensive, and of limited value. The amount of effort spent on inventory and assessment should reflect the relative importance of each resource, the sensitivity to and expected magnitude of effects from project proposals, and the level of confidence in existing data.

The inventory and assessment process gives managers an excellent opportunity to share and validate information. Integrated management requires that people with different training, skills, and backgrounds work together (Garcia 1989). People come to the table with different sets of values, professional languages, and modes of communicating. A successful integrated planning team learns a common language and builds trust of and respect for one another. The process of identifying needed information, developing a database, and assessing the validity of available information gives the team the opportunity to develop the skills necessary to work together successfully. Typically, information is fragmented among different organizations, is collected for other purposes, and may conflict (Yaffee 1997). Working together to develop mechanisms for sharing and evaluating information during the inventory phase will get the integrated management process off to a successful start.

Consider Social Values

As described in detail in Chapter 11, social values can be varied and complex. They also help set the context and relative priorities of the various natural resources in the project area. Inventory and assessment of the natural resources will be much more effective if focused by identifying social values early in the planning process. Considering social values and concerns throughout the planning process can reduce the potential for costly legal, political, or jurisdictional battles. Therefore, people's uses and values for an area should be one of the first things inventoried and assessed.

When an integrated approach is used, the first step is usually to identify the key stake-holders; those people who have an interest in the area being planned. Some stakeholders can be easily identified, because they have been very active and visible in the past. Other groups may take longer to identify. But there are many techniques to identify key players. A good starting point generally is a "geographic screen"; i.e., ensure that all affected and adjacent ownerships are represented. One type of geographic screen is to ask those initially involved: "Who else should be here?" When the same names come from several sources, you can be reasonably sure that most of the key interests are identified. This is known as the "snowball" technique (Goodman 1961). Agencies are generally aware of people who have a strong stake in a given project area, so talking with individuals from different agencies and other landowners will also help identify interest groups. There are many techniques for getting the word out (newspapers, web pages, notices). One simple and often forgotten technique is to visit the area and observe who uses it.

In the Greenwoods scenario outlined above, a single landowner was initially involved in the project. However, the landowner was interested in working with other groups of people to develop a management plan that would more effectively address all the resources found on the property. The landowner first contacted an extension forester who in turn recommended additional contacts. The landowner next consulted with both a wildlife biologist and a fisheries biologist from the state department of natural resources. The wildlife biologist suggested that the hydrologist from the municipal water department would also be interested because the stream on the property was a tributary to the domestic water intake. The hydrologist suggested working with a local canoe club that uses the stream. From other projects and community works, the landowner knew that other potentially interested groups included the Audubon Society and Trout Unlimited. All these contacts led to the establishment of an informal team of resource professionals and interested citizens working on a management plan for Greenwoods Forest. This is a good first step. Involving multiple interests, having representatives from many disciplines, and having a clear decisionmaking process are important components of integrated resource management (MacKenzie 1997).

Once key players have been identified in the project, it is essential to develop a list of concerns (issues) about the project area (referred to as "scoping" by federal agencies using the National Environmental Policy Act (PL 91-100)-(NEPA) process). Identification of key issues will help managers focus plan efforts and will direct future resource inventories. Techniques for identifying key issues may include everything from mailings, interviews, and surveys to highly structured workshops (Moore 1987). The time spent on identifying key issues should relate to the degree of interest (or opposition) you expect in the project (Daniels et al. 1996). It's better to spend time up-front, identifying concerns and interests, than to deal with them later when people are upset because they feel they have been excluded from the process.

In our example project, several concerns were identified. The wildlife biologist was concerned about the bald eagle nest located near the stream as well as the importance of riparian areas for a variety of wildlife species. The fisheries biologist identified protection of the Blue Ribbon Trout Fishery as very important and, for all the reasons outlined in Chapter 9, highlighted the importance of coarse woody debris. The municipal water district hydrologist was concerned about protecting drinking water quality. The canoe club president wanted to maintain a quality recreation experience for canoeists and had concerns about trees falling and blocking passage in several places. The landowner wants to maintain timber productivity and provide for long-term sustainable harvest.

At this point, two long-term goals were added to reflect the presence of an eagle nest and a previously unrecognized high level of canoe use on Shady Creek:

4. Shady Creek provides a high-quality canoeing opportunity.

5. Potential bald eagle nest trees are present along Shady Creek.

Inventory Natural Resources

An intensive survey of all biological resources in an area is usually unnecessary and impractical. One way to focus an inventory is to consider surveying only those resources that are most critical or sensitive, those that are most important in meeting the goals, or those that most relate to issues identified during up-front public involvement.

In our example, Shady Creek is identified as a Blue Ribbon Trout stream and the landowner's goal is to maintain this status. Factors that influence the status of a Blue Ribbon Trout stream include existing trout populations (number, size, species, distribution), water quality (temperature, sedimentation), riparian vegetation (stream shade, supply of future coarse woody debris) and in-stream habitat structure (amounts and location of coarse woody debris). When planning a project that could affect the trout populations in Shady Creek, it would be important to include a survey of these resources.

Inventory of natural resources consists of two phases: (1) assessment of existing information and (2) on-the-ground collection of information. Existing sources of information — GIS, maps, files, databases or something as simple as personal experience may preclude the need for further survey work on the ground. However, on-the-ground inventory work may be needed to fill in gaps in existing data or resolve conflicting information from different sources. On-the-ground inventory work can range from an informal walk-through of the area for very simple resource information needs to intensive sampling (Figure 13.2). Sampling intensity should be related to the significance of the issues being addressed and to the sensitivity of the resources being sampled. One important consideration is that all people involved in sharing information should agree that the quality and extent of information is acceptable. Previous chapters discuss some of the actual tools and techniques used to inventory certain resources.

Multiple spatial scales should also be considered in resource inventories. Some information may be adequate at the project scale, while other information must be considered at a larger, landscape scale. In our example, the landowner was able to work with the state fisheries biologist to determine that a maximum summer temperature of no more than 62°F is reasonable for trout streams in that region. In previous survey work, the biologist found that the maximum summer water temperature in Shady Creek is 59°F. Therefore, a project design that involves no more than a 3°F rise in summer stream temperature seems to meet trout water quality parameters. However, to be certain, the landmanager needs to consider other past and future management activities both upstream and downstream of this project such as the timber harvest planned on other ownerships in the next two years and the effect of lack of shading in the upstream pastureland. Other activities may result in a cumulative rise in temperature that is unacceptable. In an interdisciplinary process, other resource issues would be considered in a similar manner.

Develop Alternatives

Alternatives are simply different ways to reach the same desired end, as defined by the long-term goals. The foundation for developing sound alternatives for a project is a clear understanding of how the project relates to the stated long-term goals. In our example, the reason for the project is the deteriorating jack pine that does not meet the stated long-term goal for forest health. A project could just as easily be aimed at achieving any of the other long-term goals not being met. For example, unacceptably high water temperature or sediment levels in Shady Creek could lead to a project for restoring the water quality needed to meet long-term goals for fish habitat and drinking water supply.

In an integrated management approach, alternatives offer a way to display varying levels of emphasis on the resources of the project area. Most projects include actions to mitigate the adverse effects of a project. An integrated approach seeks ways for the project to actively move toward all the long-term goals rather than just mitigate effects which might move away from those goals. Alternatives display different approaches, effects, benefits, and costs so that they can be compared.

In the Shady Creek example, an integrated approach was used to identify ways to accomplish a variety of resource goals. Although a project was originally conceived to harvest timber, other opportunities identified were incorporated into one or more alternative approaches. These include improving canoeing experiences, providing large dead snags and trees for eagle nesting, and providing future coarse woody debris for fisheries and water quality.

Alternatives were developed to display tradeoffs in timber production related to providing streamside buffers and long-lived trees along Shady Creek for shading and long-term supply of coarse woody debris. Alternatives also explored specific ways to add coarse woody debris and provide for canoe passage. The objective of this alternative was to produce pools and rapids that benefit both fish habitat and canoeing.

Consider the Capability and Limitations of the Ecosystem

Considering the inherent capabilities and limitations of the land is an essential step in determining whether alternatives are feasible and practical. For example, different sites are capable of growing some tree species but not others; water chemistry may limit the ability to support aquatic life; the presence of wetlands may satisfy habitat requirements for some wildlife species but not others; some areas may have soil or topographic conditions that limit activities completely or seasonally; and some ecosystems may have been stressed sufficiently by previous activities to limit any further disturbance. Equally important to the biological or physical capability of the immediate area is recognition of its relationship to surrounding lands. For example, goals of solitude may only be achievable if adjacent lands have similar objectives.

Address Issues and Concerns

A primary reason for exploring alternatives is to address the issues and concerns that surface from interested parties. Each alternative should move the area toward the long-term goals for the project but will use a different approach to address the issues. In general, if nobody is concerned about your proposal, there is no reason to consider alternatives (except for federal agencies, where a no action alternative is required under NEPA).

In our example, the planning team developed several alternatives that addressed most of the concerns and suggestions for change that came through the integrated, collaborative approach to project design. Alternatives modified harvest near the stream and enhanced erosion control to protect water quality. Alternatives also maintained shade and provided coarse woody debris to improve fisheries. All alternatives were modified to protect the eagle nest because that was required by the Endangered Species Act. The landowner decided to look at alternative rates for establishing future tree species composition to address concerns that sudden change may have undesirable impacts on the birds that use the riparian area. While the ultimate species composition was established as part of the desired future condition, moving toward that goal at a faster or slower rate would have different effects.

Evaluate Effects and Trade-Offs

Once alternatives are developed, they should be assessed to determine their effects. Effects can be positive, such as improving fisheries or providing timber products or negative, such as degrading water quality or creating barriers to recreation. Usually, there will be a combination of both positive and negative effects. Effects can be local and tied directly to the site at a project scale or they can be indirect, perhaps resulting in changes beyond the immediate area. For example, increased sedimentation or temperature could affect local fish habitat and be felt far downstream.

Effects can be immediate and short-term, such as the immediate change to wildlife habitat from harvesting timber, or they can be gradual and long-term such as changes in vegetative type. Anticipated long-term changes resulting from the project should be consistent with the desired condition for the area. It is important to recognize that short-term actions may seem to contradict a long-term goal but may still be a necessary step in reaching the goal. For example, heavier than normal timber harvest in a riparian area might be required to establish the long-lived species desired for long-term coarse woody debris.

The effects of one project may add to past or future actions to result in changes far exceeding that of any individual project. An analysis of these cumulative effects would include the past and anticipated effects of other land-use activities in the area. In our example, this would include assessing effects on stream temperature and sedimentation of past and future timber harvest and the past clearing of pasture along Shady Creek.

Alternatives are developed so that the decisionmaker and the public can fully understand the various possible approaches, the short and long-term effects, and the tradeoffs associated with taking different courses of action. Since most actions will have both positive and negative effects, the final decision on what to do will reflect the combination of tradeoffs the decisionmaker believes to be best.

Check Back with Affected People and Groups

Integrated management for riparian areas requires collaboration with many people, agencies, and groups. For Greenwoods Forest, collaboration began by interested parties discussing the desired conditions for the area and objectives for the project. Alternative approaches were developed that addressed issues, resulted in different environmental effects, and were feasible given the capabilities and limitation of the land.

After developing alternatives and before proceeding with further analysis or final decisionmaking, it is best to check back with those who have participated in planning thus far or will be affected by the project. This contact helps ensure that the factors considered in making the decision are the most important factors and that alternatives are adequate for addressing the significant issues.

Even if people do not agree with the final decision, they are more likely to accept it if they had ample opportunity to be heard and get involved. Federal and state agencies, as well as landowners, have a responsibility as good neighbors. When the decision is made, you must be able to show you have listened, considered the results, explored the options, and have good reasons for selecting a particular course of action.

Selection of an Alternative for Implementation

Many decisionmaking processes, both informal and formal, can be used to select a plan of action from a list of possible alternatives. Formal processes may involve one of the many techniques available for rating, weighting, and selecting the best approach. Informal processes may involve nothing more than an intuitive comparison of alternatives. Regardless of the process or approach used, comparison of alternatives should include consideration of how each alternative relates to the goals for the area, issues and concerns identified during scoping, and information gathered about stockholder values and desires.

In our Greenwoods Forest example, the landowner selected an alternative with the following features:

- Harvest a total of 100 acres of jack pine immediately, focusing on the areas experiencing highest mortality. Harvest the next highest priority 100 acres in 10 years and the final 100 acres in 20 years.

- In the harvest areas, leave buffer strips along Shady Creek. As planned, buffer strip density on the north bank will be much lower to facilitate establishment of intolerant, long-lived species that will provide large logs for coarse woody debris in the future. Buffer strip density on the south bank will be higher to provide shading for temperature mitigation.

- Fell some large trees into Shady Creek to enhance fisheries habitat and provide stream bank stability. Consider the need for safe canoe passage and the possibility of creating pools and rapids to enhance canoeing.

- Leave three large trees or snags per acre within 100 feet of Shady Creek for potential eagle nest trees and future coarse woody debris.

Monitor and Adapt

Monitoring (see Chapter 16) is a key step in implementing a project. It is important to know whether project objectives are being met and how well the project is helping achieve the desired future condition. Even the best plans sometimes do not achieve the desired results. Adaptive management is an approach that embraces this reality to continually improve results by "adjusting management actions in light of new information" (Interagency Task Force 1995) from monitoring, research, or other projects being implemented (Figure 13.3). In simpler terms, adaptive management is "learning by doing" (Walters and Holling 1990). Management decisions are viewed as experiments subject to modification rather than fixed and unbending final rulings. Adaptive management recognizes the limits of knowledge and experience and helps us move toward goals in the face of uncertainty (Interagency Task Force 1995).

In our Greenwoods Forest example, a monitoring plan was developed in collaboration with several of the interested parties including the Municipal Water District, the state, the county, the local canoe club, and a local fishing club. In fact, the monitoring plan was developed in such a way that it could be used for future projects on any of the ownerships along Shady Creek. Monitoring parameters were focused on key issues such as sediment, water temperature, riparian vegetation, and coarse woody debris. Monitoring of Shady Creek after harvest indicated that stream temperature did not change noticeably. However, smaller jack pine logs felled into the stream to meet coarse woody debris recommendations were washed downstream during high spring flows while larger jack pine logs remained. Under adaptive management, buffer strip densities might remain about the same for future projects, while coarse woody debris guidelines might be modified to rely solely on larger logs.

Conclusion

Forest resource management has become more important and complicated as the pressures, interests, and demands of society have increased. This is even more true of riparian resources because water moves — it can travel across ownership boundaries and through many ecosystems, affecting and being affected by many factors as it does so. In addition, the importance of water is evident to more people than other, less-visible resources. Everyone understands the value of abundant, clean water. Integrated management is an effective way to address complex, sensitive resources and resource management issues. An integrated approach to management with broad involvement of specialties and interests should not be considered a barrier or burden to achieving landowner goals. Rather, it offers a way to achieve both sound resource management and widespread acceptance of management activities.

Figure 13.2 Inventory of natural resources is an important part of project planning; it may mean installing a water temperature recorder (top), or measuring the depth of sand covering a brook trout spawning riffle (bottom) (both in Minnesota).

Figure 13.3 Whether prescribing timber harvest in Minnesota (top) or evaluating a stream ford in Ohio (bottom), assessing and improving management practices is the object of adaptive management.

Chapter 14

Ecological Principles for Riparian Silviculture

Brian J. Palik, John C. Zasada and Craig W. Hedman

"The goal of silvicultural management in riparian management zones should be to provide the natural ecological functions of riparian vegetation...." Gregory, 1997

We value riparian forests for diverse things, perhaps placing greater value on them than on any other type of forest. For instance, riparian forests offer recreation opportunities that are unparalleled by other types of forest. Riparian areas are important for flood mitigation and water quality control. Many riparian forests provide critical habitat for species that require the land/water interface in their life cycle, as well as travel corridors and refuges for species that use it occasionally. We also value riparian areas for the productive forests and unique timber species they support. The special place of riparian forests in our value system is a result of the equally special place of riparian forests in the landscape — at the interface between land and water. This landscape position imparts unique ecological attributes to riparian forests in the form of linkages between land and water (Chapter 2) and is responsible for the associated array of values we place on them.

Historically, riparian forest management has disregarded the natural links between land and water (Swanson and Franklin 1992). Ecologists refer to these as functional links, because each one serves an important function in sustaining aquatic ecosystems and in the life history of the plants and animals found in riparian areas. Functional links describe the flow of material, energy, and nutrients (e.g., fallen trees, limbs, leaves, nutrients in the plants, nutrients in the water, the water itself, sediment, and sunlight) between riparian forests and aquatic ecosystems. More recently, managers are placing greater emphasis on incorporating a functional perspective into the silviculture of riparian areas (Berg 1995). To do so requires an appreciation of three basic ecological principles of riparian forests. First, riparian forests vary in composition and structure, requiring equally diverse silvicultural approaches. Second, the functioning of riparian forests depends closely on stand structure and species composition.

233

Figure 14.1 Some examples of riparian forests in the Lake States and Georgia. Clock-wise from the top: a lakeside conifer peatland in northern Minnesota; a black ash-dominated floodplain along a small stream in northern Minnesota; a hardwood bottomland grading into longleaf pine in southeastern Georgia; a mixed stand of paper birch and conifers on a lakeside slope in northern Michigan; and a northern hardwood forest along a headwater stream in northern Minnesota.

Finally, riparian forests are ecotones (they are not easily delineated stands of uniform structure and function).

We explore these principles in this chapter, pointing out their implications for riparian area management. We then present examples of how a manager may integrate functional objectives with other objectives in silvicultural prescriptions for riparian areas. We close by suggesting some guidelines for riparian silviculture that incorporate this ecological understanding.

Riparian Forest Variability

Riparian forests vary greatly in their biotic and abiotic characteristics, and the landscape settings where they occur. They are not simply floodplain forests next to perennial streams. Rather, they occur adjacent to a variety of aquatic and wetland ecosystems and on a diversity of geomorphic surfaces, including terraces and slopes (see Chapter 2).

Variation in Space

Riparian forests occur next to rivers and streams, but also lakes and open-water wetlands. They support a diversity of plant communities, not just bottomland forests (Figure 14.1). Unfortunately, many managers, and natural resource scientists still hold that riparian forests are simply river floodplain communities (see Zentner 1997).

Often, management approaches for riparian forests ignore their natural variability by using conventional, one-size-fits-all practices. Consider the scenario shown in Figure 14.2 that is based on real-world examples common in the eastern United States. As the figure illustrates, an important first step in applying a silvicultural prescription is to first delineate stand boundaries. In riparian settings, this amounts to drawing a line at some distance from the aquatic system to delineate a riparian management area (Figure 14.2a). We refer to this as *the magic line* because managers may lose sight of the fact that this line does not always reflect the functional boundary of a riparian forest. In practice, politics more than science decides the location of this line which is one reason it is often the same distance from the water (e.g., 30 m) regardless of the characteristics of the riparian area.

An all-too-common silvicultural prescription is to designate the riparian management area as a no-cut buffer (Figure 14.2b). This is a legitimate management alternative in some situations. However, when pursued out of uncertainty about potential impacts, a no-cut buffer simply limits management options and opportunities. These opportunities include not only management for obvious features, such as desired commercial species and timber products, but also enhancement and restoration of riparian functions (see next section). Our point is that no-cut buffers do not accommodate the natural range of variability in riparian forests, including differences in potential composition and productivity. These buffers ignore the fact that disturbance is a natural part of riparian systems (see next section), and they provide minimal flexibility for meeting diverse management objectives.

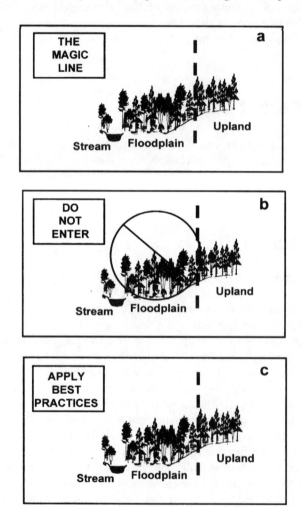

Figure 14.2 Stylized depiction of a riparian forest showing the placement of a riparian management-boundary (a) and two common alternatives within the designated management areas: a no-cut buffer strip (b) and application of Best Management Practices (c).

An alternative to the no entry scenario is the application of Best Management Practices or BMPs within the designated riparian buffer (Figure 14.2c). Riparian BMPs protect water quality, and sometimes aquatic and terrestrial habitats, from negative impacts during and after harvest. Typically, BMPs are based on residual basal area and soil disturbance requirements. Table 14.1 provides some examples of BMPs for several eastern states.

Table 14.1 Best management practices for riparian forests in selected eastern states

State	Stand Characteristic	Recommended Practice	Degree of Ecosystem Specificity
Minnesota[1]	Management Area Width	50 ft 100 ft	Non-trout waters Trout waters
	Residual Basal Area in Riparian Mgt. Zone	25 ft^2/acre 60 ft^2/acre 60 to 80 ft^2/acre	Intol. trees adjacent to any waters Tolerant species non-trout Tolerant species trout waters
	Forest Floor Disturbance	Mineral soil exposure less than 5%	None
Pennsylvania[2]	Management Area Width	At least 50 ft	Increase 20 ft with each 10% increase in slope
	Residual Basal Area in Riparian Management Zone	Maintain at least 50% residual crown cover No cutting within first 10 ft	None Seasonal ponds and sinks
	Forest Floor Disturbance	Avoid disturbance Avoid ruts deeper than 6 inches within 200 ft of water	None Seasonal ponds
Tennessee[3]	Management Area Width	25 ft	Increases 20 ft with each 10% increase in slope
	Residual Basal Area in Riparian Mgt. Zone	Maintain 50 to 75% residual crown cover	None
	Forest Floor Disturbance	Avoid disturbance	None

[1]MDNR 1995, (1999 changes offer variable widths based on stream width or open wetland size. For even-age mgt. on non-trout waters widths are 50 or 100 ft, and uneven-age mgt. widths are 50, 100, or 200 ft. On trout streams and lakes the even-age mgt. width is 150 ft, and the uneven-age mgt. width is 200 ft).
[2]Forested Wetlands Task Force 1993.
[3]Tennessee Department of Agriculture 1996.

Riparian BMPs are not silvicultural prescriptions. Unfortunately, managers sometimes give riparian BMPs undue weight by using them as the core of a silvicultural prescription rather than incorporating them as elements into a comprehensive silvicultural plan. BMPs should not form the basis of silvicultural prescriptions. They typically vary only minimally to accommodate changes in slope, or the presence of important fish populations, or perhaps even major differences in shade tolerance among desired commercial species (Table 14.1). They do not include the full range of activities that must be part of a comprehensive prescription. Applying BMPs indiscriminately and uniformly within some predetermined management area ignores the diversity of riparian forest types and management objectives.

Variation in Time

Riparian forests vary in composition and structure over time because of natural and human disturbances. The role of natural disturbance in regulating the amount and spatial pattern of successional stages in watersheds is an important management consideration. A detailed review of natural successional dynamics in riparian forests is beyond the scope of this chapter. However, several points are important to keep in mind when planning and implementing riparian silvicultural operations.

First, remember that, just as they do in other types of forests, various natural disturbances affect riparian forests to reset the successional clock (Chapter 5). Consequently, within and among watersheds, natural disturbances create a spectrum of stand age classes and compositions; not just old, late successional forest. For example, consider the area of river bottom forest in the Lakes States during the mid-19th century. During this period, there were approximately 1.8 million hectares of river bottom forest in Michigan, Minnesota, and Wisconsin (Frelich 1995). However, old-growth apparently made up no more than 55% of the total; the rest was younger and presumably composed of some earlier successional species.

Remember also that the disturbance regime (the type, severity, and return interval) differs greatly within and among watersheds and geographic regions. Consequently, managers must evaluate disturbance regimes on a site- and watershed-specific basis. Armed with this knowledge, it may be possible to create similar stand-age and compositional patterns with silviculture (Oliver and Hinckley 1987). Aside from simply copying nature's patterns, there may be sound ecological reasons for creating a diversity of stand age classes within a watershed. For instance, the capacity for riparian vegetation to absorb nutrients from water moving to an aquatic system (see next section) may depend upon the stage of forest development. Periodic and dispersed disturbances, which foster new vegetative growth, help to maintain uptake efficiency (Lowrance et al. 1984). This suggests that silvicultural prescriptions that maintain some stands, or portions of stands, in a vigorous growth stage may enhance the filtering capacity of a watershed.

One of the important ways that management alters plant composition and successional pathways in riparian settings is through changes in hydrology. Hydrologic factors that alter species composition include seasonal variation in water table depth, maximum depth to the watertable during the growing season, depth and duration of flooding, the source of water (ground, surface), drainage patterns, depth of organic soil, and texture of the mineral soil. Many plants that grow in riparian areas have adapted (through their physiology, anatomy, morphology, or phenology) to the particular hydrologic regime of a system. McKevlin et al. (1997) provide an excellent review of the literature on this topic. A key point to remember is that even small to moderate changes in hydrology can lead to significant changes in plant species composition. The consequences of these changes may be favorable or unfavorable, depending on management objectives. To avoid undesirable consequences, managers should understand the hydrology of a site in as much detail as possible to minimize disturbances from single and cumulative silvicultural treatments. In some cases, it may be necessary to return the hydrologic regime to a pre-disturbance condition to achieve desired species reproduction and growth.

Riparian Functions and Stand Structure

The second principle is that structure and composition of stands link riparian forests to water through ecological functions (Table 14.2). Silvicultural approaches for maintaining or restoring healthy riparian forests should incorporate methodologies that sustain these functional links to water. We briefly discuss these functions in the next section. Other chapters in this volume contain greater detail.

Table 14.2 Riparian forest functions and stand structural characteristics that affect functions

Functional Attribute	Structural Feature
Aquatic coarse woody debris recruitment	Species composition; live and dead tree density; tree age and size
Aquatic particulate and dissolved organic matter input	Species composition; stand density
Water temperature and light regulation	Canopy heterogeneity; stand density; species composition
Bank stability	Tree age and size; tree density; abundance of understory vegetation
Regulation of sediment, nutrient, and organic matter movement or uptake	Stand density; coarse woody debris abundance; species composition; forest floor and understory vegetation
Terrestrial habitat for riparian species	Forest composition; tree age and size; coarse woody debris; understory vegetation

Coarse Woody Debris

Coarse woody debris input to streams is probably the best recognized functional link between riparian forest stands and water. The riparian source area for debris in surface waters is the zone defined by the tallest trees growing along the bank (Fetherston et al. 1995; Gregory et al. 1991). In practice, most trees come from closer distances, e.g., nearly 50% from within 3 ft (Murphy and Koski 1989). The actual distance varies with factors that influence tree height: e.g., site productivity, species composition, and stand age. The source distance may exceed tree height if secondary transport of the dead tree occurs, such as in floods or with mass soil movement.

Forest stand characteristics that influence stream debris include tree size distributions and canopy composition. For instance, the longevity of coarse woody debris in streams is directly proportional to bole diameters (Murphy and Koski 1989; Webster et. al. 1992) because larger stems decay at slower rates than small stems (Harmon et al. 1986) and large logs are less mobile in the steam than are small logs. Reductions in tree diameters in the woody debris source zone, through shortened rotation lengths or conversion to short-lived species, will affect characteristics of debris in the water body. Log decomposition rates also differ among species (Harmon et al. 1986), as does the susceptibility of streamside trees to mortality from disturbance (Palik et al. 1998). In concert, these factors influence addition and depletion rates of debris among riparian forests of different compositions.

Particulate Organic Matter

Canopies of riparian forests produce the coarse particulate organic matter, primarily leaf litter, that is an important energy base for food webs in aquatic ecosystems. Leaf litter input is particularly important in shaded areas because primary production is low (e.g., low-order streams (Vannote et. al. 1980) and seasonal forest wetlands (Barlocher et al. 1978)). Additionally, particulate organic matter is an important energy source in large, unshaded rivers through deposition on their floodplains and eventual transport into the active channel (Johnson et al. 1995).

Some riparian management guidelines suggest that most particulate litter entering river channels originates within one-half tree height of the bank (e.g., FEMAT 1993). There are little published data for any region of the country to support or refute this contention. Data we have collected suggests that the one-half tree height rule of thumb does not always hold (B. Palik unpublished). Figure 14.3 shows autumn litter input to headwater streams in Minnesota after a summer harvest. The streams flow through (1) uncut forest, (2) a 100 ft wide uncut buffer with adjacent clearcuts in the upland, or (3) a 100 ft wide, selectively cut buffer (with the adjacent upland clearcut) having a residual basal area of 35 ft^2/acre. The data indicate that one-third of the biomass of litter in the stream comes from distances beyond 100 ft. This distance exceeds the mean maximum tree height for the system of approximately 72 ft. The data also show that litter input to the stream the first year after selective cutting of the buffer is even lower than in the uncut buffer. However, litter inputs to the stream should rise as leaf area in the cut forest recovers.

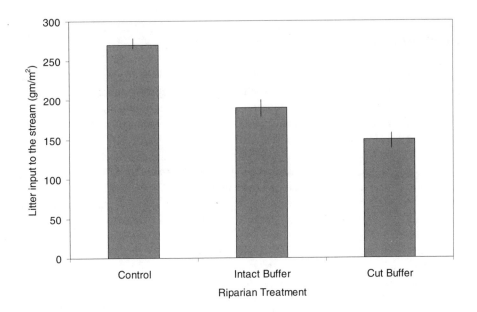

Figure 14.3 Coarse particulate organic matter input to small streams in northern Minnesota under differing riparian forest management conditions. The streams are surrounded by uncut forests (control), a 100 ft wide uncut buffer with the adjacent upland clearcut (intact-buffer), or a 100 ft wide buffer (with adjacent upland clearcut) that is selectively cut to a residual basal area of 35 ft^2/acre (cut-buffer). Litter was collected from September through November 1997. Values are means (n = 3± 1 standard error).

The quality of leaf litter entering a water body is also important for stream functioning. Riparian canopy composition has profound effects on litter quality. For instance, litter of alder (*Alnus* spp.), which can be abundant on floodplains of streams in many regions (e.g., Swanson et al. 1982; Pautou et al. 1992), increases turnover of organic matter due to high foliar nitrogen content (Cole et al. 1978). Differences in litter quality and processing rates can affect aquatic invertebrate communities (Webster and Benfield 1986) largely by determining the timing of resource availability to litter processors (Cummins et al. 1989).

Forest composition also determines the phenology of litter additions to water. Pulsed seasonal inputs of litter occur when deciduous species dominate the riparian forest. In contrast, evergreen species provide a more constant litter supply. This suggests that managers should maintain mixtures of evergreens and deciduous species where they occur naturally in managed riparian forests.

Solar Radiation

Canopies of riparian forests control the amount and quality of light reaching surface waters. Changes in light environments through alteration of canopy structure are of greatest concern in smaller aquatic systems where shading is usually high. Light regulation has important consequences for primary production and aquatic animals (Tschaplinski and Hartman 1983; Gregory et al. 1991). For example, macrophyte production increases with decreased shading of the water. Riparian canopies also regulate water temperature (Barton et al. 1985) which has important consequences for fish populations through effects on primary productivity (Swanson et al. 1982) and dissolved oxygen concentrations (Ringler and Hall 1975).

Cartographic models, based on stream orientation, riparian slope, vegetation height, and solar angle and direction, provide insight into the influence of riparian forests on stream shade and water temperature. One such exercise identified buffer requirements needed to maintain stream temperature at pre-harvest levels. Shade-strip widths ranged from 0 ft on the north bank of east-west running reaches to 115 ft on steep south sides of northeast running reaches (Dick 1989). Another study found that maximum shading of small streams occurs when the width of riparian buffers on the south side of the stream approaches 150 ft (Steinblums et al. 1984). While this distance may buffer the stream itself, it does not necessarily provide microclimatic buffering in the riparian forest by protecting it from adjacent upland influences, e.g., large adjacent clearcuts (Brosofske et al. 1997).

Bank Stability

Vegetation growing on streambanks and lakeshores binds soil which reduces surface erosion and sediment runoff to surface waters. Riparian plants also buffer banks from the erosive force of flowing water (Smith 1976). Large trees with expansive root systems should be better at preventing bank erosion than small trees. However, woody vegetation is not the only consideration for bank stability. For instance, roots of herbaceous vegetation may reduce erosion rates of streambanks (Dunaway et al. 1994). This argues for considering the entire riparian plant community in silvicultural prescriptions.

Regulating the Movement of Nutrients, Sediment, and Organic Matter

Riparian forests regulate the movement of water between terrestrial and aquatic ecosystems, and along with it, nutrients, sediments, and particulate organic matter. Riparian forests affect material movement in several ways. Coarse woody debris in active stream channels traps suspended sediments and particulate organic matter. Floodplain forests also trap sediments and particulate organic matter outside the active channel during over-bank flows (Swanson et al. 1982; Grubaugh and Anderson 1989). The boles of trees, as well as coarse woody debris on the forest floor, may be key impediments to sediment movement in floodplains (Swanson et al. 1982; Harmon et al. 1986). The sediment trapping function of floodplains suggests that this entire geomorphic feature be included in riparian delineation and management guidelines. Too often, Best Management Practices fail to specifically consider the functional links operating on river floodplains (Gregory 1997).

Non-floodplain forests trap sediments in overland flow before they enter the aquatic system. Tree boles, coarse woody debris, herbaceous vegetation, and litter all trap sediment in overland flow. The first 200 to 300 ft of forest probably traps most suspended sediments in overland flow (O'Laughlin and Belt 1995). However, this distance increases with slope of the forest land. The composition of riparian stands also affects cycling and transformations of dissolved nutrients. Riparian trees contribute dissolved nutrients to aquatic systems through leachates from litter (Fisher and Likens 1973; Fiebig et al. 1990). As discussed earlier, species vary in the nutrient content of their litter. Additionally, forests on floodplains are more effective denitrifiers than forests in higher geomorphic positions (Gregory et al. 1991). Soil moisture plays a role in this, but forest composition also is important. For example, denitrification rates are much higher in alder-dominated floodplains than in conifer-dominated systems (Gregory et al. 1991).

Habitat

Structural features of riparian stands (e.g., rotting logs, snags, vertical canopy layering, large trees with cavities, etc.) provide habitat for a variety of organisms, just as they do in non-riparian forests. The added benefit of adjacent water makes the riparian setting a particularly important habitat for many species. Riparian species include both obligate (must live in this habitat) and facultative (use it by preference occasionally) users. Birds provide the best examples of obligate species. Bald eagles (*Haliaeetus leucocephalus*), for instance, rely on water as a source of fish and require large trees in riparian forest for nesting and observation. Wood ducks (*Aix sponsa*) also are obligate riparian users. They feed in water but need large cavities in trees for nesting. The eastern gray squirrel (*Sciurus carolinensis*) is an example of a facultative riparian mammal. While found in many habitats, gray squirrels use hardwood stringers along ephemeral streams in upland pine and pine-hardwood forests (Fischer and Holler 1991).

Sustaining Riparian Functions Through Silviculture

Silvicultural prescriptions for riparian forests must incorporate up-to-date knowledge of forest-water interactions directly as a principal objective, not as a secondary application of BMPs. In some cases, this may be possible using traditional silvicultural for a given forest type that occurs near the water's edge. Multi-cohort and crop-tree systems (Perkey et al. 1993) for northern hardwoods are examples. In contrast, conventional single-cohort systems, which leave little mature structure, are not appropriate in many riparian settings — at least they are not appropriate to use throughout the riparian area or throughout a watershed (see next two sections). The distinction is that multi-cohort systems maintain tree cover and mature stand structure in the riparian area and, consequently, maintain the functional links to water that trees provide. Single-cohort systems break these links when they leave little or no mature structure.

Riparian Areas as Ecotones

The final principle to remember is that riparian forests function as ecotones between land and water (Naiman et al. 1988). Ecologists have been studying riparian ecotones for decades, but managers rarely consider the ecotone concept when delineating and managing riparian forests.

Three types of ecotones are of concern in riparian forests: ecotones of structure, ecotones of function, and ecotones of impact. The structural ecotone reflects changes in abiotic and biotic characteristics from the water's edge to the upland forest. Various factors change across this ecotone, including geomorphology, soil features, disturbance regimes, and plant communities (Figure 14.4). Not all riparian ecotones are as structurally complex as the one shown in Figure 14.4, but even structurally simple riparian ecotones consist of at least two intergrading forest ecosystems that differ in geomorphology, soil texture and drainage, plant community composition and structure, and natural disturbance regimes.

The functional ecotone reflects the fact that the ecological links between riparian forests and water are not equally important at all distances along the gradient, we show some rules of thumb for these distances (adapted in part from O'Laughlin and Belt 1995)(Figure 14.5). However, the exact distances at which functional links are no longer important are difficult

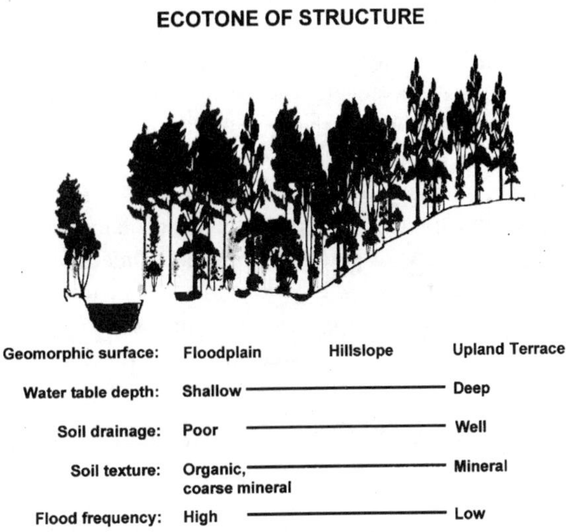

ECOTONE OF STRUCTURE

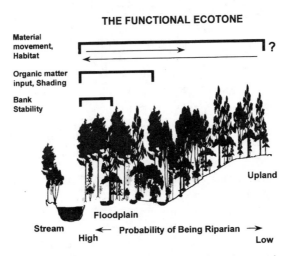

Geomorphic surface:	Floodplain	Hillslope	Upland Terrace
Water table depth:	Shallow ———————————— Deep		
Soil drainage:	Poor ———————————— Well		
Soil texture:	Organic,———————————— Mineral coarse mineral		
Flood frequency:	High ———————————— Low		

Figure 14.4 Diagram of a riparian forest ecotone showing changes in abiotic and biotic characteristics from the water's edge to the upland.

THE FUNCTIONAL ECOTONE

Material movement, Habitat ?

Organic matter input, Shading

Bank Stability

Upland

Floodplain

Stream   ←— Probability of Being Riparian —→
High   Low

Figure 14.5 Diagram of a riparian forest ecotone showing changes in various ecological functions from the water's edge into the upland. Material movement includes exchanges of water, sediment, and nutrients between the aquatic and terrestrial systems. Organic matter input includes coarse debris, particulate and dissolved organic matter.

to determine with certainty. It is instructive to think about the functional ecotone in a probabilistic sense. The probability of a forest having riparian function is great immediately next to the water. This probability decreases away from the water as various functions become less important. The point where the probability of being riparian reaches zero may be difficult to determine with certainty. It is usually not so difficult to determine when the probability is low, at least for many of the important riparian functions. The field key for delineating riparian areas in Chapter 2 incorporates this probabilistic idea.

The final riparian ecotone is one of potential impacts to functions (Figure 14.6). The idea here is simple: the importance of riparian functions increases with decreasing distance to the water. Consequently, the potential to degrade these functions also increases closer to the water. For example, the probability of soil transport to a stream after site preparation is

ECOTONE OF IMPACT

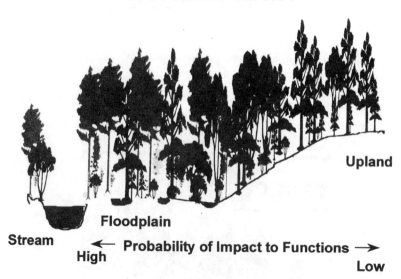

Figure 14.6 Diagram of a riparian forest ecotone showing changing potential for management to negatively affect functions.

higher within 10 ft of the channel than it is at 70 ft from the channel. The potential to use silviculture to enhance riparian functions also increases closer to the water's edge. For instance, thinning in the riparian forest may increase growth of residual trees, thereby hastening an increase in stream shading and large debris contribution (Gregory 1997).

Obviously, the potential to use thinning to accomplish these objectives decreases with distance from the water.

The probabilistic models of function and impact are the conceptual basis for riparian management buffers. Unfortunately, there is little scientific data to support the use of specific buffer widths. Nevertheless, the basic idea is sound: the gradient in riparian function loosely corresponds to the gradient in potential impact. Consequently, the farther one is from the water, the lower the probability that alteration of the forest will affect riparian functions.

Integrating Functional Objectives into Silvicultural Practice

How can we use our ecological understanding of riparian forests to improve silvicultural practice? The first step is to delineate riparian management boundaries ecologically, by recognizing that riparian systems are ecotones of structure, function, and impact. Riparian area delineation is a challenging exercise, particularly when based on functional characteristics. The on-the-ground approach presented in Chapter 2 of this volume provides insight into dealing with this delineation.   The second step is to prescribe site-specific practices that protect or enhance riparian functions along the ecotone while meeting other management objectives. At this step, managers need to consider two important points. Always remember that the starting point for all silvicultural activities is the prescription. The prescription outlines the silvicultural system for the stand as well as a method for monitoring results (Nyland 1996; Zasada 1995). The prescription covers a long period (e.g., one or more rotations) and, consequently, the prescription itself must be flexible. It is a first approximation that is subject to alteration and refinement as more information becomes available. Remember that monitoring in riparian forests must include the effects of silvicultural practices on riparian functions.

A major purpose of the prescription is to ensure that all activities are complementary and based on current knowledge and technology. In other words, the practice of silviculture in riparian forests should anticipate the future and prevent problems rather than respond to problems as they develop (Wagner and Zasada 1991). Recognizing the important functional links between forests and aquatic systems is an important tool for developing silvicultural approaches in riparian settings.

Another consideration, when developing silvicultural prescriptions, is to minimize cumulative effects of individual activities. Cumulative effects are "the impact on the environment which result from the incremental impact of the action when added to other past, present, and reasonably foreseeable future actions (Gosselink et al. 1990)." Cumulative effects can result from seemingly minor actions taking place over an extended period; collectively, these actions have significant negative impact. To avoid or minimize cumulative effects, the prescription must coordinate activities occurring at different times during stand development. In the riparian setting, this means understanding how to avoid small, but cumulatively significant, impacts to riparian functions.

Examples of Site-Specific Silviculture along the Riparian Ecotone

Some examples help to illustrate our ecotone concepts. First, consider the stylized riparian area depicted in Figures 14.7 and 14.8. In Figure 14.7a, the riparian management boundary occurs at 100 ft from the edge of the stream, following guidelines similar to those in Table 14.1. Figure 14.7b depicts a common harvest scenario. The forest inside the boundary remains uncut, while the adjacent area outside the boundary is clearcut to regenerate a single-cohort stand. The management boundary may bisect the ecotone at a point where the forest still has important riparian functions, i.e., across the floodplain. Also, the harvest has just two contrasting potentials for effects on functions: low effects to the inside and high effects to the outside.

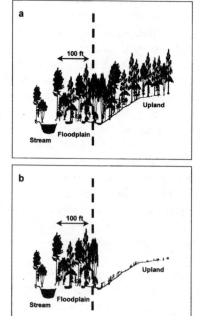

Figure 14.7 Stylized depiction of a riparian forest showing (a) placement of a riparian management boundary in the middle of the riparian ecotone, and (b) contrasting silvicultural activities on each side of the boundary.

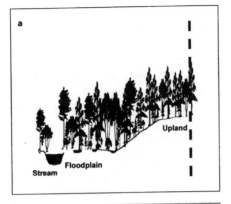

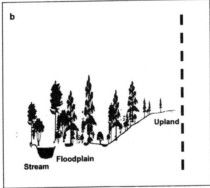

Figure 14.8 Stylized depiction of a riparian forest showing (a) placement of a riparian management boundary line at the upland end of the riparian ecotone and (b) a gradient of silvicultural activities and potential impact to riparian functions along the ecotone.

Now consider the situation in Figure 14.8a. The figure shows an adjusted location for the riparian management line in the same riparian forest. Its new location is at a distance from the water where potential impacts on riparian functions are low. Again, the field-based key for delineating riparian areas in Chapter 2 helps to locate this line. An advantage of the key is that it accounts for variability in riparian lateral extent among different areas (e.g., the location of the line varies among types of riparian areas). Figure 14.8b depicts a potential post-harvest stand structure in the riparian area. The residual stand structure reflects a silvicultural prescription that includes maintenance of riparian functions and sustained use of the timber resource as objectives.

Note several important items in the Figure 14.8b. First, the intensity of management decreases from the upland to water, paralleling the ecotone of impact to functions. Second, structural complexity of the residual stand increases along the same gradient, reflecting increasing probability of riparian function. Third, timber resource objectives change along the gradient, from more intensive, single-cohort management to less intensive, two- or multi-cohort management. This is only one of many possible scenarios. An alternative approach may emphasize increased intensity of management, favoring single-cohort structure along a greater proportion of the ecotone. In this case, the manager must recognize the potential for greater impact to riparian function. When weighing these alternatives, managers must keep their watershed disturbance perspective in mind. More intensive management along a greater portion of the ecotone is not necessarily a bad thing when restricted to only a portion of the watershed.

Harvest and residual stand structure are only two considerations in a silvicultural prescription. The same gradient approach applies to other practices that managers use in existing forests, including type and intensity of competition control, seedbed preparation, and pre-commercial and commercial thinning. In some situations, additional species may be desired or, if none exist, all species must be planted or seeded. Standard silvicultural texts (e.g., Nyland 1996; Smith et al. 1997) discuss appropriate silvicultural methods for artificial regeneration, but whatever the prescription, it should be site-specific along the ecotone, as opposed to the more traditional approach of applying a uniform prescription throughout the management unit. There may be an economic incentive for the ecotone approach. The larger volume of timber removal at the upland end of the ecotone may help compensate for income lost by reducing the volume of timber cut closer to the water. This may be particularly important for restoration of riparian areas. In our experience, loggers show greater willingness to operate in the riparian forest when the timber sale includes larger upland volumes, either through increased timber removal or increased upland sale area. In other words, it may make sense economically to treat riparian forests concurrently with adjacent uplands when forests in the latter are being thinned, fertilized, or prepared for planting.

Another example illustrates our idea of site-specific silviculture along the riparian ecotone. Consider the hypothetical riparian forest in Figure 14.9. Figure 14.9a looks down on the uncut forest. The different shaded circles represent crowns of different tree species. The dark band on the left represents a stream channel, and the dashed line on the right delineates the ecologically determined riparian management boundary. The silvicultural objectives for this forest include maintenance of riparian functions and sustained use of the timber resource. Figure 14.9b depicts one possible post-harvest stand structure. We divide the riparian ecotone into three areas that differ in intensity of harvest, decreasing from upland to stream. The specific timber management objectives also change along the ecotone. In the portion of the ecotone farthest from the stream, the timber objective is to regenerate a single-cohort stand of intolerant, early succession species. In the middle portion of the ecotone, maintenance of mature stand structure receives greater weight because of its importance for riparian function. In this portion of the ecotone, multi-cohort management is used to favor

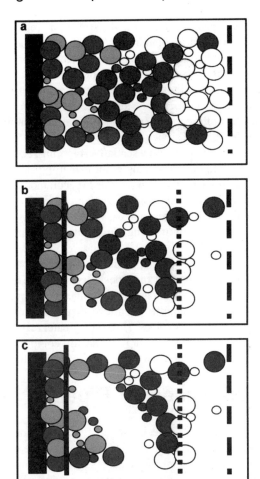

Figure 14.9 A riparian forest showing different harvesting patterns along the riparian ecotone. Shaded circles represent the crowns of different tree species. The thick band to the left is a stream channel; (a) the uncut forest showing the riparian boundary line placed at the upland end of the ecotone (dashed line on the right); (b) the riparian forest is cut along a gradient of intensity reflecting single-cohort management on the right, multi-cohort management in the middle, and uncut forest nearest the stream. Residual basal area in the middle of the ecotone is dispersed by cutting many small gaps; (c) similar to b except that residual basal area in the middle part of the ecotone is clumped to open a single large gap. In both a and b, mature stand structure and protection of riparian functions increase closer to the stream.

generation of more tolerant species. Nearest to the stream, no removal occurs, recognizing the potential to greatly affect riparian functions at this location along the ecotone.

Figure 14.9c depicts an alternative management scenario for the same forest. The difference from the previous frame lies in the middle portion of the ecotone. In both scenarios, an equal amount of residual basal area remains in the middle of the ecotone. However, the spatial arrangements of the residual trees differ. In Figure 14.9c, one large opening is cut in the middle of the ecotone (i.e., the residuals are clumped), as opposed to the many small openings in 14.9b. The silvicultural prescription calls for creating a two-cohort structure (with the two cohorts being distinct spatially), favoring regeneration of less-tolerant species than in the previous figure. The main point is that the silvicultural prescription is variable (i.e., site-specific) along the riparian ecotone to accommodate changing emphasis on different objectives. Both Figures 14.9b and 14.9c depict scenarios that may comply with BMPs that require a certain amount of residual basal area and minimal soil disturbance. The difference lies in the approach for meeting timber management objectives while still satisfying the intent of the BMPs.

A real world application of our spatial approach is in the Southeastern United States. In this region, where mixtures of hardwoods and pines often dominate terraces and slopes along small streams, a manager may favor harvest of large pines and leave some residual mature trees to satisfy Best Management Practices. A uniform distribution of these residual trees can inhibit regeneration of shade-intolerant pines. However, greater regeneration may be possible by clumping residuals, i.e., creating larger gaps, while still complying with BMPs.

Ecological Guidelines for Riparian Silviculture

Riparian areas are the physical, biological, chemical, and functional focal point of many forested landscapes. As such, they are integral to maintaining the health of both terrestrial and aquatic ecosystems. In many regions, managers already value riparian areas for their ecological importance, particularly for nonpoint source pollution control and wildlife corridors and refugia. As our chapter illustrates, there are other ecological reasons to value riparian areas, including organic matter input to water, shading of streams and small wetlands, and bank stability.

Managers also value riparian areas for their timber resources. Many watersheds contain riparian forests that are highly productive because they occur on nutrient-rich floodplains. Some riparian forests contain valuable commercial species; for instance, lakeside eastern white pine (*Pinus strobus*) in the upper Lake States. In some regions, a large percentage of the commercial forest base is riparian simply because of an abundance of water. In Minnesota, for example, 10% of commercial forests occur within 200 ft of a lake or steam (Laursen 1996).

Timber managers rarely ignore forests that are productive, abundant, or contain valuable species. Riparian forests are no exception. Their income potential is difficult to ignore—especially for private landowners that often own forests associated with lakes and rivers. Because of this potential, associated aquatic ecosystems are vulnerable to ecologically damaging silvicultural practices — practices that degrade the important ecological linkages that exist between riparian forests and aquatic ecosystems.

The fundamental question then is what kind of silvicultural practice is appropriate within the boundaries of a riparian area? At a minimum, these practices should protect or enhance the functional connections that exist between riparian forests and aquatic systems. Our goal in this chapter is not to present detailed prescriptions for the many types of riparian forests found in the Eastern United States; rather, our intent is to present a conceptual approach for incorporating important ecological principles into silvicultural prescriptions in a more comprehensive way than most agency BMPs do. The approach can be applied in most types of riparian forests. We summarize this approach with four ecological guidelines for riparian silviculture.

- **Guideline 1: Understand the characteristics of your system.**
  Recognize the type of riparian area you are in. What are its current, historical, and desired future conditions? This understanding must apply equally to watersheds. What are the characteristic natural disturbances affecting riparian forests in the watershed? What are the current, historical, and desired compositions and age structures of the watershed? What is the current and desired hydrologic regime for the site? How has past management altered hydrology and, consequently, species composition? Remember, do not treat all riparian areas as if they have similar structures and responses to disturbance.

- **Guideline 2: Include maintenance or restoration of riparian functions as a principal silvicultural objective.**
  Keep the important riparian functions in mind: organic matter input to water, shading, bank stability, regulation of sediment, water, nutrient movement, and wildlife habitat. Remember that traditional silvicultural systems for a particular forest type may be fine for meeting timber management goals but may not protect riparian functions.

- **Guideline 3: Remember that Riparian forests are ecotones.**
  Remember that riparian forests are ecotones of structure, function, and have potential for impact to functions. Also remember that the probability of a forest being riparian increases with decreasing distance to the water's edge, as does the potential for impact to these functions.

- **Guideline 4: Apply site-specific comprehensive silvicultural prescriptions in riparian areas.**

  Avoid applying prescriptions uniformly across the entire riparian ecotone. Be creative when integrating Best Management Practices into silvicultural prescriptions. Remember, there are diverse ways to implement practices that protect riparian functions (e.g., clumped versus dispersed residual basal area). In other words, how you accomplish riparian protection is just as important as the protection itself. Remember that riparian silvicultural prescriptions are long-term planning tools that should minimize the potential for cumulative impacts on riparian functions. Further, they are flexible to accommodate changing conditions and new information acquired through the monitoring of ecological functions and timber resources.

Which silvicultural option would serve the riparian functions and values for this lakeshore site with over mature white birch and scattered conifers and other hardwoods?

Chapter 15

Harvesting Options
for Riparian Areas

James A. Mattson, John E. Baumgras,
Charles R. Blinn and Michael A. Thompson

"My objective is not to stop the ax but to guide it so the forest is used wisely."
Gifford Pinchot — as quoted by Max Peterson, Chief, U.S. Forest Service, 1985.

As the chapters in this book demonstrate, forested riparian areas provide many important functions and values, including wildlife habitat, recreation, water, timber production, and cultural resources. The high soil moisture and nutrient availability in these areas make them highly productive sites for plant and animal life, including trees, and this, coupled with the fact that many of these areas have gone unharvested due to difficult terrain and wet soil conditions, makes them a valuable source of wood products. Much of the current debate about riparian areas is focused on the level of timber harvesting that should occur within these areas.

Riparian management areas are defined to recognize that management practices may need to be modified to protect and enhance the diverse functions and values of the riparian ecosystem. They are not intended to be zones of management exclusion, although the best prescription may be to exclude active management in some cases. Minimizing disturbance is the primary objective when developing strategies for timber harvesting operations in riparian areas. Limiting disturbance helps reduce the potential for erosion, and minimizes impacts on shade and other terrestrial and aquatic habitat factors.

Harvest Planning for Riparian Areas

Timber harvesting includes felling, processing, extracting, sorting, loading, and hauling of timber products. Harvest operations usually require that haul roads, landing areas, and skid

255

trails be constructed and maintained to access the resource. All these operations have the potential to negatively impact riparian resources and should be carefully planned to protect the riparian area functions (Meek 1994). Many states now have operational guidelines, commonly called Best Management Practices (BMPs), that are designed to protect water quality and a limited number of other resources (MI DNR 1994; MN DNR 1995; WI DNR 1995). The guidelines for locating roads, skid trails, landings, buffer strips, and water crossings are particularly pertinent to harvesting operations in the riparian area.

Landowner objectives, terrain, and season of operation are also significant factors that need to be considered when planning harvesting operations. A well-developed harvest plan using appropriate BMPs will lead to harvesting operations that protect water quality and other riparian functions and values, remove forest products efficiently and profitably, promote sustainable forest growth, and maintain aquatic biodiversity. It is also important to meet with operators on-site before harvesting to make sure provisions of the harvest plan are well understood. Active supervision during operations is critical to ensure that the plan is being properly implemented.

Each site and harvest operation is unique and should have a separate harvest plan. The need to consider each site individually is illustrated by an example from the Monongahela National Forest where past harvesting practices and activities on adjacent private land resulted in stream sediment at or above critical levels (Baumgras 1996). Harvesting operations that produced just a small amount of sediment may have harmed populations of native brook trout. Although harvesting with conventional systems could comply with state water quality BMP requirements, the unique attributes of this site required alternative harvesting systems to protect riparian resources. Consequently, much of the wood on this large timber sale was yarded with helicopters.

Access

Forest roads that are poorly located, constructed, or maintained are one of the largest sources of nonpoint source pollution from forest management activities (Patric 1976). Roads on steep slopes, erodible soils, or stream crossings hold the greatest potential for degrading water quality. Effective road construction techniques minimize the disturbance to the natural flow of water over the landscape and ensure the structural integrity of the road embankment (Kochenderfer and Helvey 1987; Swift 1984c). The goal is to provide a simple road structure of adequate strength to support heavy vehicle traffic while providing drainage structures for water to pass through the road corridor at normal levels. Choices about road construction standards and maintenance activities will be influenced by site characteristics, the value of the resources served, the season of use, and the intended duration of use of the road. State BMP manuals generally contain road building recommendations.

Operating equipment in or near perennial or intermittent stream channels may add sediment directly to streams (Lynch and Corbett 1990; Martin and Hornbeck 1994). Stream crossings that are poorly located or constructed may erode streambanks. As roads approach a stream crossing, proper road drainage is critical to avoid depositing sediment into streams. Stream crossings must be designed, constructed, and maintained to safely handle expected vehicle loads and to minimize stream disturbance.

A variety of environmentally and economically acceptable temporary or portable stream and wetland crossing options can be used during timber harvesting and hauling operations (Blinn, et al. 1998). Increased awareness of options that reduce damage to streams and wetlands can help minimize the cost of protecting these valuable resources. State BMP manuals include sections on stream crossing options that are permitted in that state and make recommendations for their proper installation and use. Before installing a stream crossing, contact the appropriate water resource regulatory agency in your state to determine if permits are required.

Buffer Strips

Many state BMPs recommend the use of filter and shade strips around water bodies (MI DNR 1994; MN DNR 1995; WI DNR 1995). The main purpose of a filter strip is to provide a zone of infiltration to slow down the water and to help keep sediment and other polluting agents from washing into the water body (Castelle et al. 1994). Activities conducted within the filter strip must minimize the exposure of mineral soil to help ensure that vegetation is continuously present on the site. The recommended width of the filter strip varies with slope length and percent. A shade strip helps maintain moderate water temperatures by minimizing the warming of the water body, reducing the warming of overland flow into the water body, and moderating temperatures in the filter strips (O'Laughlin and Belt 1995).

The amount of timber left within a shade strip depends on site conditions and the management goals. Many states have adopted minimum operational widths with minimum tree stocking levels and special water protection measures. These Riparian Management Zones (RMZs) are usually narrower than the variable width Riparian Area defined in Chapter 2. Managing shade-tolerant species generally results in a higher residual tree volume. Filter and shade strips can be a valuable source of coarse woody debris for the terrestrial and aquatic systems. They can also provide energy to the water body, serve as a visual screen, and provide cover and travel corridors for wildlife species. Where RMZ management goals include a desire to establish long-lived species, minimum soil disturbance levels may be exceeded to provide adequate regeneration. Many long-lived species, which are important sources of coarse woody debris and downed logs, require mineral soil exposure to naturally regenerate. Check with the appropriate organizations before deviating from the recommended guidelines; sedimentation is still a concern even where minimum disturbance levels of filter strips are purposely exceeded.

Terrain

Riparian areas are sometimes thought of as low-lying flat areas adjacent to water bodies. However, the terrain can vary considerably within a riparian area. In mountainous areas, stream courses will be found in the bottom of valleys or gullies, the sides of which can often be very steep (Hornbeck and Reinhart 1964). In steep terrain, especially sites with a relatively dense network of small intermittent streams, harvesting, road building, and skid trail construction, even located far from larger perennial streams and more typical riparian areas, can still contribute to stream sedimentation. Therefore, protection of aquatic habitats in mountainous terrain may require careful planning and conduct of forest operations on sites not normally classified as riparian.

Even in gentler terrain typical of the Midwest, many streams and rivers have created deeply cut river valleys and channels that have very steep slopes. Steep terrain in riparian areas will increase the potential for erosion, sedimentation, and mass movement of the hillside. Bank instability can be a serious problem in some areas, especially where residual basal area is low (e.g., clearcuts) and the residual root mat breaks down before the new stand can become established. Slope and natural surface roughness are the main factors determining how far sediment and other contaminants are transported to the water body. The steeper the terrain, the further away from the water body that impacts from operations can be felt, making it necessary to consider terrain when defining the riparian area. The combination of wet and steep ground on the same site can lead to significant negative impacts on the site if harvesting operations are not carefully planned and conducted.

Seasonal Effects

Seasons of the year have a significant impact on harvesting operations in riparian areas throughout the Eastern United States. Precipitation in all parts of this region varies over the course of the year, making soil moisture levels vary considerably from season to season. This causes the bearing capacity and trafficability of the soil to vary by season. In many areas, winter or spring flooding can be extensive. Vernal ponds also have wet soils adjacent to them requiring caution with equipment operation. Aust (1994) summarized recommendations for improving the operation of harvesting systems on wet sites:

- Recognize difficult site conditions — Use available topographic, soils, and vegetation maps, on-site inspection, as well as GIS, to identify difficult site characteristics before beginning harvesting operations.

- Plan better — Use better harvest planning techniques before operating on a site. Time the harvest on very wet sites to occur during dry or frozen periods that permit equipment operation. Lay out designated skid trails on the drier, higher strength soils. If stream and/or wetland crossings are required, follow state BMP guidelines. Paint the boundaries of the RMZ so that operators can identify them.

- Use existing equipment better — Use existing equipment and technologies properly to minimize the impacts of harvesting on wet sites. Low ground pressure technologies, such as wide tires, dual tires, bogies, tire tracks, and tracked machines, can be used to reduce impacts. A common problem with these technologies is the tendency of some operators to build larger loads and extend their operation into areas they would not normally traffic with conventional equipment. Without proper planning and control of the harvesting operation, this can result in more serious impacts than would have occurred with standard equipment.

- Develop new technologies — As new technologies are developed, their potential application to sensitive sites needs to be evaluated. Aerial systems, tracked machines, and other low ground pressure concepts have been developed that may be lower impact alternatives for harvesting on wet sites. Also, cut-to-length technologies offer an option for conducting low impact harvesting within these areas. These alternative technologies will be discussed further later in this chapter.

Temperature regimes vary greatly over the Continental Eastern United States. Ground in the northernmost areas may be frozen over part of the year, while ground in the southernmost areas may never freeze. Areas in between have the potential for frozen ground, but occurrence is highly variable. Even where air temperatures remain below freezing for extended periods, deep snow can reduce frost penetration. Packing the snow may be required in these situations to produce or maintain frozen ground conditions (Blinn and Dahlman 1995). Harvesting operations in the region have to be highly mobile and flexible to take advantage of dry and frozen ground when it does occur.

Operational Site Impacts

Many riparian areas have a gradient of site conditions ranging from upland conditions away from the water body to conditions typically characterized as wetlands near the water. With the exception of unstable soils on steep slopes, the most serious constraints placed on selecting and operating harvest systems (except possibly for unstable soils on steep slopes) are the conditions occurring closest to the water body where high soil moisture is the predominant consideration. Extensive research has studied the impacts of forest harvesting on wetland functions and values, particularly in the Southern United States where extensive wetland forests contain large volumes of high-value hardwood timber (Aust 1994; Aust et al. 1995). Much of this work can be drawn upon to discuss appropriate harvest systems for riparian areas of the Eastern United States in unfrozen ground conditions.

The interactions between machines, soil, and soil moisture are the primary determinants of site disturbance during forestry operations (Reisinger and Aust 1990). Soil moisture governs the effect of machinery traffic on the soil. Dry soils generally have sufficient strength to resist most compaction and rutting. Moist soils are weaker, making them vulnerable to compaction. Compaction increases soil bulk density and decreases the size of large soil pores, resulting in poor soil aeration and drainage. As moisture content increases water fills the soil pores, reducing soil strength and shear resistance. When trafficked, very wet soil ruts rather than compacts. Rutted soils have slow rates of soil aeration and drainage and can serve as a conduit through which sediment can move toward a water body.

Potential impacts of harvesting on wetland soils were reviewed by Aust (1994). Studies have shown that compacted and rutted skid trails slow internal water movement in both moderately well drained soils and poorly drained soils. However, poorly drained soils already have low rates of hydraulic conductivity before being compacted or rutted. Harvesting effects on water quality for these wetland soils were found to be minimal as long as forestry BMPs were properly applied.

In several studies, compaction and rutting were found to reduce site productivity. Loss in site productivity can result in a reduction in the type, amount, and structure of terrestrial biodiversity, leading to direct economic losses over the life of the stand. Severe rutting on sites may have an immediate economic effect for the operator. The difficult operating conditions result in decreased harvest system productivity, lowering profitability. Also, machines subjected to severe operating conditions require more maintenance and repair than similar machines operating in good conditions, further reducing profits.

Site disturbance by harvesting equipment can also harm cultural resources, such as cemeteries, archeological sites, traditional-use areas, and historic structures. The displacement of soil during harvest operations can break fragile artifacts, disrupt human remains, and make interpretation of the artifacts difficult when they are moved from their original setting. Erosion resulting from soil disturbance can do the same and may destabilize building foundations.

Overview of Equipment Systems

Harvesting systems for riparian areas need to be able to perform the same functions as systems operating on upland areas. Felling, delimbing, bucking, in-woods transport, sorting, and loading all need to be accomplished efficiently and economically. Operating in the riparian management zone requires special considerations of soil and water impacts when selecting and operating the harvesting system. Avoiding the direct operation of equipment within the RMZ is the most effective way to minimize impacts. If possible, operational activities such as loading, fueling, and maintenance should be located outside the RMZ. Roads and skid trails should be located outside the RMZ whenever practical.

Ground-based harvesting systems are still the most commonly used because they have a definite productive and economic advantage over alternative systems available at this time. However, ground-based systems have the greatest potential for site disturbance and associated degradation of site and water quality.

Carrier Vehicles

All mobile harvesting equipment has some sort of vehicle base that serves as the platform for mounting the operating hardware and provides the mobility of the unit. It is the interaction between the carrier vehicle and the ground that causes most of the impact to the site. The characteristics of carrier vehicles are generally independent of the operating equipment they carry, and thus can be discussed independently of the functions they perform in the harvesting operation.

Low ground pressure vehicles are generally recommended when operating in RMZs. Reducing the pressure applied to the soil limits compaction and rutting, which can decrease plant growth, adversely affect surface and subsurface water movement, and accelerate erosion. Rubber-tired carrier vehicles for harvesting machinery are still by far the most popular option because of their proven performance, productivity, lower purchase and operating costs, and flexibility. However, the negative side of conventional rubber tired machines is their propensity for rutting and site damage.

Many studies have looked at the use of rubber-tired forest harvesting machines, primarily skidders, and documented the rutting potential of these machines (Koger et al. 1984; Sirois and Hassan 1985; Murosky and Hassan 1991). Reisinger and Aust (1990) reviewed several studies that investigated the feasibility of using wide and dual tire set-ups on skidders and found that most studies concluded that these tires do have the potential to provide better flotation and improved traction. However, use of these tires may still damage the site if used too long into the wet season. Skidders equipped with dual tires have been found to be a more acceptable and less expensive alternative to wide tires (Rummer et al. 1997). Contractors also have the flexibility of removing the outer tires when soil conditions are dry enough to operate without the dual tires.

Before the development of rubber-tired skidders, tracked machines were the most commonly used ground skidding equipment. Lower ground pressure, increased traction, and reduced site disturbance are still advantages of tracked machines. Disadvantages include lower productivity and higher operating and maintenance costs. Tracked machines used in logging are typically machines designed for construction uses and modified for forestry uses. Typical crawler tractors and excavators are the machines generally adapted to forestry uses. Many crawler tractors, together with a winch and logging arch, are seen in use as skidders, quite often on steep terrain where the inherent stability of the crawler tractor is an advantage, or on wet sites where the lower ground pressures aid trafficability.

Excavators have commonly been adapted to use as feller-bunchers. The inherent stability of tracked excavators, the long reach of the boom that minimizes the need to move the machine to each tree to be felled, and the low ground pressures exerted by these machines make them natural choices for feller-bunchers. The success achieved with modified construction excavators in forestry uses has led to the design and development of a new generation of tracked rotating platform machines specifically for forestry (Figure 15.1).

Figure 15.1 Typical feller-buncher utilizing a tracked rotating platform carrier designed for forestry applications.

An inherent problem with this machinery is the wide tail (back end) of the rotating platform that extends outside of the tracks, making it difficult to operate in narrow corridors. Recent developments have led to "zero tail swing" machinery where the entire rotating platform is able to stay within the width of the tracks.

The low speeds and high maintenance costs of steel track crawler tractors led to the development of flexible track machines intended to increase speeds and lower costs while maintaining the advantages of tracks. Stokes (1988) found that flexible track machines were commonly used in skidding applications in wetlands in the South. Skidders with flexible steel tracks generally have a front-drive sprocket (in contrast to the rear-drive sprocket rigid track machines) that reduces track tension. This allows the track to conform with uneven ground conditions and provides better traction and lower ground pressures. High initial and maintenance costs have limited the use of these machines, but the concept is still viable. Major manufacturers have introduced tractors with rubber tracks. These machines are currently being used in agriculture, but the potential for forestry use is significant.

The development of forwarder technology to a high level of sophistication in Scandinavia over the last two decades has led to the availability of another alternative for carrier vehicles in forest harvesting machines. The forwarder chassis has become the carrier of choice for several harvest machines, notably harvesters that use the forwarder chassis for mobility and will attach a harvester head to the boom for felling and processing. The trafficability of the forwarder chassis, particularly those versions that have double-axle bogie wheel assemblies either on just the rear or on both front and rear axles, is extremely good (Figure 15.2). The

Charles R. Blinn

Figure 15.2 Double-axle Bogie assemblies on a clam bunk skidder.

low ground pressure, low speed, and high load capacity of this combination make it especially well suited to harvest operations on sensitive sites, such as riparian areas.

Felling Equipment

Mechanized felling is used for a larger percentage of wood production each year; the notable exception to this trend is hand felling large timber on steep terrain. Mechanized felling methods are safer, more productive, and permit better integration with the other components of modern, generally fully mechanized, harvesting operations. Various designs of shears, chain saws, and circular saws have been incorporated into commercially available felling/bunching heads. For harvesting near recreational areas or home sites, the felling head should be selected to minimize the noise created in the felling operation, as well as the likelihood of the felling head throwing debris a significant distance from the machine. A shear or chain saw unit will generally operate more quietly with less chance of throwing debris than a circular saw unit.

The most significant consideration for mechanical felling in riparian areas is the carrier vehicle of the unit. Rubber-tired feller-bunchers are generally considered to be faster moving machines with lower initial and operating costs than other alternatives. However, rubber-tired feller-bunchers generally have fixed felling heads that require them to drive to each tree to be felled. This need to drive to each tree and the need to move rapidly to maintain production rates raise the possibility of significant disturbance across the entire site when a rubber-tired feller-buncher is used on soft ground. Residual stand damage is also more likely as the machinery moves back and forth through the stand.

Tracked swing-to-tree feller-bunchers are preferred when site disturbance is a major consideration, such as in a riparian area. Although more costly and slower moving, this type of feller-buncher can fell several trees from one location, reach into sensitive areas without entering them, and pile wood away from the water body, all of which will help minimize the area impacted. These machines can also be quite productive because travel time is minimized. Some tracked feller-bunchers have been developed to operate on slopes up to 50%. The lower ground pressure they exert and the smaller percentage of the site being trafficked are significant advantages for these machines in riparian areas. However, reduced soil disturbance may make it more difficult to establish species that require soil scarification.

Extraction Equipment

Rubber-tired skidders are still the most common means of moving wood from the stump to a landing. Cable skidders and grapple skidders are both widely used. The cable skidder, in particular, has significant advantages on difficult sites because it can be kept out of sensitive areas and its winch can be used to pull wood to the machine. This feature can be used to limit the area trafficked, thereby reducing impacts. A cable skidder can also drop its load when it bogs down in difficult conditions, then move ahead and winch the load through the difficult area.

Cable skidders tend to be narrower than grapple skidders, making them more maneuverable and reducing impacts to residual trees. They are also better suited to staying on designated trails and working in steeper terrain. Alternatives to wire rope are being examined to help overcome problems associated with operator fatigue when pulling the cable over distances longer than about 50 feet.

Grapple skidders are generally more productive than cable skidders when operating conditions are favorable. Their advantage comes from the operator not having to get down from the machine to secure the load. Grapple skidders also tend to come in much larger sizes, which under the proper operating conditions, leads to greater production rates. In either case, the trafficability of the skidder in difficult conditions can be greatly improved by the use of dual or wide tires as discussed earlier. Grapple skidders are not recommended for use on steep slopes because of stability and safety concerns.

Skidding, by definition, involves dragging trees or logs across the ground. Where site disturbance is a major concern, forwarding may be a good alternative. Forwarding consists of carrying the wood, usually in the form of cut-to-length logs or pulpwood, out of the stand on the back of a machine. Site disturbance from the wood dragging on the ground is eliminated, a cleaner product is delivered to the landing, and landing size requirements are relatively small. On the negative side, a forwarder can create greater ground pressure, which increases the potential for soil compaction and rutting. An additional disadvantage is that conventional forwarders traffic more of the area than a cable skidder. This effect is somewhat offset by the forwarder's ability to move the same volume of wood in fewer trips because of larger average load size.

The soft ground and steep terrain in riparian areas, can make aerial extraction systems a viable possibility. Skyline cable yarding is an extraction system uniquely suited to harvesting timber on steep slopes and unstable soils (Baumgras et al. 1995). In its simplest form, this system uses a yarder with a tower and two cable drums. One cable is a skyline that is strung between the tower and the far end of the site where it is anchored to a holding point such as a large tree or stump. A carriage runs along the skyline and carries the mainline with chokers out to the logs. Logs are pulled into the yarder by the mainline. The main advantage of this system is that heavy machines do not traverse the site, and the need for skid trails is reduced. Steep terrain in riparian areas can be beneficial to the operation of a cable yarding system, providing the change in elevation needed to get good deflection, or lift, on the cable system.

Helicopters are also a technically viable harvesting system for riparian areas. The system is very expensive to operate, but causes virtually no site disturbance. High timber values or high values on other resources such as water quality or aquatic habitats are required to justify helicopter logging, and the planning and logistical support for the operation are critical. Balloons, another option, have not been widely examined. Also, walking machines are under development and may soon be used in the woods. Because extraction and transport equipment can cause both on-site and in-channel resource damage, its use requires diligent planning.

Processing Equipment

Traditionally, processing functions such as delimbing, topping, and bucking were done with chain saws, either at the stump or at the landing. The trend toward mechanization has resulted in machinery to handle these functions. Except for cut-to-length harvesters that process harvested trees at the stump, most of the mechanical processing equipment is designed to work at the landing. Delimbers, slashers, and whole-tree chippers are intended to be landing-based machines with the felled trees brought to them from the stump area, generally by skidders. Some combination of feller-bunchers, skidders, roadside delimbers, and slashers is the most common full-tree harvesting system currently in use. A major advantage of this type of system for harvesting sensitive sites is that much of the operation is done at the landing rather than the stump area, which is the area most vulnerable to site disturbance. Disadvantages are that full-tree skidding can increase site and residual stand disturbance and results in nutrients being taken off-site. Full-tree skidding also removes most sources of coarse woody debris from the riparian area.

The other common configuration for mechanizing processing functions is in combination with felling in a harvester. A harvester combines the functions of felling, delimbing, topping, and bucking into one machine. A harvester can be built onto a rubber-tired or tracked carrier, rigidly mounted to the front of the carrier, or boom mounted on the carrier. Regardless of the configuration of the harvester, the result is harvested wood that is delimbed, cut-to-product lengths, and piled in the stump area. A significant advantage of harvesters for working in sensitive areas is their ability to deposit the delimbing residue they generate in front of the machine and thus build a slash mat to travel on. This capability can greatly reduce site disturbance and rutting. Forwarders are generally used to collect the processed wood and transport it to the landing or roadside where it can be piled or loaded directly onto haul trucks.

Harvesting Systems

Regardless of the individual pieces of equipment involved, the total harvest system needs to be able to function as a whole, producing wood products in an economically efficient manner while protecting the diverse functions and values of the ecosystem (Stokes and Schilling 1997). With the continuing trend toward increased mechanization, most harvest systems used in the near future will likely be some variation of the full-tree system or a shortwood cut-to-length (CTL) system.

The full-tree system consists of feller-bunchers, skidders, processing equipment such as delimbers, slashers, and whole-tree chippers, haul trucks, and support equipment. Selection of specific equipment will depend on several factors such as species and size of trees being harvested, size of harvest areas, products, production levels, terrain, soil type, and ground moisture. General advantages of full-tree systems include (Gingras 1994):

- Full-tree systems can efficiently handle a variety of tree sizes.

- The individual machines in full-tree systems are generally mechanically simpler, which leads to less down time.

- Owning and operating costs are generally lower for full-tree systems because less labor is required and production is high.

- Operation of the machines in a full-tree system is simpler, thus requiring less operator training and quicker attainment of maximum productivity.

The CTL system consists of pairs of harvesters and forwarders matched to their productive capacity. Selection of harvester specifications, such as cutting capacity, can be further refined to match the machine with the available stands to be harvested, optimizing use of the technology. General advantages of the cut-to-length systems include (Richardson and Makkonen 1994):

- CTL systems can efficiently work small tracts because there are only two machines to move between jobs, and minimal landing space is required on small sites.

- The CTL system is well suited to partial cutting because it processes the harvested trees to shortwood lengths at the stump, minimizing damage to the residual stand.

- The CTL equipment works well in wet areas and on sensitive sites because of its capability to work on a slash mat it produces as it moves through the stand.

- Forwarders can economically work over longer distances because of the larger loads they can carry, reducing the needed road network.

- Well-defined skid trails do not have to be created. The trails that are used can be narrow and meandering.

- Decking areas smaller than typical landings can be used along the roadside.

- The CTL system facilitates product sorting and merchandising.

Special Equipment for Riparian Areas

Working in many riparian areas requires crossing streams or wet and weak soils. The crossing should minimize the potential for sedimentation as well as the obstruction of waterflow and the movement of aquatic organisms. Temporary structures are most often used to facilitate these crossings and protect the environment. Blinn et al. (1998) provide a review of these technologies, which can be divided into temporary stream crossings and wet area crossings (see Chapter 6 for culvert sizing recommendations that allow spawning fish to pass).

Temporary Stream Crossings

Avoid crossing intermittent and perennial streams whenever possible because these locations are the most vulnerable to harvesting impacts on water quality and aquatic habitats. Where avoidance is not practical, the number of crossings should be kept to a minimum. Stream crossing options include fords, culverts, and temporary bridges. A permit may be required in some states to cross some or all perennial and/or intermittent streams. Check with the appropriate regulatory or natural resource agency before establishing any stream crossing.

Fords

A ford is a crossing where vehicles drive directly through the stream. They are inexpensive to install and can result in minimal impacts when constructed, used, and maintained properly. Fords should be used only for short-term or infrequent access. A location with a low flow, a firm base of rock or gravel, and a gentle slope with low-banks is best for a non-frozen ford. Placement of suitable base materials may be necessary if no adequate natural base exists. A trench should be excavated to hold the added base material so it does not extend into the normal channel cross section area, constrict flow, and induce bank erosion at and below the ford site. Sediment from excavation is minor compared to the everyday sediment produced from blocking the channel cross section. Use at least small cobble material or material larger than that found on point bars (an indication of what the streams will carry at bankfull flows). Logs, brush, and soil should not be used in ford crossings because these materials may be carried downstream during high flow periods, causing sedimentation and/or obstructions in the stream channel. Fords should not be constructed during fish spawning, incubation, or migration.

Culverts

A culvert is a structure (usually a large pipe) that allows the stream to flow under the road. Culverts are recommended for some locations where a ford is not, such as streams with steep banks, streams with a soft base, or areas where frequent or extended access is required. The primary disadvantages of a culvert include sedimentation during placement and removal, the potential for the culvert and fill to wash out during high stream flows, and the need to frequently maintain the culvert to avoid blockage. Care must be taken to size the culvert properly to accommodate normal high-water levels (see Chapter 6) and to ensure that it extends to the toe of the fill slope. Cover the culvert with at least 12 inches of road base (24 inches is needed to ensure ponding water upstream of the road for the 25 to 50-year events and avoiding road washout). This depth minimizes damage to the culvert during maintenance of the road surface and distributes the weight of passing vehicles, thus preventing crushing. During installation, compact fill properly to minimize seepage of water through the subgrade. Place the culvert deep enough in the stream channel (about 1/6 of the culvert diameter) to prevent perching which can obstruct fish migration.

Temporary bridges

Temporary bridges provide an excellent means of crossing a stream without disturbing the stream channel (USDA FS 1997). They have been built of timber (native logs or lumber), polyvinyl chloride (PVC) or high density polyethylene (HDPE) pipe bundles, snow or ice, steel, pre-stressed concrete, reinforced used truck trailers, and used railroad flatbed cars. We recommend only the use of bridges that have been engineered to safely carry the intended load. Bridges are commercially available in a variety of lengths and carrying capacities and should last for many years. Although some of these bridges are expensive, the fact that they are reusable brings the cost per use down considerably.

Bridges should be placed on appropriate abutments to help level the structure, to minimize disturbance to the streambank, and to make removal easier (MNDNR 1995). PVC or HDPE pipe bundle crossings function best with geotextile underneath. Geotextile is a fabric mat that allows water to drain through, provides additional support to the crossing, and provides separation from the soil, facilitating removal. A frozen streambed or a snow and ice bridge may work well in the winter in some areas.

Temporary Crossings for Areas with Weak or Wet Soil

Many access roads and skid trails must cross soils with poor bearing capacity (e.g., peat and wet clay). Waiting for the soil to freeze or dry is often not an option. The hauling and placement of fill to stabilize these areas can be expensive and may negatively impact them.

Several alternative methods for crossing these areas are available. Most of these options function best with a nonwoven geotextile under them.

Corduroy is a mat formed from logs, sawmill slabs, or brush placed across the wet area. It is a common method for crossing soft soils because the material used is normally abundant and found on-site. Expanded metal grating can sometimes provide adequate support for crossing soft soils, and it is relatively cheap, reusable, and easy to handle. Other reusable crossing options include wood mats, wood planks, wood pallets, and tire mats. Most of these options can be built from local materials or are commercially available. Although some of these options are expensive and bulky, all are reusable and provide summer access at a much lower cost than building a permanent road.

For any of these options, the surface should be flat (maximum-grade of 3%) and free of high spots (e.g., stumps and large rocks). The length and width of the various options should be sized to meet anticipated loads, soil strength, and installation equipment. On very weak soils with low bearing strength (e.g., muck and peat), the option used may need to be longer and wider than usual to spread the weight over a larger surface area. Most options are best suited for use with hauling and forwarding operations because skidding can cause movement and increased wear of the crossing material.

Conclusions

Forested riparian areas are highly productive and a valued source of wood products and many other resources. Harvesting operations in riparian areas must protect and even enhance the resource functions and values of the riparian zone. Minimizing disturbance is the primary objective when developing strategies for harvesting operations in riparian areas. Operational guidelines, commonly called best management practices (BMPs), have been developed in most states to provide practical guidelines for protecting water quality. BMPs may also provide some level of protection for many other riparian functions and values. Well-developed harvest plans that apply BMPs and timber sale provisions are needed to protect riparian resources while recovering forest products efficiently and cost effectively. It is also important to meet with operators on-site before harvesting to make sure provisions of the harvest plan are well understood. In addition, active supervision during operations is critical to ensure that the plan is being properly implemented.

Ground-based equipment will be the basis of most harvesting systems used in the foreseeable future because of its productive and economic advantages. However, ground-based systems also have the greatest potential for site disturbance and associated degradation of riparian resources. Proper selection and use of equipment features that can lower site impacts, such as wide or dual tires, flexible tracks, and double-axle bogie wheel assemblies, will be essential. Decisions about specific equipment and mitigation strategies to apply must be made on a site-specific basis, considering site conditions, landowner goals, silvicultural requirements, economic implications, local regulations, and anticipated site and stand impacts. Opportunities to reduce impacts should be fully employed wherever possible (e.g., designated skid-trails or taking advantage of frozen or dry ground). Alternative harvesting systems, such as aerial cable systems, helicopters, balloons, and walking machines, should be further developed to take advantage of their low impact nature in an economically effective manner.

Art Elling

Operating equipment on frozen soils, whether organic or mineral, minimizes soil compaction (roadside loading of black spruce harvested from a frozen peatland in northern Minnesota).

Federal Interagency Restor. Group

Protective measures like this installation of silt fencing actively recognize the link between heavy equipment operations and fishing.

Sandy Verry

Caught and eaten! It's the fishing!

Chapter 16

Best Management Practices for Riparian Areas

Michael J. Phillips, Lloyd W. Swift, Jr. and Charles R. Blinn

"A thing is right when it tends to preserve the integrity, stability, and beauty of the biotic community. It is not when it tends otherwise."

Aldo Leopold — A Sand County Almanac

Forest streams, lakes, and other water bodies create unique conditions along their margins that control and influence transfers of energy, nutrients, and sediments between aquatic and terrestrial systems. These riparian areas are among the most critical features of the landscape because they contain a rich diversity of plants and animals and help to maintain water quality and terrestrial and aquatic habitats (Hunter 1990; Gregory et al. 1991). These fragile areas are easily disturbed, and caution is needed whenever forest management occurs within them. Riparian areas are often linear features of variable width that have high edge-to-area ratios but generally occupy only a small part of the landscape. However, the linear nature of riparian areas means that resource managers and loggers, either through active management or the need to gain access to a site, will invariably come into contact with these features. Therefore, the proper management tools are needed to maintain the functions of riparian areas and minimize disturbance to the terrestrial and aquatic systems.

Best Management Practices (BMPs) are developed to prevent or minimize the adverse impacts of forestry activities on water quality while permitting the intended forest management activities to occur. They serve as the cornerstone for most state water quality protection programs (NCASI 1994; Phillips 1995; Stuart 1996). Possibly the earliest effort to establish BMP guidelines was the "Criteria for Managing the National Forests in the Appalachians" (1971) by Regions 8 and 9 of the USDA Forest Service. The development of BMP programs has been a collaborative effort among state agencies and organizations (both public and private) and federal agencies to identify practices that reflect the particular physiographic, economic, technical and political considerations of each state. Monitoring has shown nationally that compliance with BMPs is relatively high (Hook et al. 1991; Adams and Hook 1993, 1994; Henson 1995, 1996; and Carraway et al. 1998). However, by definition, BMPs were designed to protect water quality, not the other functions and values of riparian areas. We need to move beyond BMPs based solely on water quality to address these additional functions and values.

All state BMP programs recognize the importance of retaining some form of riparian management zone (RMZ) with management options that minimize impacts to the water resource (Vermont Department of Forest, Parks, and Recreation 1987; Kentucky Division of Forestry 1998; Maryland DNR 1992; Alabama Forestry Commission 1993; Florida 1993; Tennessee Division of Forestry 1993; South Carolina Forestry Commission 1994; Georgia Forestry Commission 1995; Kittredge, and Parker 1995; Minnesota DNR 1995; Wisconsin DNR 1995; Cassidy, Aron, and Trembly 1996; Virginia Department of Forestry 1997; and North Carolina DENR 1998). Achieving BMP compliance in these areas generally requires greater care, reduced physical intrusion (e.g., skid trails, roads, equipment) into the riparian management zone, and often reduced levels of harvest (e.g., thinning, uneven-aged management) or no harvest at all. It also requires preharvest planning that considers the landowner's management objectives. The BMPs discussed in this chapter pertain to lands where silvicultural or other forest management activities are planned and conducted. This chapter will describe the management issues of concern, water bodies that are addressed by traditional BMPs, RMZ options, and approaches to the development of RMZ guidelines that move beyond BMPs and address issues other than the protection of water quality.

Management Issues of Concern

Any forest management activity adjacent to or intruding into the riparian area has the potential to negatively impact water quality, terrestrial and aquatic habitat, and other riparian functions and values. These activities include timber harvesting, mechanical site preparation, pesticide application, prescribed burning, fire line clearing, insect and disease control, road construction and maintenance, and recreational development. The concern that has received the greatest attention is nonpoint source (NPS) pollution, which originates from diffuse sources across the landscape. NPS pollution contributed from any particular area may be small or insignificant, but can create water quality problems when combined across the landscape. Nationally, forests occupy approximately one-third of the land base, but forest management is credited with contributing only 1 to 5% of NPS pollution in assessed waters of the United States (U.S. GAO 1991; Kochenderfer et al. 1997). Because forests can help decrease NPS pollution, groups of trees are sometimes planted adjacent to waterbodies where agriculture is the predominant land use (NRCS 1996). Sediment is the principal pollutant associated with forest management operations (Pardo 1980; Golden et al. 1984). Sediment originating from the construction and use of logging roads and skid trails generally exceeds that from all other forestry activities (Megahan 1972; Patric 1976). Stream crossings are the dominant feature where roads make the major contribution of sediment to water bodies. State BMPs have traditionally been designed to reduce and trap erosion and to control subsequent sedimentation of water bodies. Research has shown that where BMPs are properly employed, significantly less erosion and sedimentation occur (Black and Clark 1958; Hewlett and Douglass 1968; Swift 1984a, 1984b, 1986; Kochenderfer and Helvey 1987; Burroughs and King 1989; Briggs et al. 1998). Other sources of NPS pollution that are of concern include fertilizers, pesticides, fuels and lubricants, organic matter and nutrient leaching, and thermal impacts from removal of vegetative cover.

Forest management activities can also impact other riparian functions and values such as cultural resources (e.g., logging camps, cemeteries, burial mounds, artifacts); streamflow and water quantity; forest soil productivity; terrestrial and aquatic habitat type, structure (e.g., inputs of coarse woody debris), and amount; bank stability; recreation; aesthetics; and rare, threatened, and endangered species. Forest management activities can have multiple impacts on the resource if BMPs or other forest practice guidelines are not followed (Table 16.1).

Table 16.1 Potential impacts to riparian areas from forest management activities

Forest Management Activity	More Erosion and Sedimentation	Chemical Pollution to Water	Slash Disposal	Temperature Increases	Soil Compaction &	Habitat Type, Structure and Amount Terrestrial	Aquatic	Cultural Resources	Forest Soil Productivity
Timber harvest	X	X	X	X	X	X	X	X	X
Site preparation	X	X	X	X	X	X	X	X	X
Pesticide use		X		X		X	X		X
Prescribed burning	X	X		X		X	X	X	X
Fire line construction	X	X	X		X		X	X	X
Roads	X	X	X	X	X	X	X	X	X
Recreation	X	X	X	X	X	X	X	X	X

Water Bodies Addressed

Riparian areas exist around all water bodies. The water bodies of concern for forest management include lakes, perennial streams, rivers, intermittent and ephemeral streams, vernal and autumnal pools, nonopen-water wetlands, and open water wetlands. A necessary question is whether all water bodies should be or can be practically managed to the same standard of importance in terms of riparian functions and values. When developing riparian guidelines, a decision must be made about the degree of protection given to each of these water bodies. For example, lakes, perennial streams, rivers, and open-water wetlands are likely to receive a higher standard of protection than nonopen water wetlands, intermittent or ephemeral drainages, and vernal and autumnal pools. This varying level of protection may occur due to the relative presence/absence of each type of water body within the landscape, their ease of identification (e.g., ephemeral streams and vernal and autumnal pools may be difficult to identify during dry periods or under a cover of snow), the perceived importance of each water body, and their associated functions and values.

The terminology for management within riparian areas is variable and often describes the function of the practice or identifies the type of water body. Common terms include streamside management zone, RMZ, filter strip, riparian buffer, and shade strip. A streamside management zone (SMZ) is a designated area that consists of the stream itself and an adjacent area of varying width where management that might affect water quality and aquatic habitat is modified. The SMZ is not a zone of management exclusion; instead, it is a zone of closely managed activity. Despite the name of the zone (i.e., streamside), the associated guidelines are also applied to open water bodies in many states that apply SMZs. Florida, for example, uses the SMZ acronym to mean special management zone (Florida 1993). Within the SMZ, the use of practices such as filter strips and shade strips are commonly prescribed. The RMZ performs the same functions as the SMZ but is named to be inclusive of other types of water bodies.

A buffer is a transitional area between two different land uses that mitigates the effects of one land use on the other. A filter strip is an area of land adjacent to a water body that provides for infiltration of surface runoff and traps sediment and associated pollutants and may be a designated function for land within or around SMZs and RMZs. A key aspect of a filter strip is that rutting, compaction, and exposure of mineral soil are minimized to permit the filter strip to function. A shade strip is an area of land adjacent to a water body where sufficient timber or other vegetation is retained to provide shade that maintains temperatures within the normal range. In this chapter, RMZ will be the preferred term.

Options in the Riparian Management Zone

Many of the BMP recommendations for operating within or adjacent to the RMZ (e.g., skid trail location, water bar spacings) are fairly consistent between states. One difference, however, is in specifications for RMZ width and harvesting restrictions. How wide should the RMZ be? Should the RMZ be a fixed width or variable width based on site conditions? How much harvesting is permitted within the RMZ?

The criteria for establishing effective RMZ widths and acceptable management restrictions within the RMZ have been heavily debated among resource professionals and the concerned public. Some will argue that the wider the RMZ, the greater the protection given to riparian functions. At some point, increasing the width of the RMZ and imposing more restrictions on management will conflict with economic considerations, the landowner's management objectives, and issues of property rights. For resource management agencies and public lands, the "wider is better" approach will likely have more appeal where economic objectives for management are not dominant. Many resource management agencies apply wider RMZs than are required by state guidelines for water quality protection to provide for "other" riparian functions and values. This does not imply that agencies are not concerned about economics or that these wide RMZs are BMPs for water quality protection, but rather that the agencies' management is or may be more encompassing than that of private landowners. However, economic considerations, particularly for private lands, may result in a minimal RMZ width. It could be difficult to convince nonindustrial private forest (NIPF) landowners to maintain a wide RMZ with many management restrictions where riparian edge is a significant portion of small tracts.

The width of the RMZ can be selected in two ways: (1) reserve a fixed width or standard width that may vary based on slope or water body type; and (2) establish a variable width based on specific site conditions (e.g., composition, age, and condition of vegetation; site geomorphology; animal and plant species present on the site; watershed-level issues; adjacent land use; sensitivity of the site to disturbance). Some examples of recommended buffer widths are given in Table 16.2. Most RMZ guidelines recommend minimum widths in the range of 50 to 100 feet.

The advantage of fixed-width RMZs is that they are easily applied and monitored for compliance. Applying a fixed-width RMZ does not require a knowledge of ecological principles. Therefore, they may be more easily applied without management assistance. A disadvantage of fixed-width RMZs is that they are based on a narrow set of site conditions that may not commonly exist. Compliance monitoring of the implementation or application of fixed-width RMZs, while easily accomplished, does not indicate protection of riparian functions. A fixed distance may result in a RMZ that is not wide enough to protect these functions. In other cases, it may be more than what is needed to protect them. However, applying a minimum fixed width may protect some of these functions. An argument for fixed-width RMZs for NIPF landowners is that we are currently reaching a relatively small

Table 16.2 Examples of RMZ widths and harvest restrictions

State	Water Type	RMZ Width-ft.	RMZ Harvest Restrictions
Maryland (1992)	Perrenial streams and wetlands	50-250 w/ slope %	Maintain 60 sq. ft. BA/ac. Even distribution.
Wisconsin (1995)	Lakes and nav. streams	100	0-50 - no harvest 50-100 - maintain 60 sq. ft. BA/ac.
	Nav. Int. streams	35	Maintain 60 sq. ft. BA/ac. Even dist.
North Carolina (1998)	All water bodies (critical areas)	50-250 w/ slope % (75-300 w/ slope %)	Selection harvest up to 25% of existing canopy
	Ephem. streams		Same, no bare ground
	Intmit. streams		Same, and < 41% bare ground
	Peren. streams		Same, and < 21% bare ground
New Jersey (1995)	Open water, incl. intermit. streams	25-165 w/ slope % & erosion hazard	Any harvest system can be used if the integrity of the soil surface is maintained.
Pennsylvania (1993)	Streams, lakes, ponds, wetlands	25-165 rds, lndngs 50-165 harvest area	Maintain 50% BA. RMZ width double in municipal watersheds
Vermont (1987)	P. strms & lakes	50-110+ w/ slope %	Light thinning or selection harvest
Ohio (1992)	Shade strips (peren. streams)	25	No cut or light cut
	Filter strips (water courses)	25-250 w/ slope %	Selection harvest. RMZ double in municipal watersheds
Florida (1993)	Peren. streams	35-200	Selection harvest.
	Interm. streams	35-300	Maintain integrity of soil surface
	Lakes, sinkholes	35	Selection harvest.
Minnesota (1999)	Trout waters	150-200	Maintain 60-80 sq. ft. BA/ac.
	Nontrout waters Even age mgt.	50-100 w/ stream width or lake size	Maintain 25-80 sq. ft. BA/ac.
	Nontrout waters Uneven age mgt.	50-200 w/ stream width or lake size	Maintain 80 sq. ft. BA/ac.
	Filter strips (p.& i. strms, lks, o.w. wetlands, ponds)	50-150 w/ slope %	Maintain integrity of forest floor (also around seeps and springs)

percentage of these landowners with professional assistance (Gathman et al. 1992), and most of the landowners and loggers are not likely to have the expertise to evaluate the site-specific conditions and determine the appropriate width of the RMZ. However, the percentage of landowners receiving technical assistance may be increasing due to education and incentive programs, such as the Stewardship Incentives Program. When that technical assistance is used, BMP compliance will likely increase (Henson 1996).

Compared to fixed-width RMZs, variable-width RMZs are more apt to be applied on public and private industrial lands. Making the right decision about width requires trained judgment. Larger ownerships maintain technical staff who can evaluate the specific site conditions and identify the appropriate width of the RMZ. Variable-width RMZs allow flexible management decisions based on ecological and landscape principles, specific site conditions, intensity of adjacent land use, and the need to maintain and protect identified functions. Making the decision will likely require site visits that collect detailed information about the site and surrounding landscape. This means that deciding on the RMZ width will be more costly and time consuming. However, this approach will likely provide more protection to water quality and other functions of the RMZ.

Once the RMZ is defined, the management activities within the RMZ, such as harvesting levels, are considered. Harvesting in the RMZ can be conducted so as to leave the residual basal area scattered uniformly across the RMZ or harvest so that a higher proportion of the uncut trees are left adjacent to the water body. Whatever the harvesting regime, residual trees must be windfirm to resist blowdown and maintain functions of the RMZ. The bulk of protection for water quality, aquatic habitat, and riparian functions occurs closest to the water body and diminishes with increasing distance from the water body. The trees closest to the water body will provide large woody debris for both terrestrial and aquatic habitats, promote bank stability, provide shade, maintain water temperatures within the normal range, provide detritus to the water body, and address some recreational and aesthetic concerns. The outer portion of the RMZ may be more suited to addressing terrestrial biodiversity issues while providing additional protection to the inner zone (e.g., windfirmness). This management approach is embodied in the USDA Natural Resources Conservation Services buffer strip interim conservation practice standard for the Conservation Reserve Program (USDA NRCS 1996). Because the different types of water bodies have both overlapping and differing functions and values, RMZ guidelines may vary by type of water body.

Tree orientation is another consideration for harvesting within the RMZ. Within one mature tree length of the water body, retain trees leaning towards the water body because they will eventually fall into it and provide large woody debris for aquatic habitat niches. Trees leaning away from the water body are preferred for harvest. However, one reason not to remove all of trees leaning away from the water body is that they will also provide terrestrial biodiversity habitat once they fall down. In addition, maintaining a mixture of conifers and deciduous species in a multi-layered canopy is desirable for the maintenance of plant and animal diversity.

Determination of RMZ widths and forest management restrictions is influenced by economic considerations. Ability to harvest riparian species with economic value is necessary to encourage continued landowner commitment to maintaining these areas. For the landowner and logger, there are real costs associated with BMP implementation in riparian areas. Some of these costs include opportunity costs of not harvesting, reduced stumpage payments due to increased harvesting costs, or the added expense of installing a particular type of stream crossing or culvert (Ellefson and Miles 1985; Dissmeyer and Foster 1987; Lickwar et al. 1991). Recognizing that guidelines may affect operating costs and efficiency, guidelines need to be flexible enough to protect water quality and other riparian functions while providing for an economically viable operation. There are also indirect costs to agencies and organizations associated with implementing various programs (e.g., technical assistance, education-extension, monitoring compliance, regulatory enforcement). For NIPF land, incentive programs need to encourage compliance with BMP recommendations and other practices within RMZs. These programs can include tax incentives, financial incentives (cost sharing), educational-extension programs, technical assistance, voluntary guidelines that provide flexibility, and some form of regulation.

While it is relatively easy to identify and quantify BMP implementation costs, few studies have tried to quantify benefits. Many benefits may accrue to society as a whole or to individual groups within society (e.g., hunters, bird watchers), off-site or in the future, and some benefits may be associated with species that do not have a current market value. Also, it is difficult to develop appropriate production and price relationships for many of the benefits. This lack of information about BMP benefits makes it difficult to conduct complete economic analyses of riparian management zones.

Best Management Practices

Before land management activities begin within a RMZ, the landowner or manager needs to identify their management objectives for the area within the RMZ. Once these objectives have been identified, the landowner or manager should plan their BMPs. Planning is, in itself, a Best Management Practice and is probably the most important BMP. It is an opportunity to identify site-specific needs, landscape-level concerns, potential problems, conflicts, and to select mitigating activities that prevent impacts, modify the intensity of impacts, or improve pre-existing poor conditions. Because the RMZ is identified as a zone of special concern, it is important to consider where and how much disturbance is appropriate, the size of the RMZ, the location and type of mitigating practices, and the best season for the activity. Planning helps to identify the risks and costs of management activity in the RMZ. If possible, the RMZ should be visited to identify sensitive areas such as stream crossings; special vegetation or animal habitats; cultural resources; sensitive habitats, troublesome topography, soils, or geology; and potential sources and routing of sediment. During the visit, existing disturbances such as old roads can be identified so that planning may include appropriate remedial activities or future use if existing sites are properly located and stabilized.

BMPs specific to the RMZ fit into three categories: those that limit disturbance, those that exclude pollutants, and guidelines for river and stream crossings. State BMP guidelines exclude or control activities that disturb the soil such as roads, skid trails, log landings, boat landings and other recreation sites, site preparation, and fire and fire lines. Some guidelines exclude the use of machinery such as tractors or skidders, requiring the logger to pull logs by cable from equipment set outside the RMZ. Other guidelines limit the number, size, or location of trees that may be cut in order to reserve trees for shade strips, aesthetic view protection, or coarse woody debris.

BMPs are designed to mitigate or prevent adverse impacts due to sediment movement, water temperature shifts, changes in streamflow, input of chemicals, organic debris, solid waste disposal, and habitat alteration. Where soil is disturbed and erosion occurs, fine soil particles can be deposited as mud in water bodies or transported as suspended sediment in turbid waters. Larger particles add to the bedload volume and may fill stream channels and reduce lake volume. Excess sediment reduces water quality and covers the normal substrate of water bodies.

Reduction of vegetation density in the RMZ can lead to increases in summer water temperatures and possible reductions in winter temperatures, thus altering the habitat for aquatic organisms (Swift 1983). Temperature change is directly related to the location and amount of vegetation removed (Kochenderfer et al. 1997). Where reducing the density of forest cover increases soil moisture storage, nonstorm stream flow rates can increase (Swank et al. 1988). Road drainage will add to stormflow volume if there is a direct surface connection with the stream system or if the drainage becomes part of the subsurface stormflow.

Most pollution control BMPs focus on techniques to exclude sediment from the water body. These include controlling erosion at its source, trapping sediment in natural or constructed barriers, dispersing storm water and its suspended sediment away from the water body, and avoiding or controlling situations that initiate debris slides. Various guidelines cover methods of draining storm water from a roadway such as waterbars on closed roads, broad-based dips, the sloping or shaping of roadbeds, and ditch line drainage. Guidelines protect water quality by recommending against the disposal of storm water into the water body or its RMZ.

BMPs control chemical pollutants by restricting vehicle or equipment maintenance, trash disposal, and pesticide application and equipment servicing within the RMZ. Chemicals used in the RMZ may move to and pollute the water body. They include spilled fuels and lubricants from vehicles and earthmoving or logging equipment, residues washed off pavements, and pesticides and fertilizers carried by wind drift or storm waters. Trash and solid waste disposal in the RMZ also may pollute the subsurface and surface water. The wetter soils characteristic of riparian areas increase the probability of spills impacting subsurface water. Excess organic debris, such as slash disposal, can add nutrients to the RMZ, which may be either beneficial or polluting.

The aquatic habitat may be physically altered by stream crossing structures, which can block the upstream migration of fish and amphibians. Undesirable practices include road locations inside the RMZ where construction has moved the channel or mined the substrate of the water body for gravel. Problem stream crossings must be identified early and remedial actions taken to eliminate or minimize their impact.

Stream and river road crossings are the major disturbances (e.g., decreased bank stability, increased erosion, and subsequent sedimentation) in the RMZ. BMP guidelines specify the culvert size and placement, and use of bridges, fords, or other type crossings. A management guideline may specify that crossings are removed when intermittent use roads are closed, effectively disturbing the stream an additional time. Portable bridges offer a low-impact alternative for such roads. BMPs to reduce stream crossing impacts include using brush barriers below road fills, putting erosion-preventing materials and plantings on fills at crossings, covering of fills with geotextiles, using geogrids, temporary wood mats, tire mats, or wood planks to stabilize soils at crossings, and putting rock in ditch lines and fords. Stream and wetland crossing options and some of their associated impacts have been summarized (Blinn et al. 1998).

When the specific BMPs to be used in the RMZ have been identified, the landowner or resource manager should ensure that the appropriate lines are marked to identify the edge of the RMZ. They should also mark the trees to be left and identify skid trail and water crossing locations (Figure 16.1). In addition, the landowner or resource manager should include the restrictions as appropriate language in contracts and meet with the contractor and logger on site before harvest entry begins. Timber sale administration is critical to ensure that protection standards are met.

Other Riparian Considerations for Guideline Development

As noted earlier, BMPs were designed to protect water quality. However, functions not generally covered in most BMP guidebooks are receiving increased attention including protection of cultural resources, maintenance of travel corridors and habitat structure for wildlife, maintenance of unique habitats, and maintenance of soil productivity. Some of these situations are covered by antiquities and endangered species regulations. Protection of these other functions and values should also consider landowner objectives where voluntary programs have become established. The survey and planning associated with management of the RMZ provide opportunities to identify and protect these resources. Forest management that considers these, and other issues, provides a more comprehensive approach to protecting the broad array of riparian functions and values.

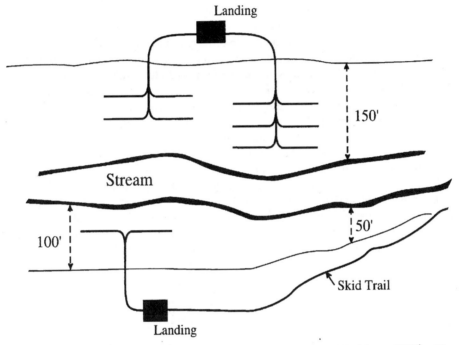

Figure 16.1 Skid trails and landings in riparian areas (Blinn and Dahlman 1995). The limit (width) of the RMZ is noted in feet from the water's edge.

Because of differences in terrain or vegetation between a riparian zone and the surrounding landscape, the RMZ may be a travel corridor for some animals, including humans. BMPs may indirectly protect these uses, but where such corridors are identified, specific practices can be created to serve that special need.

Riparian areas historically constituted much of the landscape that various cultures used (i.e., settlement location, economic activities, social activities, and spiritual activities) (Emerson 1996). Protection measures for cultural resources include excluding the land containing the cultural resources from the sale area; clearly marking the boundaries of the cultural resources when included in the sale area; keeping roads, skid trails, and landings away from the cultural resource area; and measures reducing soil disturbance in cultural resource areas.

Riparian areas have high plant diversity, both horizontally and vertically from the water's edge, which contributes to the high diversity of animals that live in these areas (Hunter 1990). However, riparian area diversity may be reduced in some situations by past land-use. Measures to protect and maintain wildlife habitat in riparian areas include use of silvicultural systems other than clearcutting, retention of mast, retention of cavity trees, retention of slash and downed logs, and retention of conifers to provide food, cover, and nesting sites. While the retention of standing and downed materials provides unique habitats for wildlife, the potential for insect and disease infestations as well as operator safety need to be considered.

Maintaining soil productivity is a key to maintaining riparian benefits on a sustainable basis. Soils within riparian areas are generally wetter than soils in adjacent upland areas. Issues of concern for soil productivity include compaction, rutting, erosion, loss of organic matter, depletion of available nutrients and nutrient reserves, and reduction of soil fauna and flora. Many of the measures in existing state BMPs protect against negative impacts to soil properties in riparian areas. Important protection measures include matching tree species to site, minimizing exposure of mineral soil in riparian areas, using harvesting techniques that promote retention of slash and debris within the riparian area, using equipment that is suited to the site and the size of the material being harvested, minimizing intrusion of equipment into riparian areas, confining equipment to designated trails, and using corduroy (e.g., logs, brush mats) or other suitable materials (e.g., geotextile with reusable wood mats, wood planks, tire mats) to construct temporary crossings on weak or wet soils (see Chapter 15).

Approaches to BMP Development

In developing NPS pollution control programs, states need to design BMPs that respond to local needs and conditions and protect water quality; while the U.S. Environmental Protection Agency retains the responsibility for program review and oversight. State BMPs are developed and implemented through basically two approaches: voluntary or regulatory programs. Most of the states have opted for the voluntary approach, which assumes that BMPs are first developed and then promoted through education and technical assistance. Workshops, demonstration areas, and brochures are critical to program implementation (Teeter et al. 1997). Landowners are encouraged to use loggers and operators who have a working knowledge of state BMPs. In many states, BMPs are then monitored for their use (Hook et al. 1991; Adams and Hook 1993, 1994; Phillips et al. 1994; Holaday et al. 1995; Henson 1996; Carraway et al. 1998; and Adams 1998). Monitoring information is then used to modify the BMPs, to improve BMP efficacy, and to target future education efforts and technical assistance.

Many regulatory programs use mandatory controls and enforcement strategies enacted in state forest practice rules or water quality statutes (Ellefson and Miles 1985). These programs use some combination of plans, permits, and prescriptions. Harvest operations are often reviewed by the responsible state agency. Implementation of these BMPs is required based on site-specific conditions, and penalties may be levied to ensure compliance.

In most states, acceptable compliance is judged by the installation of a specific set of voluntary or required practices without regard to how effective the practices are in an individual case. The North Carolina program may be unique because a wide range of practices is suggested for use, but acceptable performance is judged solely on the landowner's ability to protect aquatic resources regardless of BMP methods selected (NC DENR 1998).

States have taken two basic approaches in developing guidelines and practices: (1) through an agency followed by public review and (2) through involvement of all concerned stakeholder interests, which may be followed by public review. Development of BMPs through an agency approach is problematic. These BMPs are generally strong on science but often weak on practicality and flexibility. Economics may not be considered to the degree that many land management organizations would desire. Often the time needed to develop these BMPs is relatively short compared to the long period of implementation because there will be resistance from interest groups not involved in development of the guidelines. Without stakeholder involvement in guideline development, there is little opportunity to build the trust among affected groups that is critical for effective implementation of any BMP program.

The involvement of stakeholder groups in the development of BMPs and other forest practice guidelines requires more patience and time. Once agreement is reached, however, implementation will likely be more rapid and effective since there is a greater probability that the interest groups have bought into the product produced. The BMPs or other forest practice guidelines developed by consensus are more likely to reflect a balance of science, practicality, and economics. There is also greater likelihood that trust will develop among the many stakeholders involved in BMP development, which is necessary for successful program implementation.

Conclusion

Most states have developed and published BMPs that address water quality and have begun programs to encourage and evaluate their implementation. The forestry community within each state must continue to support education efforts, technical assistance, research, and other implementation efforts to ensure improvement in the adoption and use of BMPs. At the same time, it is important that the forestry community move forward holistically in addressing other key issues to maintain its credibility in dealing with important resource concerns within the RMZ.

Sandy Verry

Water quality has long been the emphasis in forestry BMPs, like using temporary bridges to prevent sedimentation in streams, but new BMPs include habitat considerations, like leaving legacy patches for this boreal owl.

Tom Nichols

Many states have adopted mining BMPs as well as forestry BMPs to prevent this type of habitat destruction where the stream is used as a road for mining access.

Federal Interagency Restor. Group

Chapter 17

Monitoring the Effects of Riparian Management on Water Resources

Gordon W. Stuart, Pamela J. Edwards,
Keith R. McLaughlin and Michael J. Phillips

"Trust but verify."   President Ronald Reagan, October 23, 1986, Springfield, MO

This chapter presents information on determining the effects of riparian forest management and use on water resources. The focus is on water resources because much of the public concern about forestry and many of the legal requirements relate to water. We are also limiting the discussion to streams. Lakes and reservoirs present a different set of monitoring challenges. Furthermore, while forestry ranks low as a pollution source for streams, it ranks even lower as a pollution source for lakes (USEPA 1994).

Monitoring activities are conducted by numerous agencies and groups. Key monitoring steps have been identified by several authors, including MacDonald et al. (1991):

- Identify the purpose of monitoring and the questions to be answered.
- Gain the approval of management for the cost and necessary time commitment.
- Use existing information and knowledge to design the monitoring effort.
- Select measurable parameters that can document the expected response.
- Test and adjust the design to achieve the desired level of accuracy.
- Consider ways to gain acceptance of your findings.
- Prepare a monitoring plan including quality assurance and data analysis.
- Prepare a report of the findings and recommendations.

Ponce (1980) and MacDonald et al. (1991) defined several types of water resource monitoring. These types include baseline, trend, implementation, compliance, effectiveness,

287

project, and validation. The attention here is on project monitoring to determine if the desired objectives are being met during forest management or use (MacDonald et al. 1991). These objectives include these:

- Avoid nonpoint source (NPS) pollution from timber harvest.
- Avoid NPS pollution from forest uses such as recreation or grazing.
- Enhance functions of riparian forests, including aquatic habitats.
- Buffer the runoff from non-forest land uses.
- Increase the floodplain deposition of sediment.

The water resource aspects of riparian forestry are not limited to the use of Best Management Practices (BMP) for pollution control. Other aspects include reforesting cleared land, improving floodplain conditions, and enhancing fish habitat.

Many people look at how activities are conducted. Less attention is paid to the actual results of the activities in terms of hydrology, water quality, and biology. The quote, "Trust but verify," is appropriate for monitoring. Too often the results are assumed.

There are many techniques for measuring accomplishments in riparian management, including measures of erosion, morphology, habitat, BMP compliance, tree planting, water quality, and landowner plans. There are many sound theories about how streams respond to these measures. But monitoring the results in terms of stream biology or beneficial uses provides real-world verification of our concepts and assumptions. Finally, if a monitoring effort is to be successful, it must be acceptable to the intended audience.

This chapter will provide an outline to help you achieve these monitoring goals. We focus on aspects of monitoring that supplement those already available in published guides. For more detailed information, see the recommended reading section at the end of this chapter.

Identify the Monitoring Objectives

Setting clear monitoring objectives produces the best results. Too often monitoring efforts simply measure a lot of things with no clear idea of why they need to be measured. Both the questions to be answered and the audiences to be informed are important. These dictate the required level of precision. Monitoring for self-assurance, a professional paper, or legal action requires different levels of precision. Questions may include these:

- Are the BMPs adequate to avoid impairment at the municipal intake?
- Are the BMPs adequate to protect stream biology?
- Was a cold-water fishery restored by reforestation?
- Is water quality impaired by recreational stream fording?

Measuring change is not an end in itself. The significance of a change must be explained. Pollution is defined as human-caused impairment of beneficial uses (Water Quality Act of 1987). Changing water quality or habitat within a range that does not affect fisheries, public water supplies, or other uses may not be meaningful. The threshold at which change becomes impaired must be based on local conditions.

Expected Responses and Sources of Variation

Forest management is only one of the factors that influence water resources. Many forested watersheds are affected by natural characteristics, historic land uses, and current land use. These also affect their response to management. Therefore, it is important to identify on-site baseline conditions before interpreting the changes due to forest management.

Natural Characteristics

Landscape factors of climate, geology, and vegetation have a controlling influence on water resources and biology. Broad physiographic delineations, e.g., Omernick 1987, indicate the general condition of water resources over large areas.

Local factors also influence streams. For example, Ponce et al. (1979) demonstrated how geology changes stream chemistry in West Virginia, and Bailey and Hornbeck (1992) documented how the bedrock sources of glacial till influence water quality in New Hampshire. Dunson and Martin (1973) documented the downstream effects of acid bog drainage on fish populations in Pennsylvania.

Past and Present Land Uses

In the east, historic mining and agriculture commonly influence present day chemistry and sediment loads in forested streams. For example, Clover Run in West Virginia had an elevated sediment load 60 years after farm abandonment (Kochenderfer and Helvey 1987).

Forestry is just one of the current land uses on many forested watersheds. Agriculture and urban land uses frequently are intermingled. These uses affect hydrology, water quality, channel stability, and stream biology (Bolstad and Swank 1997; Richards and Horst 1994). Stream channelization for home sites, highways, and emergency flood work is common throughout the east. Channelization reduces the complexity of channel morphology and impairs stream biology (Karr and Gorman 1975). Forest land uses also include recreation, grazing, and other activities that may need to be considered.

Changes to Expect from Forestry Activities in the Eastern U.S.

Research results show that most forestry-related changes are temporary and localized. Some changes such as increased streamflow can be beneficial; other changes such as increased sediment are considered harmful when they impair beneficial uses.

Common changes

Increased streamflow becomes measurable when basal area reductions exceed 25% over an entire watershed (Hornbeck et al. 1993). Most of the increases occur during summer low flows, when it is most beneficial. Stream biology benefits from the increased wetted perimeter that is available for biological activity (Sweeney 1993).

Harvesting that stimulates regeneration and increases the number of woody stems on a floodplain increases floodplain roughness and influences sediment deposition (Arcement and Schneider 1989). Aust et al. (1991) found a long-term gain in sediment deposition on harvested floodplains.

Streamside forests provide organic material to the stream community. Harvesting that reduces the supply of large woody debris by removing all the large trees along streams can harm aquatic habitats, particularly on streams with sand and gravel substrates (Karr et al.1983). Detritus, small organic material, is a key ingredient in the food chain (Sweeney 1993). Intensive harvesting on large areas may increase summer flows to a point that detritus is flushed downstream.

Harvesting that removes both the overstory and understory shade and exposes streams to direct sunlight will increase water temperature (Pierce et al.1993). These increases can be harmful to cold-water fish and invertebrates (Sweeney 1993). Ground temperatures in cutting units are increased by removing the overstory, but this does not increase stream temperature when shade strips are left along the stream (Pierce et al. 1993).

Sedimentation is the most common cause of water quality impairment from harvesting (USEPA 1973). Eroded soil may reach streams and impair uses if BMPs are not used. But present-day timber sales also can reduce sediment delivery by rebuilding and stabilizing old roads.

Harvesting, especially in New England, increases nitrate concentrations (Hornbeck et al. 1987; Pierce et al. 1993; Fenn et. al. 1998). Nitrates were significantly higher below clearcuts and thinning on the Mill Brook timber sale in New Hampshire (Stuart and Dunshie 1976). None of the nitrate increases exceeded water supply standards or negatively affected stream biology. The population of macroinvertebrates increased on the Mill Brook sale, but this change was attributed to increased streamflow, not increased nitrogen. Noel et al. (1986) found increased biological activity in water within the clearcuts reported by Pierce et al. (1993). They attributed this to increased light and temperature.

Less common changes

Cumulative effects are possible from excessive harvesting on a watershed. Concentrating too many clearcuts, partial cuts, or new roads on a watershed increases the risk of sediment delivery to streams and channel instability. However, most cutting units are small and scattered because forest ownership is highly fragmented (Birch 1996). Peak flows may increase during either the growing or dormant season (rain or snowmelt peaks) as shown in studies from New York, West Virginia, and Minnesota (Schneider and Ayer 1961; Rothacher in Hornbeck et al. 1977; Verry et al. 1983). Hornbeck, et al. (1997) show that dormant season flood flows are not greatly affected by harvesting, but bankfull flows which are not floods, but strongly impact in-channel conditions, can double in size (see Chapter 6; Verry 1986). Other research from small research basins has not shown increases in peak flow rates (Edwards and Woods 1994). Furthermore, the cumulative effect of mixed land uses on various ownerships may over-ride individually small changes from small harvesting units.

Timber harvesting does not generally increase dissolved phosphorus. Monitoring in the White Mountain National Forest in New Hampshire, and the Chippewa National Forest in Minnesota showed no significant increases in phosphorus from timber harvesting (Stuart and Dunshie 1976; Verry 1972). This is part of the reason for the limited impact on lakes.

Planning a Monitoring Project

Riparian management or use is not easy to monitor. Adams and Hook (1993) surveyed 177 timber sales and found 57 with perennial streams, but only 27 of these were found suitable for research. While research and monitoring are not synonymous, selecting the project area and sampling sites is critical to success.

A substantial amount of information that can be helpful to your monitoring effort may already be available. Check with the U.S. Geological Survey, USDA Forest Service, universities, state agencies, forest industry, and private organizations.

Sampling Design

Sampling, by definition, seeks to measure a small part of something larger. Statistical methods indicate the number of samples you need to take to have confidence in your results. Statistical methods help identify which changes are real versus those within the sampling error. MacDonald et al. (1991) and Dissmeyer (1994) provide more detailed information on sampling design.

What to sample

Streams have a complex web of chemical, physical, and biological parameters. It may be best to use the most direct, most reproducible, and most economically feasible indicators of results. Often a few parameters or observations can document the attainment of water resource objectives, such as:

Parameter	Objective
• Turbidity	protecting municipal water supplies
• Macroinvertebrates	controlling adverse stream sedimentation
• Fish community	controlling adverse temperature and habitat changes
• Overland flow	buffering non-forest land uses
• Woody stem density	increasing floodplain roughness
• Streamside trees	controlling temperatures and providing woody debris
• Bacteria	controlling health hazards from recreation use

How to sample

Grab samples or spot checks work well for many parameters. Continuous recorders or automatic stream gauges often are not necessary, however, they can be useful or necessary for assessing parameters such as turbidity, which must be sampled during storms. Others such as temperature must be sampled during low summer flows. For detailed guidance on measuring various parameters, see the publications listed in the recommended reading section.

Where to sample

Typical forest activities affect local water resources and small streams. Most research results are measured in or at the edge of cutting units on 30 to 100 acre watersheds. Monitoring on the White Mountain National Forest could not detect changes on watersheds over 2,000 acres (USDA FS 1972). Look for changes on watersheds that are no more than four times the size of the area harvested or under concentrated use.

Sample at locations that are most likely to demonstrate a response and where the change is most critical. To document change or impairment, sample close to the activity where a beneficial use such as a fishery exists. To document possible downstream impairment (e.g., a municipal water supply downstream) sample at the water intake. Downstream impairment should not be assumed when a change is measured in a cutting unit.

Sampling points layout
Single points or paired points will provide some answers. Both can detect trends or changes before and after management. Paired points can separate the effects of management from other changes, but neither may provide statistical verification of results.

Multiple points are more effective for monitoring. These points can be above and below the activity or between a control and treated sample point. Unfortunately, the typical site has limited opportunities for multiple sampling points. At a minimum, chemistry can be sampled above and below the use or treatment; biology and habitat can be sampled many places above and below. A comparable stream can be sampled at similar upstream and downstream points.

Comparable sample points
The baseline, reference, or untreated control points must be similar to the stream in question. Differences in watershed size, watershed history, geology, storm occurrence, stream shading, stream gradient, substrate composition, embeddedness, and other factors can mask the effects of harvesting or use. Watershed size and stream order alone are not good indicators of comparable conditions (Montgomery and Buffington 1993).

Local sample points are best for making comparisons, but long-term results from research watersheds, U.S. Geological Benchmark stations, and environmental agency reference watersheds may provide useful information. Be sure to use information from sites that have similar geologies, watershed uses, and channel conditions. Be aware that reference sites are generally picked as the best possible sites and they may not represent the baseline condition at your more typical site.

Comparable points are most critical for biological monitoring. Substrate composition has a controlling influence on macroinvertebrate populations (Hynes 1970). Substrates frequently change dramatically between stream reaches on the same stream due to differences in geology and/or stream gradient. Mason (1982) found close relationships between types of stream reaches and macroinvertebrates in New Hampshire (Table 17.1). Consider the following factors when selecting comparable sample points:

- Watershed or stream size
- Geology
- Land use history
- Channel modification

- Existing sources of pollution
- Canopy shade
- Substrate composition
- Shared storm events

Sampling techniques
Stratified random sampling works best for both biologic and chemical sampling. Water quality parameters can be grouped by stormflow or baseflow. Biologic or habitat survey data can be grouped by similar stream segments, habitat types or other parameters.

When to sample
To characterize pretreatment conditions, try to begin monitoring at least one year before the activity begins. This will help define the natural differences between sample points. In general, the longer the pretreatment sampling period, the easier it is to distinguish between natural variations and management effects. Information from long-term monitoring points may be helpful.

Continue monitoring at least 2 years after the activity ends or until the key parameters reach baseline levels. Research has consistently shown that water quality changes are measurable during the sale and for a short time after the harvest. On a worst case harvest with no road planning, no BMPs, and no road closure, turbidity levels recovered to baseline levels in 3 years (Reinhart et al. 1973).

Changes due to timber sales can be delayed in slide-prone areas of the Pacific Coast Range (MacDonald et al. 1991), but eastern landscapes have demonstrated less risk of delayed changes from landslides. In the East, post-harvesting changes have occurred when roads, for short-term timber harvesting, are left open for recreation use without maintenance or upgrading with long-term design features.

There are some questions about long-term effects on channel structure. Remember that many stream channels have been disturbed by early agriculture clearing or forest harvesting activities as well as large floods. If there is no measurable impairment of beneficial uses, lingering changes in stream geomorphology may not be significant. Where ongoing uses are being monitored, sampling should be done during a representative period of that use.

Stormflow sampling
Water quality, especially turbidity, changes during storms. Sampling may need to include storm periods to get a complete picture of changes due to management or use. The influence of other land uses also will become more pronounced during storm runoff. Bolstad and Swank (1997) reported increased levels of turbidity, nitrogen, and bacteria during stormflows and attributed much of this to overland flow from pastures, roads, and other uses.

Shifts in weather conditions
Before and after sampling results are frequently complicated by weather cycles. Pretreatment and post-treatment sampling may be done during different weather conditions. The best approach for handling these differences is to present all the data and explain how the change

in weather may have influenced results. The localized nature of many storms is another reason for keeping treatment and control points close together, so they share the same events.

Timing is important in biological monitoring

The time frame for sampling macroinvertebrates following a possible impairment is limited. Macroinvertebrates migrate downstream (Hynes 1970), which can affect the results of sampling above and below an activity such as a stream crossing. There is a general downstream drift over a period of months, but the diurnal drift will affect results of short-term forestry changes. Chisolm and Down (1978) found that benthic communities recovered within months after sedimentation ended from highway construction in West Virginia. Herricks and Cairns (1974) eliminated macroinvertebrates by acidifying stream reaches in Virginia. They reported full recovery to pretreatment populations in 19 to 28 days and attributed this to drift.

How many samples

An important aspect of reaching valid conclusions is taking enough samples to characterize the parameters of primary interest to the project (e.g., sediment, temperature), and then to determine if there is a change due to management or use. MacDonald et. al. (1991) discuss sample size and suggest that 30 to 40 samples may be necessary for attaining statistical significance, but that 10 samples may be adequate depending on cost considerations.

Biological monitoring

Biological monitoring uses fewer samples and relies more on indices and ratios (Plafkin et al. 1989). The number of samples used in riparian habitat evaluations is based on the sampling approach and statistical probability (Platts et al. 1987).

Habitat units are not suitable for monitoring the response of streams to human activities because the measurements are too subjective and the impairment from human activities may not be apparent in habitat characteristics (Poole et al. 1997). Monitoring for changes in the biology itself is more a direct and effective approach.

Macroinvertebrates are a popular and effective measure of water quality, and may relate strongly to substrate character (Table 17.1). They respond more rapidly than fish to adverse changes (MacDonald et al. 1991). It may also be easier to document the effects of sedimentation using macroinvertebrates than it is to sample sediment or turbidity during storms which occur at odd times of the day. Total numbers provide an effective indication of impairment due to sediment. Some of the popular procedures use indices that are more

Table 17.1 Relation between macroinvertebrates and substrate type on the White Mountain National Forest

Stream Reach	Total Number	Diversity	Stream
Stable, bouldery, shaded	2537	3.68	2-4
Stable, bouldery, exposed	1060	3.63	5
Stable, embedded cobble	204	3.46	1-3
Unstable, loose cobble	2832	3.45	3-4
Meandering, sandy	895	2.95	4
Wetland, organic	218	2.86	2

effective in cases of organic pollutants that reduce oxygen levels (Plafkin et. al. 1989). One example that shows the differential affects of sediment on biology is by Gammon (1970), who studied the effects of crushed limestone on stream biology in Indiana. Macroinvertebrate populations decreased as suspended solids increased, all taxa were reduced equally, diversity indices did not change, and the fish remained in pools until the pools filled up.

The focus in this chapter is on results. The number of samples is important for statistical determination, but biological impairment and statistical significance may not be mutually dependent. Statistical results may not fully describe a significant threshold for biological change, but biological impairment may be readily apparent using deductive reasoning (see Chapter 6 to compare the use of Fisher statistics and deductive reasoning in the scientific method).

Monitoring cost

The cost of collecting field information, analyzing the data, and reporting the results needs to be considered and approved when the monitoring plan is prepared. Costs vary greatly for different methods and degrees of accuracy. The best information may be available locally from others who do monitoring surveys or research. Management commitment to the monitoring project, including staff time and cost, is critical. Having the funds cut in the middle of a project is something to avoid.

Gaining Acceptance from Your Audience

Importance of the information

Report what is important. Public concerns are frequently expressed about threats to human health, fish kills, and water clarity. Changes in water clarity are the most noticeable to the public, but bacteria may pose the greatest health risk.

Who to involve

Involving different expertise and interests will strengthen the results by increasing its credibility and accuracy, especially on projects where precision is most important. Combining the forester's understanding of management practices, the biologist's understanding of aquatic life, and the hydrologist's understanding of geomorphology will improve the product. In his 1975 lecture entitled *The stream and its valley*, H.B.N. Hynes, Canadian Professor of Biology, stated that "one could say that some of our most important recent discoveries have been the existence of hydrologists, foresters, and soil scientists."

When precision counts

As the nature of your audience becomes more critical, more care is needed for design, sampling, and analysis. Attention to the number of samples, statistical design, and adherence to biological monitoring protocols becomes essential where court action is possible. Consult research scientists and statisticians. Splitting samples among reputable laboratories or experts is another way to validate and support your findings.

Present all the data

It is better to explain strange results than to eliminate data. Eliminating data will lower the level of public trust in you and your interpretations. Explain the changes you found or did not find as best you can, and ask a lay person to review the document for clarity.

Do not assume impairment based on appearances

Some BMP surveys have assumed pollution based on seeing soil erosion near streams or trees cut near streams. Where there are indications of pollution, some form of in-stream verification should be done to document actual impairment. Visual surveys are a good way to scope out problem areas that may warrant more detailed monitoring. They are also useful during an activity when corrective action can be taken before impairment occurs. But visual appearance is not an accurate measure of impairment.

Know the Audience

- Public acceptance requires trust.
- Management acceptance requires relevance to institutional goals and business decisions.
- Regulatory acceptance requires accuracy based on the use of standard methods.

Gain Public Attention

When Maine's Governor signed the law limiting dioxin releases, he said the proof would be in the fish, and he jumped in the Kennebec River. Your state governor may not participate in your monitoring project, but you may be able to gain some public attention and support by having a field day, stream walk, insect count, or fish survey.

Explain the Conflicts with Preconceived Ideas

Monitoring results must overcome some serious barriers to acceptance. The public hears repeated messages about the bad effects of forestry. There is also persistent belief in watershed folklore (Sartz 1969; Verry 1986). Citing research results that are consistent with yours may be the best way to overcome out-of-date concepts.

Make the Results Accessible

The expense of measuring success, warrants advertizing the results. Furthermore, providing public access to the data so that individuals can evaluate the data and form their own opinion leads to better understanding than relying on third-party interpretations. Putting the results on the Internet is one way to make them accessible.

Frequency of Monitoring Projects

Repeat monitoring projects at your own discretion for self-interest. Where public debate necessitates increased monitoring, a few well-chosen public demonstrations may be more effective than many in-house projects.

These demonstrations should be of actual activities rather than of artificial demonstration areas where the practices may not be needed or may not work. They should be repeated within a geographic area once or twice a decade. Monitoring results have a short half-life in the public memory, and regulatory agencies tend to focus on results since passage of the most recent legislation.

Close

Verifying the real effects of management leads to improved performance and builds trust. The proof is in the results.

Summary

- Set clear objectives that can be tested.
- Know what changes to look for.
- Be able to isolate the activity or use from all other variables.
- Prepare a plan and have the support of management.
- Use comparable sampling points.
- Use the most direct measures of results.
- Use appropriate levels of precision.
- Plan the monitoring project with audience acceptance in mind.
- Monitor frequently enough to retain public awareness and credibility.
- Build trust.

Recommended Annotated Readings (full citation in references)

On-the-ground procedures
Evaluating the effectiveness of forestry best management practices in meeting water quality goals or standards by Dissmeyer (1994). This publication provides guidance to forest managers on developing plans to monitor and evaluate Best Management Practices. Both on the ground and in-stream procedures are included.

Guide for selecting Manning's roughness coefficients for natural channels and flood plains by Arcement and Schneider (1989). This publication provides field procedures for measuring floodplain roughness based on the number of woody stems.

In-stream methods — water quality/biology
Estimating total fish abundance and total habitat area in small streams based on visual estimation methods by Hankin and Reeves (1988). This publication provides procedures for estimating habitat features and the number of fish in habitat units, which are calibrates by more accurate measurements.

Fisheries techniques by Murphy and Willis (1996). This is a compendium of methods and techniques used by fisheries biologists and others to sample fish habitat, fish populations, and other aquatic organisms.

Monitoring guidelines to evaluate effects of forestry activities on streams in the Pacific Northwest and Alaska by MacDonald et al. (1991). This publication provides guidelines to assist people in developing water quality monitoring plans for all types of forest land uses. A wide range of parameters is discussed.

Rapid bioassessment protocols for use in streams and rivers by Plafkin et al. (1989). This publication provides a practical reference for conducting cost-effective biological assessments. Three levels of macroinvertebrate and two levels of fish protocols are provided.

Restoring Life in Running Waters by Karr and Chu. (1999). This book is a clear discussion of using biologic indices to evaluate the health of aquatic systems. It dispels the criticisms of multi-metric indices and tells why they are needed and how to use them.

Standard methods for examination of water and wastewater by the American Public Health Association (APHA 1989). This book provides the acceptable standards for laboratory and field tests of water quality parameters.

In-stream methods — morphology
Applied river morphology by Rosgen (1996). This book provides technical concepts on natural stream stability for channel restoration based on morphological characteristics.

Methods for evaluating riparian habitats with applications to management by Platts et al. (1987). This publication provides comprehensive coverage of habitat evaluation techniques both on the ground and in the stream.

Better trout habitat by Hunter (1991). This is a western guide to stream restoration. While several of the conditions are different from eastern forested riparian areas, it provides a narrative explanation of the methods of Platts et al. (1987) and includes embeddedness.

Field comparison of three devices used to sample substrate in small streams by Grost, et al. (1991). Three methods of substrate sampling are compared. The shovel method proved least expensive, least cumbersome, and provided acceptable results when compared to the excavated-core and freeze-core methods.

Stream channel reference sites by Harrelson, Rawlins, and Potyondy (1994). This illustrated guide provides detailed instructions on establishing reference sites including bankfull indicators and pebble count procedures.

Federal Interagency Restor. Group

Monitoring for water quality has long been used to evaluate land management practices, but monitoring can also include, invertebrates, habitat condition, sediment amount, erosion rate, vegetation condition, trail condition, fish communities, and recreation facility condition.

Soil Conservation Service

Natural Resource Conservation Service

Shaping of the channel bank on the eroded banks of the Winooski River in Vermont in 1938 (above) and the use of brush mats and pole jetties stopped the excessive bank erosion for over 57 years (same area in 1995 below).

Chapter 18

Riparian Restoration

Russell A. LaFayette, Jerry Bernard and Donald Brady

"Eventually, all things merge into one, and a river runs through it. The river was cut by the world's great flood and runs over rocks from the basement of time. On some of the rocks are timeless raindrops. Under the rocks are the words, and some of the words are theirs."
Norman Maclean — A River Runs Through It, and Other Stories

After many years of neglect, riparian areas are finally getting the attention they deserve. Inventories of riparian area extent and condition routinely point to a need to restore hundreds of thousands of acres around streams and wetlands across the nation. We are now able to see where their condition is diminished, and to plan objectively for a desired condition that meets our restoration goals and social values. Interest, personnel, energy, and money are finally moving toward accomplishing the complex task of restoration. Across the country, efforts to restore watersheds and riparian areas in urban, agricultural, and forest settings are underway by citizen groups and all levels of government (Forsgren 1997;Wood et al. 1997; and USEPA 1997).

As with any large effort undertaken over a wide area by many people and levels of expertise, some succeed, while others do not. Why is this the case, and how can we increase the likelihood of success in future work? What can we learn from past efforts to ensure higher rates of success in the years to come?

Evaluations of projects have consistently highlighted several steps, processes, and features that promote success. Some have three components: (1) technology/science, (2) economics, and (3) social/values, others work through a planning process, often having five general phases: (1) overall planning, (2) project design, (3) implementation, (4) monitoring and evaluation, and (5) adaptive management and maintenance. Either approach will aid organization, as other groupings might. However, regardless of artificial groupings, successful restoration generally has most, if not all, of a common series of features.

One thing is true regardless of the project size, complexity, ownership, or location. Each project has its own unique features that restoration specialists must deal with. At the same time each project can benefit from the experiences of others, learning from their efforts and maximizing the possibility of success for each new project. We hope the following observations will lead to greater success in future riparian restoration projects across the nation.

Define and Articulate Clear Visions, Goals, and Actions

Defining clear visions, attainable goals, and specific actions are keys to riparian restoration success. Visions, goals, and actions must be carefully developed, written down, and assessed periodically to ensure that all are reasonable and achievable. Setting overoptimistic or unreachable targets will undermine success and deflate the enthusiasm of all involved.

An overall vision statement sets the stage for the process. While necessarily broad and sweeping, the statement must have punch and content, and be short and to the point. Develop the vision in close cooperation with project stakeholders so all key participants share the same overall vision of success. Differing visions of the eventual outcome will result in unachieved expectations and disappointment. Work with partners to set the overall vision for the work to come.

Specific goals define how to attain the overall project vision. Make goal achievement measurable, definable, and compatible with future monitoring. Specificity that includes a measure to test against is the key. "Increase streambank vegetation by 20% in stream reach #5" and "Eliminate 25 miles of unpaved roads in the Smith Creek subwatershed" are examples of specific goal statements. "Improve the trout fishery in Jones Creek" says too little and defies measurement over time.

Specific actions are the heart of riparian restoration. They relate back carefully to one or more specific goals and thereby to the vision statement. Actions are the work that achieves success. Depending on the scope of the project, the package of actions may include only a few specific small projects or dozens of projects of varying size. As with the vision and goals, write down specific actions but in more specific detail. Say carefully what the action will entail. Include specific location, estimated time, required materials, estimated cost, personnel needs, and similar detail. For example, a project to plant willow cuttings by hand along a section of streambank should specify:

- The length, width, and location of the project.
- The density of plantings.
- Type and source of stock planted.
- Specific harvest, handling, and storage instructions.
- Number of crew members.

- Number of types and tools needed.
- Transportation requirements.
- Time required to do the work.
- Monitoring and maintenance needs.
- Estimated total cost of the project.

A more complex project, such as the obliteration of an unneeded road segment, might have several phases and include multiple operations. Specificity keeps project workers on track and costs in line. The action plan should allow for adjustments if on-site changes are needed but should not allow for undisciplined overruns of time and materials.

Obtain the Most Effective Leadership

Leadership throughout the life of the project is absolutely essential to success. Leadership may be shared, as through a steering committee, or it may come from a dedicated individual. In either case, consistent leadership throughout the many phases of the project, from its inception through monitoring, provides the solid foundation needed to make a project a success.

Leadership does not necessarily need to be the same individuals throughout the project. Leadership may change hands several times as long as a smooth and thorough transition takes place that provides continuity for the long-term vision and specific goals for the project. Indeed, changes in leadership during the life of long-term projects will be more common than having the same leaders throughout the project. This is especially true where projects use a steering committee approach. Consistency provides the basis for success. It breeds confidence in participants at all levels.

Leadership has two key components: (1) commitment, and (2) the ability to empower others. Most restoration projects, especially large or complex ones, require strong commitment to enable the leader to work through the many problems and conflicts that inevitably arise. Staying power, coupled with a long-term consistent vision of the desired outcome, is essential.

Good leaders empower others on the team to make the project a success, and train their own successors. Empowerment and building trust at the smaller scales works just as well as on larger projects. No leader can track and manage all the details of a complex project. Nor should they. Leadership entails the ability to trust others to accomplish work through proper decisionmaking and shared responsibility. Moving leadership to the most appropriate levels of an organization breeds trust and commitment throughout the project. Selfish leadership styles frequently fail to build support and trust needed for project success. Complex projects with many partners, landowners, and action items require dispersed leadership and trust throughout the ranks. Even small, less-complex projects may require overall project leaders plus local level leaders to implement treatments.

Select and Develop Local Project Coordinators/Leaders

Local watershed coordinators provide the tie to the resource, and most projects will profit from a local watershed-level coordinator. While the overall project may be managed by an individual or steering committee at a location away from the restoration actions, having local leadership that knows the country and the work force helps translate overall visions, goals, and action plans into quality on-the-ground work. Even the best and most thoughtful plans fail if local leadership is missing or ineffective. Leading projects at the local level takes special skill and determination. Agency and other overall restoration managers must identify, select, and develop those skilled local project leaders, and then nurture their skills for continued and high-level leadership, passing their skills to others, growing new local leaders. Debates rage over whether leaders are "born" or "made." Both types exist. Seek out the naturals and get them on board. Find people with leadership potential and help them learn the needed skills.

Ensure Environmental, Economic, and Social Values are Compatible with Technical Requirements

The best technical solutions cannot ensure the success of projects that ignore key environmental needs, economic demands, or the mix of social values unique to the watershed. While much of the interest in riparian restoration lies with improving our technical knowledge and skills in areas such as assessment, implementation, and monitoring, most seasoned restoration professionals know that few projects, if any, succeed on technical merit alone. Indeed, the environmental, economic, and social values interwoven with the watershed and riparian area are equally important. Few watersheds are owned and managed by a single entity. Almost all watersheds, particularly larger ones, have a complex mixture of ownership, economic needs, and social values. And all watersheds drain somewhere, leaving that "somewhere" on the receiving end of practices performed in that watershed.

Successful riparian restoration takes into account and works within the context of technical, environmental, economic, and social requirements and limitations. The requirements of many environmental laws (e.g., Clean Water Act, Clean Air Act, National

Environmental Policy Act, Endangered Species Act, and National Historic Preservation Act, etc.) including many state and local laws, plus the responsibilities of many agencies, provide many constraints, but also provide important opportunities. Successful restoration integrates these complex needs throughout all of the many steps that lead to long term project success, from the initial vision, through planning, and design, and implementation, to monitoring, evaluation, and maintenance.

Because of this complexity, each riparian restoration project has its own needs. While most restoration projects can follow similar basic steps, restoration managers must address the unique characteristics of each individual project. The mix of solutions that work well in one watershed may not be the proper mix for the adjacent watershed. Watersheds are as individual as people. Each has similar general needs, but each has its own set of special needs as well. Successful restoration projects recognize and deal with these special needs. Failure to address these special characteristics frequently leads to dramatic failures. Similarly, projects that recognize these "nontechnical" needs must include the proper mix of technical solutions. Success results from a proper blend of all these components.

Recognize the Power of Partnerships

For many reasons, successful riparian restoration efforts include a wide variety of strong partnerships. Projects need support: political, emotional, physical, and financial. Support comes from partners.

Due to the mixture of ownership, environmental complexities, economic needs, and social values, it is easy to find one or more strong opponents to nearly any watershed or riparian project. Partnerships forged from a strong base of supporters can overcome opposition and push onward to successful project implementation. Failure to build partnerships leaves even sound projects open to the attacks of a few vocal or well-financed opponents.

Build and nurture partnerships throughout the many phases of a riparian restoration project. Ideally, partners provide the original sponsorship to get the project up and running. They remain part of the project to promote monitoring and make sure proper maintenance takes place as needed. Partnerships, of many types, are key to project success. Some partners provide funds to do the needed planning and implementation. Others contribute labor or supplies.

Some partners bring special or unique skills to a project that may otherwise be missing, such as local archeological knowledge or a historical perspective. Others may know local plants and other key elements to successful design. Some may have skills at monitoring treatment effects. Others may undertake the often thankless task of maintaining a completed project so it continues to function as planned. Some are skilled fund-raisers that rally support and keep the economics of the work in order. Nearly anyone in the subject watershed, and many others outside the work area, can bring the enthusiasm, skills, and support needed to make the project work. Many agencies, such as the USDA Forest Service, have programs like the Challenge Cost-Share that matches partners with projects to leverage success. Others, like the Environmental Protection Agency, offer grants for particularly important project work.

Many organizations now specialize in providing partners for good projects. Several agencies and private organizations publish lists of likely partners, contacts for potential work, or they bring parties together. Several key publications have insights into building partnerships with hundreds of phone numbers, fax numbers, e-mail addresses, and mailing addresses to make the necessary connections (USEPA 1997). Obtain and use these invaluable tools to build your base of committed supporters.

Use the restoration process to increase the knowledge of partners about watershed and riparian management. While partners provide a source of funding and labor, they also are the continuing advocates of needed future restoration work. Help your partners learn about riparian processes and functions. Help them understand the interaction of watershed management and riparian health. Knowledgeable partners are the future of watershed and riparian management.

In short, partners equal support. Partners equal success. Find, educate, train, and nurture your partners. You need them as much as they need you. Make the connection.

Use and Integrate the Best Available Science to Develop Restoration Plans

Riparian restoration knowledge exists to satisfy almost every need of the restoration specialist. While some specialized research will help address specific areas of knowledge, the real challenge for most of us is assembling and using methods workable for our specific project. Let's look at some of the key concepts.

Use the Best Information Available

Modern technology has made most of this information readily available to most all restoration specialists. Literally thousands of references are accessible so analysts, planners, and designers can learn the most recent information available for their use. The Internet provides a wealth of sources to access easily. Libraries and agency offices have paper copies of thousands of articles, symposium proceedings, and workshop results. Video reference materials give specific instructions about processes and projects. Agency handbooks contain the results of many years of experience (USGPO, 1998). Conferences around the country hosted by professional organizations, universities, and societies facilitate exchange of the newest information and experience. Tours of existing projects hosted by local groups provide hands-on experience to real projects. Membership in professional societies and user organizations brings a wealth of knowledge. The real challenge is having the time and wisdom to sort through the available knowledge and find the information best suited to your specific needs and situation. Make the best use of the existing and emerging knowledge base to give your project the best chance of success.

Perform a Watershed Analysis

The riparian area is a direct reflection of the watershed it drains (Heede 1980; DeBano and Schmidt 1989; LaFayette and DeBano 1990). Understanding the condition of the riparian area requires a basic knowledge of the contributing watershed. We cannot simply walk a riparian area, see problems, and prescribe treatments. The process of watershed analysis will provide invaluable information to the riparian restoration team about processes occurring throughout the watershed, so correct cause-effect relationships can be deduced and problems addressed (Ziemer 1997). Several effective watershed analysis methods are available and range from the simple to the complex (Montgomery et al. 1995). Make sure you use the appropriate analysis method for your area. Only through an understanding of the watershed will you make the right choices for the riparian area.

Understand Cause vs. Effect Relationships

A proper watershed analysis will likely clarify the cause-effect relationships in a watershed (Hanes and LaFayette 1987). Incredible amounts of time and money have been spent fixing symptoms of problems rather than the actual cause. Active gullies, abnormally meandering streams and related cutbanks, upland vegetation growing where riparian vegetation should grow, and similar problems are almost always symptoms of a larger cause. Something caused that gully to form. Something threw the stream out of equilibrium and initiated bank erosion. Something allowed those upland plants to encroach on the riparian area. Attacking the gully, bank erosion, or inappropriate vegetation will most likely fail if the real cause of the problem is not addressed. Beware the urge to "do something" when you see a situation that appears to be a problem. Do a proper analysis. Find out what is behind the apparent problem. Then find a way to fix the real problem. Many times it will prove more difficult to address the real problem: poor livestock management, a poorly managed road system, inappropriate timber harvest methods, a popular but poorly placed recreation facility, or a similar sticky situation. Remember: If you lack management control of the watershed, you cannot manage the riparian areas. You will waste your time and money, and you may even make things worse, if you persist with the riparian treatments anyway!

Use the Proper Scale

Working at the proper scale is seemingly simple but actually difficult to do. Restoration projects often tend to be too ambitious or don't do enough. More common is the case of not looking far enough outside the work area to understand the real cause and effect relationships. We recommend looking not only at the watershed contributing to the riparian area of concern, but also at the next higher level of watershed to see if management or other activities in the larger watershed are affecting your project watershed. While the level of analysis of the larger watershed may be less intense than that of the subject watershed, it will provide many clues to what you may find in your more intensely analyzed watershed.

Let History be Your Guide

Understanding the watershed and its riparian area requires an understanding of the current function of the watershed as well as the historical function of the system that led to it. Many seemingly unexplainable findings in a watershed analysis become very clear when seen from a historical perspective. Digging into the history of a watershed is often a fascinating and fun part of the analysis. Be a watershed detective. Find out whatever you can about your "client" watershed. You are solving a mystery that only history can help you understand.

Many tools exist to help you conduct a productive search. Look up old photographs from file records, newspapers, reports, archives, family records, and any other place. Both aerial and ground-level photos will help. Dig through historical records in libraries, county courthouses, agency records, and old reports. Interview long-time residents and record their recollections. You'll meet some very interesting people. Find early species lists of plants and animals. Review and assemble streamflow records to assess historic streamflows. Excavate soil profiles to look for buried soil horizons, fire evidence, flood deposits, and similar features. Search out nearby or similar relic areas that show you how your area could look at its best. Find out all you can about the area to help you understand why it functions like it does today and also how it may once again function properly.

Treat the Uplands and Riparian Area as a Unit

Remember the intimate relationship between the riparian area and its contributing watershed. Treating only the riparian area may suffice in some cases but will not in others (LaFayette and DeBano 1990). Where activities or problems in the watershed continue to cause problems in the riparian area, treat both the watershed and the riparian area. Treatment of just the riparian area will likely fail if the cause of the problem isn't addressed. Treating the watershed may lead to riparian recovery and allow you to forgo investments in the riparian area, but the channel recovery time may be longer than you want to wait as the system readjusts. It may be best to treat both uplands and riparian areas simultaneously to get the most rapid results. Use "passive" treatments such as changes in land management practices, and if appropriate, "active" treatments such as gully treatments, erosion control measures, or vegetation establishment. If past or current management is the real cause of the problem, make sure you adjust management before you invest in active treatments. Use a full suite of treatments for the entire system; don't rely on one or two familiar but limiting treatments.

Remember that riparian restoration revolves around keeping water on and in the land longer. Healthy riparian areas exist in a dynamic equilibrium of water, soil, and vegetation as influenced by climate (Heede 1980). Successful restoration dampens the oscillations in the natural system and promotes slow and gradual evolution of watershed processes.

Don't be Seduced by a Specific Tool or Method

Many of us rely on a small tool kit of analysis and treatment methods. This is especially true of treatments. Just because you like hammers doesn't mean all problems are nails. Let the needs of the land tell you what will work best. That willow treatment you had good luck with in the past may not work again if channel substrates are inappropriate. Those gabion baskets you like may not be the best treatment of a gully headcut in fine-textured soils. The seed mix you like may not work everywhere. Beware of exotic species that may overwhelm native species and spread across the landscape, such as kudzu and multiflora rose. Have an open mind and tailor the treatment to the needs of the watershed and riparian area. Confirm that a technique performs the desired function before adopting it.

Beware of "Hot" Trends

Every few years a new trend in watershed and riparian restoration becomes popular in the literature. Excited demonstrations pop up everywhere. But remember, there are few truly new ideas in the world. Be a skeptic. Analyze new methods before you spend much time or money on them. Like clothing styles, restoration methods come and go. The good ones become valuable tools. The poor ideas fail and disappear. Keep on top of new methods, but screen them carefully before you add them to your bag of tricks. Beware enthusiastic salesmen with new toys or techniques. They can cost you or your client big money.

Use the Least Intrusive Methods

When you can choose among various methods to accomplish a restoration task, use the least intrusive and most natural method. While a concrete wall may stop that eroding bank cold in its tracks, a carefully developed planting of appropriate riparian vegetation may do just fine, work better, and cost far less. Seek out methods that blend well with the landscape, but still accomplish the task. Don't move 20 yards of soil when moving 10 or none will do the trick. Leave the site in as natural a setting as possible. Enjoy the anonymity of a well-done unobtrusive solution.

Remember: Small is Beautiful, Simple is OK

In a busy and complex world, many pressures exist to tackle large and complex projects. Unfortunately, size and complexity increase the risk of failure (Heede 1980). Watershed and riparian management and restoration are, by their very nature, complex undertakings. Many intricate and complex variables interact at the watershed scale because the watershed itself integrates all actions within it. It therefore makes great sense to resist temptation and keep projects both small and simple to reduce complexity and improve the likelihood of success. Seek out small and straightforward projects. Within a large overall project, break out and accomplish bite-size actions that combine to form larger successes. Don't try to be a hero by tackling the whole project at once. It is critical though, to first recognize the interactions at the watershed level, and then divide and conquer the situation with manageable actions.

Be Patient!

Except in rare cases, it took many years of gradual impacts to degrade a riparian area to the point of needing restoration. Except in rare cases, it will take an extended effort to restore the functions and values of that riparian area that is naturally self sustaining. Save the watershed and riparian area a piece at a time. Build on small successes and accumulate enthusiasm. Avoid the failures that deflate enthusiasm and drive away partners. Take your time and do it right. Time and money are precious. Spend them wisely.

Consider Life Cycle Efforts and Costs

Consider the total effort and cost in restoration projects before committing resources or choosing treatment alternatives. Total project costs involve more than just the design and .

implementation of treatments. Project planning maintenance and monitoring also play a role in total cost over the life of the project.

Put much effort into planning the project in as much detail as you deemed reasonable. What needs to be done in what order? Where and when will the work take place? Who will do the work? What skills are needed for quality implementation? How much will the work cost? What are the long-term funding sources to sustain an extended effort? Watershed and riparian restoration is a long-term undertaking, not a weekend lark. Know the details for an extended campaign that will culminate in eventual success.

Particularly important is the cost of long-term project maintenance. Where choices exist that provide similar treatment results, choose the design with the lowest maintenance requirements. Where possible, choose treatments that have a high degree of self-maintenance. Mechanical treatments like dams, fences and similar constructed efforts by their nature will require maintenance. Seedlings or other treatments that are flexible and have inherent resilience fare better in the long term.

Monitor, Communicate, Account for and Reward Progress

The history of on-the-ground actions and related funding reveals that most efforts go into the accomplishment of new treatments. Relatively little effort is focused on analysis and planning, less on monitoring, evaluation, and proper maintenance. Work to bring more balance between these restoration elements. Even if fewer projects get done with the available funds, a higher percentage will likely succeed over the long term. It is better to do a few projects well than fail many times.

Monitor your projects to learn what works so you can repeat successes and reduce failures. Share this knowledge with others, both successes and failures. Use newsletters, meetings, the Internet, published reports, and similar means to share your experiences and learn from others. Unshared knowledge is wasted knowledge. Shared knowledge is shared power. Shared power breeds success in watershed and riparian restoration.

Celebrate your work and give credit to all parties involved. Give special credit to the partners that help the project succeed. Partners who know their efforts are appreciated will gladly work with you again. Partners who feel used and unrecognized will find better ways to spend their time and energy. If partnerships are the foundation for success, then proper recognition is a cornerstone of that foundation.

"There will come a time when you believe everything is finished. That will be the beginning." — Louis Lamour

Summary

Watershed and riparian restoration is a combination of science and art. Nobody knows all the answers. Years of collective trial and error in all parts of the nation and world have resulted in a large body of knowledge about the key components of successful riparian restoration. Knowing why or how a restoration project has succeeded or failed, and communicating that knowledge will help others succeed. The preceding pages summarize our "real world" suggestions and findings from these many years of effort. Learn from the work of others. Adapt their successes to your projects and increase your likelihood of success. Learn from their mistakes and avoid making them on your projects. Many years of significant work lie ahead of us. Use your limited time and funds to their fullest extent and be a proud member of an honorable profession, saving the world one watershed and riparian area at a time.

Iowa State University

The Bear Creek riparian restoration project in Iowa illustrates how an over-grazed, broken-bank stream can be restored to productive fish habitat using grasses, shrubs, and trees.

Chapter 19

Dams: A Watershed Restoration Opportunity

Michael J. Solomon

"The nation behaves well if it treats the natural resources as assets which it must turn over to the next generation increased and not impaired in value."
Theodore Roosevelt

In the late 19th century many of our nation's eastern streams were cleared, straightened, and used as sluices to drive logs downstream to mills. In the early 20th century, these and other rivers across the nation were dammed to create hydropower to run the nation's industries (the first hydroelectric plant in the world was built on the Fox River at Appleton, Wisconsin in 1882). These dams, in addition to providing electricity, changed the condition of waterways and their riparian corridors. The construction of dams interrupts river processes. It blocks downstream movement of sediment and nutrients and alters ecosystems. Hydropower development often uses a "peaking" mode of operation where water is stored during the night and then released during the day when demands for electricity peak. This operation mode essentially floods the river system each day. It alters the timing of flows and often changes downstream river characteristics, both in shape, width and depth, as well as, in ecological characteristics. The generating turbines can kill thousands of fish, and the dams interrupt fish migration and movement. Dams can also impact temperature, dissolved oxygen, nitrogen, fluctuation of reservoirs, and public access.

Re-licensing of these hydroelectric dams is a one-shot opportunity to impact riparian and river management for 30 to 50 years. The Federal Energy Regulatory Commission (FERC) has extraordinary powers to administer and secure agreement on future conditions among dam owners and watershed landowners. This re-licensing window provides an opportunity for mitigating damage and enhancing the resources that have been impacted by these structures for the previous 50 to 100 years. For example, there are opportunities to change project operations from peaking to run-of-river regimes, to restore fisheries to stream segments that have been dewatered by interbasin canals, to stabilize eroding sites within or below the

315

reservoir, to provide upstream and downstream fish passage, and even to remove the dam and restore the site. Furthermore, FERC can require the dam owner to provide past data and to fund new studies required to assess conditions and plan for desired outcomes. Resource agencies and nongovernmental organizations must be prepared to accept the challenge and opportunity that re-licensing presents.

The NEPA Process

The 7-step re-licensing process that investigates possible changes in the operation of a hydroelectric facility (or removal of the dam) is cast in legal procedure that invokes the National Environmental Policy Act (NEPA) process. It can be costly, fraught with conflict, centered on legal posturing, and will take 5 ½ years or more (Figure 19.1).

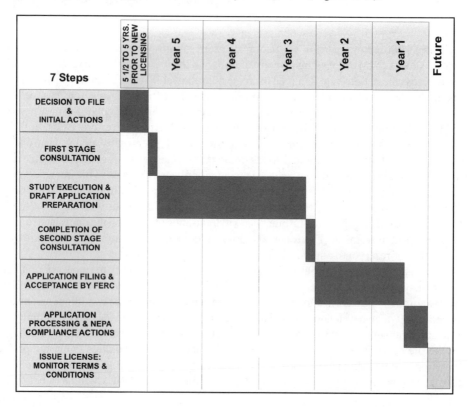

Figure 19.1 The seven steps and their duration in the FERC re-licensing process.

An alternative approach to the process, the Settlement Agreement, attempts to gain a broad consensus up-front (it may take as long as the conventional process). This approach centers on the opportunities for data collection, resource damage mitigation, and desired outcomes with less expenditure of time and money. The skillful execution of the FERC re-licensing process can seal the fate of major stream processes for half a century.

The Federal Power Act (FPA) established several important principles that apply to both the issuance of original licenses and to re-licensing. The FERC must give equal consideration to a full range of purposes related to the potential value of a river. Among these purposes are:

- Hydroelectric development
- Energy conservation
- Fish and wildlife resources, including their spawning grounds and habitat
- Recreational opportunities
- Other aspects of environmental quality
- Irrigation
- Flood control
- Water supply

If the FERC is assured that the project proposal is a comprehensive plan for developing the river, it may establish license conditions that mitigates potential damage to fish and wildlife, and protects or enhances their habitat.

For projects on Federal Agency land (e.g., a National Forest) the FPA requires that the FERC determine if a hydropower project is consistent with the purposes for which a national forest was established. In addition, in any license issued, it must include conditions the Forest Service determines are necessary for the adequate protection and use of the National Forest.

Several other authorities come to bear on the FERC process, such as Section 7 of the Wild & Scenic Rivers Act (P.L. 90-542). Existing Wild and Scenic Rivers and study rivers designated for protection under Section 3 or 5 (a) of the Act may not be used for new hydropower projects on or directly affecting such rivers. Projects indirectly affecting wild and scenic reaches may be permitted only where they do not diminish the existing wild and scenic river values. Where hydroprojects already exist that affect study or designated rivers, the effect of the license conditions on the river must be evaluated under Section 7.

The Clean Water Act (CWA) provides that state certifications shall set forth conditions necessary to ensure that applicants comply with specific portions of the CWA and with appropriate requirements of state law.

The Electric Consumers Power Act (1986) amended the Federal Power Act to require the FERC to give equal consideration to power development, natural resource conservation, fish and wildlife protection, enhancement and preservation of recreational opportunities, and environmental quality protection. This act further requires the FERC to ensure a project is consistent with federal and state comprehensive land management or watershed plans.

The Endangered Species Act requires the FERC to evaluate a project's impact on threatened and endangered species and to mitigate impacts based on the best available scientific data and economic impact. The FERC may be required to do a biological assessment to identify impacts on protected species.

The Federal Power Act (FPA) (Section 10j) requires that a license shall include conditions to protect and enhance fish and wildlife including their habitat, and to mitigate damage to their habitat affected by the development, operation, and management of the projects. The conditions should be consistent with the recommendations made by the National Marine Fisheries Service, USDI Fish & Wildlife Service, and state fish and wildlife agencies. If the FERC finds that any of these recommendations are inconsistent with the FPA or other applicable law, it must attempt to resolve any inconsistencies. Typical "10j" issues are fish screens, fish passages, minimum flows, ramping rates, reservoir draw-down limits, and wetland restoration.

Settlement Agreements

Another successful approach is for a hydropower applicant, various resource agencies, and other concerned parties such as environmental advocates, property owners, Indian tribes, and concerned citizens to develop a Settlement Agreement (SA). An SA establishes the various license conditions, mitigating activities and compensation up-front before the formal FERC review process begins. This proactive approach allows all parties to take ownership in the process and to reach balances with other concerned parties. This process can shorten the re-licensing process, build allies among the hydropower industry and various partners, and use the available financial resources for resource damage mitigation rather than legal posturing. There can be many benefits, including ownership in the agreement by all participants, less expensive studies, reduced legal costs, shortened time frames and more certainty of a desired or acceptable outcome. The FERC has supported settlement agreements in the past and encouraged partners to work on them. However, there is *no* guarantee that the FERC will accept all of the terms of an SA.

It is important to involve as many interested parties as possible. The more organizations and individuals participating in the process, the less likelihood of intervention or legal maneuvering at the end. Major organizations and agencies may persuade smaller or less influential parties, thus developing consensus more expeditiously. The more partners that can agree to and support an SA, the less likely the FERC will reject components of the agreement.

SAs can cover a wide variety of issues and may be very comprehensive documents. Issues that have been in SAs are land management plans, project operations, water quality, minimum flows, maximum flows, sediment transport, streambank erosion, Instream Flow Incremental Methodology (IFIM), historic fishery data, fish population studies, fish entrainment studies, turbine entrainment studies, toxic contaminants, fish passage, threatened and endangered species evaluations, archaeological surveys, recreational use, land management plans, and socioeconomic evaluations. The license applicant may be required to do studies in these issues or provide funding for them.

Summary

FERC re-licensing is an opportunity for resource agencies, Indian tribes, nongovernmental organizations, and individuals to influence hydropower operations and their effects for 30 to 50 years. The FERC re-licensing process requires the collection of vast quantities of data that will serve as baselines from which aquatic and terrestrial impacts will be measured in the future. The process is very structured and is shaped by a long legal history. Organizations, agencies, and individuals must understand the process to protect their interests and maximize environmental benefits. Settlement agreements are opportunities for the hydroelectric industry, together with other organizations, to negotiate a wide range of issues into a legally binding document that can be used to describe appropriate license articles by the FERC. These collaborative efforts are win-win agreements for all parties, and most important, for the water resources impacted by hydroelectric facilities.

Recommended Reading

Hydroelectric Project Relicensing Handbook, Office of Hydro Power Licensing, Federal Energy Regulatory Commission, April 1990.

Forest Service Handbook 2709.15, Hydro Power Handbook, USDA, Forest Service, Draft 1995.

Kenneth D. Kearns, Carrie Kimball Monhyan, and Anna L. West. Oct. 1995. Hydro Settlements: Providing an Opportunity for Agreement, Hydro Review.

Federal Interagency Restor. Group

Watershed restoration efforts may include changes in land use such as this fencing to limit livestock use of the riparian area.

USDA Forest Service

It's the fishing!

Chapter 20

Riparian Area Management: Themes and Recommendations

Dave J. Welsch, James W. Hornbeck, Elon S. Verry, C. Andrew Dolloff and John G. Greis

The end results of most of our management actions are reflected by the health of our rivers, streams, and lakes.   Michael Dombeck, Chief, USDA Forest Service

In this final chapter, we consider the overriding themes of riparian area management and list highlights and recommendations from each chapter.

The Water's Edge

The quote above alludes to the difference between forest management and riparian forest management. Riparian forest management is concerned with the water as well as the forest. The combination of water and forest is both a feature of the landscape, and a priority in our values that has focused resource management on riparian issues. Riparian management has passed from a history of intense exploitation a century ago, through recent decades of neglect, to the intense demand for shared decisions we see today. Protecting the essential links between land and water ensures quality water and quality aquatic habitats. Best Management Practices, which govern logging activities and tree removals in and around riparian areas, are a good example of how we design management options to reduce impacts on water quality. Silvicultural prescriptions to meet Desired Future Conditions for riparian areas give primary consideration to protection of water temperature and contributions of particulate matter and coarse woody debris to aquatic habitats. Managing for recreation in riparian areas addresses the impacts on water quality and impacts on our perception of a quality personal experience at the water's edge.

Our Values

Human values and desires shape the way riparian areas are managed. Values and desires often overlap or conflict, and there is a need to find common ground. The axiom "good stewardship is shared stewardship" is especially appropriate for riparian area management. It has given insight to complex management recommendations that consider use and the relationships among resources. Riparian areas should receive "active" management, with "hands-off" being one option.

Connected, Complex, and Fuzzy

Riparian areas are difficult to define and delineate. Physical or biological criteria seldom allow finding a location where one might comfortably say "My right foot is in the riparian area, and my left foot is not." Perhaps more than any other landscape feature, riparian areas focus our concept of landscape connections for fish and animal communities, for the interaction of forest regeneration with forest-site, and for the interaction of our personal economics with our personal recreation. Each of our disciplines (wood, water, wildlife, fisheries, recreation, farming, and transportation) finds reasons to sharpen its perception of interdisciplinary management in a landscape framework with multiple owners. This connected landscape is where land ownership and community stewardship tests the boundary between private rights and states rights. Because our riparian landscape winds through many social boundaries, and is managed with many levels of knowledge, we realize how fuzzy the boundaries really are. We need both the protection of a one-size-fits-all approach (perhaps a 2- or 3- sizes-fit-all approach), and the challenge of the custom-fit approach where we weigh our knowledge of land and water as we read the landscape and prescribe a Desired Future Condition for each riparian area we manage.

Chapter Highlights and Recommendations

1   The Challenge of Managing for Healthy Riparian Areas

- We have lived by and traveled our streams, rivers, and lakes for millennia, and the conflicts now at the water's edge are at least two centuries old: squatter's rights (private rights), forest management (regulation of forest structure on the basis of age class and acreage), flooding and sedimentation (geologic or anthropogenic), states rights (BMPs), subsidies or grants, and regional sedimentation (the landscape perspective or the site-specific guide).

- We have re-awakened to the values we find so compelling at the water's edge.

- Desired Future Condition is socially derived, and takes real, personal work to achieve.

- Broaden your competence to manage riparian areas using both your discipline and the disciplines of others.

- The acid test of our understanding is not whether we can take ecosystems apart on paper. . . but whether we can put them together in practice and make them work.

- Find the knowledge to read the land and river. When you do, we have no fear of what you will do for them.

2   Defining Riparian Areas

- Historically, we have based our definitions of riparian area on the vernacular of a particular discipline, or a particular agency.

- We review the literature and offer a definition based on how a riparian area functions: **A riparian area is a three-dimensional ecotone of interaction that includes the terrestrial and aquatic ecosystems, that extends down into the groundwater, up above the canopy, outward across the floodplain, up the near-slopes that drain to the water, laterally into the terrestrial ecosystem, and along the water course at a variable width.**

- The On-The-Ground Key offers two ways to define riparian areas for streams, rivers, lakes, ponds, and wetlands: read the land and water, or use rules of thumb based on stream width and mature tree height.

3   Diversity in Riparian Landscapes

- Riparian areas are crowded neighborhoods. Neighboring and functionally connected ecosystems within riparian areas are often dense, giving rise to greater biodiversity.

- The cause of crowding is the physiographic diversity encountered and created by rivers as they flow through the landscape. They create variation in the topography of floodplains and terraces, they mediate the space and time patterns of soil moisture, and they influence vegetation composition and structure.

- Management begins with ecosystem classification and mapping.

- Maintain or restore the physical processes that regulate riparian ecosystems.

- Put on your landscape glasses if you really want to understand and manage riparian areas.

- Manage for timber when and where appropriate.

- Buffer width depends on landscape context.

- Favor native species.

4   Classifying Ecosystems and Mapping Riparian Areas

- The size and complexity of the natural world demands that we classify and map so we can comprehend and remember what is important about Mother Nature with wet feet.

- Use the hierarchical approach to classify both aquatic and terrestrial ecosystems, and use ecological functions to group things rather than arbitrary boundaries imposed by institutions or professions.

- For aquatic classification systems, the hierarchy is split on world climate zones first, then on the similarity or difference of fish species in the community, and on the topography of watershed boundaries.

- For terrestrial classification systems, the hierarchy is split on similar world and regional climate zones, then on geologic and potential vegetation boundaries.

- The aquatic and terrestrial classification systems overlap and nearly coincide at the world climate zone and the terrestrial subsection/aquatic subbasin scales.

- Also for streams, develop Valley Segments that incorporate both physical and chemical components you can directly relate to fish, invertebrate, or mussel communities using stream width, water temperature, and alkalinity (your system's place on the acid/base titration curve), and possibly trophic status (phosphorus, nitrogen, biomass).

- For streams, use the system of Natural Stream Types to group stream reaches into common types that identify how the stream and its valley handle the energy of flowing water and sediment. Use the Types to communicate with your colleagues.

- For terrestrial units, use Land Types to group ecosystems because they relate to landscape position, geologic makeup and potential vegetation.

- Use Valley Segments as descriptors in Land Type associations or Land Types.

- Use Land Type as descriptors in aquatic types.

- With the advent of GIS and relational databases, you can map and handle information on the basis of resource hierarchal groupings, administrative boundaries, or a combination of both. How do you want to accumulate your information in a world where resource decisions are shared rather than discipline oriented?

5   Linkages Between Forests and Streams: A Perspective in Time

- Evaluate the relative contributions of your forest to streams and lakes on the basis of its successional stage: (0 to 10, 11 to 30, 31 to 80, 81 to 125, and 126 to 250$^+$ years old).

- You'll have to cut 20% of a watershed, or 25% of mature basal area to measurably increase the yield of water on an annual basis.

- If you cut the whole thing, the annual yield of water can be 40% higher than from the mature forest condition; most of this comes in the growing season when water is converted from transpiration to subsurface flow; it may take 15 years to diminish.

- Forests free of recent disturbance are the best possible vegetative cover for protecting against flooding. If your landscape is heavily cut, peak flow increases of 150% are possible (see Chapter 6).

- Intensify timber sale administration and use BMPs. Realize that sediment yield is closely related to individual storms and the condition of roads used for access. Give special attention to chronic disturbance problems at landings, roads, and stream crossings.

- Sediment yields from well-regulated forest landscapes are small, generally in the 25 to 40 lbs/acre/year range.

- Following BMPs will reduce on-site erosion.

- Site nutrient loss is highest before significant regrowth and from old stands (more than 80 or 100 years old), especially if atmospheric inputs of nitrogen and sulfur are high.

- High atmospheric inputs of nitric and sulfuric acids can denude soils of essential base cations (e.g., calcium, magnesium), especially on base-poor sands; limit biomass removal on these sites.

- Retain your stream shade. Loss of forest shade can warm streams by 9°F to 18°F unless cold, groundwater flows dominate.

- Models such as JABOWA, FORTNITE, LINKAGES, and BROOK can help you assess changes in nutrients, woody debris, forest structure, and water yield.

- Use the combination of Current Condition, Desired Future Condition, and BMPs to manage riparian forests, and manage upslope forests first because they will change riparian conditions. Consider emulating the early mature stage (80 to 125 years old) near water to provide a continuing supply of organic matter and CWD.

6   Water Flow in Soils and Streams: Sustaining Hydrologic Function

- As rainfall increases soils become saturated farther and farther from the stream's edge. This is the variable-source-area concept which states that the area that generates streamflow expands and contracts with the weather. It applies to mineral and organic soils in the watershed.

- Most of the water in forest streams is delivered as subsurface flow (within a foot or so below the surface) of either mineral or organic soils, and where deep aquifers exist, from groundwater discharging into the channel bottom.

- Depth to the water table (up 18 inches) controls the attainment of maximum vegetation height. Soil texture and organic matter content control water availability in deeper, well-drained soils, and (along with nutrients) determine the maximum vegetation height attained on these soils.

- Unless the soil is really dry or frozen, the water table must be at least 18 inches below the surface to support equipment and avoid soil compaction.

- Keep vegetation growing to its potential by keeping the water inside the soil moving — protect the macropore structure of soils so they can contain optimum amounts of air and water.

- Optimum water and air occur in surface mineral soils (to a depth of 6 inches) if soil bulk density is less than 1.3 (less than 0.1 in organic soils). In the 6- to 12-inch horizon, keep mineral soil bulk density less than 1.5. Reductions must be made to the initial bulk density values of weight and volume for soils with high rock content. These are made by subtracting the weight and volume of rocks before making the final bulk density calculation.

- Soils compacted to levels greater than above commonly reduce future biomass productivity by 20 to 30%.

- Judge water flow in streams by understanding Natural Stream Types.

- Learn what Bankfull Elevation really is and how to find it at the water's edge.

- Channels that are changing from one Natural Type to another are unstable.

- Use both deductive reasoning and Fisher statistics to evaluate watershed condition.

- Snowmelt peak flows can increase by 150% if more than 2/3 of a basin is harvested in the span of 15 years. The minimum watershed size where these increased peak flows can actually cause in-channel erosion is 1 square mile in basins with high slopes (30% or more) and 10 square miles in flat basins (slopes less than 3%).

- Restrict soil drainage networks (ditches) to less than 1/3 of a basin.

- Restrict ditched roadway area to less than 15% of the basin.

- Vigorous, strong root systems at the water's edge will minimize channel over-widening.

- Road crossings at streams are major sediment sources to the channel. Reduce fine sand inputs by blacktopping the road (to the top of the terrace slopes on each side of the road) or by surfacing with crushed rock (that will pass a 1 ½ inch sieve) in a layer 6 inches deep.

- Size round culverts at crossings by adding a foot to the bankfull channel width, or ask an engineer or hydrologist to design them for a specific site. Lay them with the same slope as the channel, and bury the bottom to 1/6 of their vertical diameter (to keep them from perching and blocking fish passage).

- In wetlands, where culverts are used underneath roads (but not in channels) use 24-inch culverts or larger, and bury the bottom half below the peat surface so the culvert will carry the everyday flows (subsurface flow).

- When deep road fills across stream valleys wash out, use a second culvert (the same size as the channel culvert) placed on top of the floodplain about halfway between the channel and the terrace slope. This "floodplain relief" culvert and the channel culvert will pass at least the 50-year flood.

7   Debris isn't so Bad

- Let everyone know that organic matter in streams is an asset to be husbanded.

- Manage the riparian area. In a natural system, change will come with or without management, wise management presents the greatest potential for watershed enhancement.

- Develop a strategy to limit "emergency cleanup" by removing only necessary amounts of woody debris in the aftermath of natural disasters.

- Manage for a component of 50- to 200-year-old trees in the riparian zone to supply near natural inputs of large wood capable of forming debris dams.

- Manage for a diversity of native species typical of riparian areas in the region to provide a natural supply of detritus that supports the needs of the macroinvertebrate community.

- Do not disrupt natural flood inputs of organic matter to streams.

- Remove trees felled into streams during logging only after consulting with a biologist or other resource specialist in charge of the timber sale.

- Leave dead and down trees in the riparian area to avoid damaging the integrity of the streambank and to enhance habitat for amphibians and other animals.

- Discourage extensive dense thickets of rhododendrons in the riparian area because they provide neither high-quality detritus nor long-term potential for large wood recruitment.

- In high-water-table floodplains, protect dense grass and sedge that bind banks and narrows channels.

8   Mammal and Bird Habitat In Riparian Areas

Landscape level considerations
- Consider variable-width and wider stream buffers.

- Limit new road building in riparian areas.

- Avoid long, linear clearcuts adjacent to buffer areas.

- Consider the potential tree species composition.

- Limit grazing activities at the water's edge.

- Limit borrow pit development.

Stand level considerations
- Favor variety and diversity.

- Manage the habitat to meet goals of typical species.

Within stand or structure considerations
- Base decisions on site potential.

- Retain higher densities of cavity trees and snags, emphasizing large diameters.

- Consider a variety of canopy closures, raptor nesting potential and perch tree potential.

- Consider softwood-to-hardwood and mast-to-non-mast basal area ratios.

- Consider increasing large dead and down woody debris potential, both in-stream and terrestrial.

- Favor multiple canopy layers: shrub thicket, grass/sedge, and herbaceous ground cover.

9   Managing for Fish in Riparian Areas

- No more than 2/3 of the watershed should be in young stands (<15 yrs old) where snowmelt may result in channel erosion and sedimentation. Where agricultural land use exceeds 30% of the watershed area, the portion of young forest should be less.

- To be effective, riparian guidelines must include small streams and dry channels.

- Design roads and trails to minimize interception and concentration of both surface water and groundwater flow.

- Design roads and trails so that bridges and culverts do not occupy low points in the road or design to divert surface runoff from the road prior to the crossing point.

- Consider designing riparian forest buffers with a width equal to at least two tree lengths for fish habitat protection.

- Consider establishing reference streams and using indices of biological integrity to evaluate streams and the impacts of management.

10   Amphibians and Reptiles in Riparian Habitats

- Maintain or restore hardwood forest canopies to maintain microclimate, leaf litter, and coarse woody debris.

- Schedule forest management activities and select equipment to avoid soil compaction.

- Severely restrict livestock's access to streams.

- Consider using buffer zones based on the seven Riparian Habitat Types and the needs of the herptofauna in question.

- Manage the aquatic edge as well as the upland edge of the riparian area, again based on a holistic assessment of the needs of the species present.

- When in doubt, manage to mimic historic disturbances that provided the evolutionary basis for natural community structure.

- Obtain a complete data set on the life histories of species inhabiting the area.

- Develop an understanding of population and seasonal dynamics and the range of threats for the species involved.

- Monitor to determine trends and to detect changes in species composition among sites.

- Hire experienced herptologists and train field personnel to conduct such studies.

11   Human Dimensions

- The human dimension is a critical and an increasingly important component of riparian management.

- For effective riparian management, it is necessary to integrate the human dimensions information early in the planning process.

- People's values, attitudes, beliefs, knowledge, and expectations are as important as plant and animal population and distribution.

- Accommodate the human dimensions by developing an array of reasonable alternatives rather than trying to "sell" your recommendation.

- A variety of qualitative and quantitative methods are available for collecting human dimensions information. These range from in-depth interviews of a small number of individuals to broad population surveys providing generalized results. Which method is chosen is dependent on the questions managers are seeking answers to.

- Human dimensions information is useful in developing management and planning options as well as in facilitating a dialogue with the public.

12   Lake Riparian Areas

- Use volunteer monitors and lake organizations in education, protection, and management programs for lake riparian areas.

- Educate lake users and lakeside landowners about the value of coarse woody debris (CWD) and discourage unnecessary removal.

- Increase habitat complexity by managing forests for a continued supply of CWD.

- Plan riparian forests in conjunction with other protective practices where trees are the natural riparian vegetative cover.

- Consider basin morphology and prevailing winds in managing riparian areas where wind stress on the surface of the lake is a problem.

- Retain lake-associated wetlands, because they are considered important sources of dissolved organic matter.

13   Integrated Management of Riparian Areas

- Social, economic and political interests along with biological considerations have added complexity to riparian management, requiring an integrated approach.

- Integrated management is characterized by collaboration, pursuit of long-term goals, dissemination and use of information, and modification of actions based on monitoring of progress (adaptive management).

- Developing integrated management plans involves setting direction, inventorying values and resources, developing alternatives, selecting a preferred alternative, and implementing.

- Management direction is the combination of long-term goals and the actions needed to achieve those goals.

- Long-term goals may be limited by laws, regulations, and organizational objectives.

- Inventory includes both assessing existing information as well as on-the-ground data collection. In both cases, efforts should reflect the value of the resource being inventoried.

- Develop mechanisms for sharing and evaluating information to facilitate collaboration.

- Considering social values at the outset can reduce the potential for legal, political, or jurisdictional battles that arise when concerned citizens and citizen groups are excluded from the process.

- Identify constituents early in the process, being careful not to confuse stakeholders and constituents that are concerned with special interests. The real stakeholders are the people who live in the watershed and drink the water.

- Consider multiple spatial scales. Some effects of resource management will go beyond the scale of the project area.

- Involve all the interests and disciplines, and establish a clear decisionmaking process.

- Develop alternatives to reach the same desired end. Evaluate the different approaches, effects, benefits, and costs associated with each. Make these trade-offs obvious to the decisionmakers and the public.

- Each alternative considered must be based on the capability and limitations of the ecosystem and must move the project toward the long-term goal.

- Document the fact that you have listened to concerns, evaluated results, explored options, and selected an alternative based on the objectives.

- Monitor progress and adjust accordingly.

14   Ecologic Principles of Riparian Silviculture

- Understand the characteristics of the system. Use your knowledge of the land and the water to apply silviculture across the ecotone from the water's edge into the upland. Consider what fits the land best as well as BMP guidelines.

- Include maintenance or restoration of riparian functions as a principal silvicultural objective. Traditional silviculture aims at producing wood, riparian silviculture aims at maintaining riparian functions. When you apply riparian silviculture, you are managing the water as well as the forest.

- Remember that riparian forests are ecotones. Consider how the values of each riparian function taper off with distance from the water, and how the need to protect the water resource intensifies approaching the water's edge.

- Apply site-specific, comprehensive silvicultural prescriptions in riparian areas. Learn how to manage with clump and patch patterns that allow a combination of single-cohort (even aged, single canopy) and multiple-cohort (uneven aged, multiple canopied) tree stands. Retain large trees at the bank, use clump and patch patterns to maximize biomass production in the second tier away from the water, and feather into single-cohort stands where desired on the upland terrace. In uneven-aged, multiple-cohort stands, consider clump and patch patterns instead of evenly spaced residuals.

15 Harvesting Options for Riparian Areas

- Route planning and equipment selection can minimize ground disturbance in the riparian zone (see our evaluations in Chapter 15).

- Wait 2 days after the day of heavy rains, so soils will drain. This allows them to convert from mud (they are at their liquid limit in this period) to the field capacity condition (gravity has drained all the water it can) when they can again support equipment.

- Use geotextiles and temporary crossing mats where weak soils must be traversed.

- Be familiar with state BMPs and apply them appropriately.

- Avoid putting sediment into stream channels. Plan road systems efficiently.

- Surface permanent haul roads with crushed rock at stream crossings.

- As a first choice, use temporary bridges to cross stream channels.

- As a second choice, use temporary culverts, pipe bundles, corduroy (pole fords in NH), or snow and ice to cross channels, but avoid pushing earth into the channel or adjacent to the bank. Remove these crossings before spring or fall peakflows do it for you.

Oversee the harvesting process
- Prepare comprehensive harvest plans. Tailor them to each site.

- Meet with the operator on-site beforehand to ensure the plan provisions are well understood.

- Supervise during the operation to ensure the plan is being properly implemented.

16   Best Management Practices

- Planning is the most important BMP.

- Mark riparian management zone boundaries, and include the operating restrictions in the language of the contract, and provide on-site administration to ensure activities comply with the contract.

- Decide the purpose and degree of protection to give to each of the various types of water bodies to be encountered.

- Involve resource specialists, industry, and concerned constituents in the development of guidelines to work toward a balance among science, practicality, and economics.

- Recognize the importance of operating with less physical intrusion and more care in riparian areas.

- The riparian area may be viewed as a zone of closely managed activity, rather than a zone of exclusion.

- BMPs applied to variable width riparian areas may be more costly and time consuming to implement, but they will likely provide greater resource protection (see Chapters 2 and 14).

- Complying with BMPs, although easier to measure on fixed width riparian areas, does not necessarily indicate protection of water quality.

- Establish a management emphasis within the *riparian area* by considering direction of lean and proximity to water as well as other factors in determining the relative riparian and aquatic value of individual trees.

- Design BMP's to prevent (or mitigate) adverse impacts from sediment input, temperature shifts, flow changes, chemical inputs, waste disposal, or habitat alteration.

- To prevent adverse impacts visit existing problem and sensitive areas within the riparian area to evaluate the location and degree of disturbance from access before operations begin, and to plan remedial work required.

- Select stream crossing methods and locations to avoid disturbing fish and amphibian migrations, deforming the bank or channel, or mining stream gravel.

- Avoid fueling, equipment servicing, and pesticide applications within the riparian area.

17 Monitoring the Effects of Riparian Area Management

- Establish an understanding of the time and cost associated with monitoring and secure the necessary commitments from all concerned.

- Identify the monitoring objective. The question to be answered and the audience to be informed will dictate the measurements to be made and the level of precision necessary.

- Measuring change is not an end in itself, the harm caused by the change must be explained.

- Follow the key steps for conducting monitoring activities by MacDonald et al. (1991).

- Understand natural sources of variation such as climate, geology and acid bog drainage, which may affect monitoring results.

- Understand the potential long-term effect of historic land use such as mining and agriculture. Elevated sediment loads have been documented as long as 60 years after farm abandonment.

- Monitor streamflow on smaller watersheds. Flow increases become measurable only from the entire watershed when more than 25% of the basal area is removed.

- Use readily available information from the United States Geological Survey, other state and federal agencies, and other sources when appropriate.

- Remember that sample points must be comparable and watershed size and stream order alone are not good indicators of comparable conditions.

- Macroinvertebrate populations are governed by substrate composition.

- Macroinvertebrate populations recover rapidly following impairment from stream sedimentation, unless the source of sediment continues to supply excessive amounts.

- Macroinvertebrate numbers are a more sensitive indication of short-term sedimentation than diversity indices or fish populations. Long-term changes can be documented with macroinvertebrate indices and fish community assemblages.

Use the following knowledge of common changes documented by past monitoring activities to develop current management alternatives
- Harvests can increase the number of stems on floodplains and increase the hydraulic roughness and sediment deposition on the floodplain. Maintain near-bank trees to prevent bank erosion.

- Harvests that increase direct sunlight to streams generally increase water temperature.

- Harvests within a tree length of the channel will reduce the supply of coarse woody debris to the stream and reduce pool development in sand and gravel channels.

- Harvesting over 25% of the basal area on an entire watershed generally increases summer flows.

- Other land-use changes in the watershed can mask and add to the changes resulting from forestry activities.

18 Riparian Restoration

- Cooperate to develop a clear vision shared by all key participants.

- Set achievable, measurable goals specific enough to test against during the monitoring phase.

- Create detailed action plans that relate back to the goals.

- Provide consistent, committed leadership that can empower others throughout the duration of the project.

- Integrate the technical, environmental, economic, and social requirements and limitations of the project.

- Nurture partnerships for political, emotional, physical, and financial support.

- Use ecological classification to identify existing condition, potential conditions, and the departure of existing conditions from potential.

- Perform a watershed analysis using the best available science and information to establish the cause of a problem and avoid treating symptoms.

- Use the most natural and least intrusive methods.

- Monitor progress and give credit to partners.

19 Dams: A Watershed Restoration Opportunity

- Use the Federal Energy and Regulatory Commission (FERC) re-licensing process to make major changes in the hydrology, channel habitat, aquatic community, and flood plain community of dammed watersheds.

- This is a one-time opportunity to impact riparian and river management for 50 years.

- FERC can require the dam owner to pay for your studies and planning.

- It's a NEPA process that can be approached with a settlement agreement that encourages partners to work together in a win-win situation.

We have no fear that in time you will read the land and the river, to manage them as a unit for many resources, indeed, we are excited about what you will do for them (adapted from Leopold 1949).

Sunset or sunrise? Voyager's National Park, Minnesota.

References

A

Aadland, Luther P. 1993. Stream habitat types: their fish assemblages and relationships to flow. *North Am. J. Fish. Manage.* 13:790-806.

Aber, J. D., J. M. Melillo, and C. A. Federer. 1982. Predicting the effects of rotation length, harvest intensity, and fertilization on fiber yield from northern hardwood forests in New England. *For. Sci.* 28:31-38.

Adams, M. D., and M. J. Lacki. 1993. Factors affecting amphibian use of road-rut ponds in the Daniel Boone National Forest. *Trans. Kentucky Acad. Sci.* 54:13-16.

Adams, T., and D. Hook. 1993. Implementation and effectiveness monitoring of silvicultural best management practices on harvested sites in South Carolina. Rep. No.: BMP-1. South Carolina Forestry Commission. Columbia, SC. 32 p.

Adams, T. O. 1998. Implementation monitoring of forestry best management practices for site preparation in South Carolina. *South. J. Appl. For.* 22:74-80.

Adams, T. O., and D. D. Hook. 1994. Compliance with silvicultural best management practices on harvested sites in South Carolina. *South. J. Appl. For.* 18(4):163-167.

Alabama Forestry Commission. 1993. *Alabama's best management practices for forestry.* Montgomery, Alabama Forestry Commission. 30 p.

Albert, D. A. 1995. Regional landscape ecosystems of Michigan, Minnesota, and Wisconsin: a working map and classification (4th rev.: July 1994). Gen. Tech. Rep. NC-178. U.S. Department of Agriculture, Forest Service, North Central Forest Experiment Station, St. Paul, MN. 250 p.

Albert, D. A., S. R. Denton, and B. V. Barnes. 1986. *Regional landscape ecosystems of Michigan.* School of Natural Resources, University of Michigan, Ann Arbor. 32 p.

Alexander, G. R., and E. A. Hansen. 1983. Effects of sand bedload sediment on a brook trout population. Michigan Department Natural Resources Fisheries Rep. 1906. Ann Arbor, MI.

Allan, J. D., and L. B. Johnson. 1997. Catchment-scale analysis of aquatic ecosystems. *Freshwater Biol.* 37:107-111.

Allan, J. D., D. L. Erickson, and J. Fay. 1997. The influence of catchment land use on stream integrity across multiple spatial scales. *Freshwater Biol.* 37:149-161.

Allen J. R. L. 1965. A review of the origin and characteristics of recent alluvial sediments. *Sedimentology* 5:89-191.

Allen, N. S., and A. E. Hershey. 1996. Seasonal changes in chlorophyll *a* response to nutrient amendments in the French River, a North Shore tributary of Lake Superior. *J. North Am. Bentholog. Soc.* 15:170-178.

Allen, T. F. H., and T. B. Starr. 1982. *Hierarchy: Perspectives for ecological complexity.* Chicago: University of Chicago Press. 310 p.

Amundson, R., and H. Jenny. 1997. Thinning. Biology: on a state factor model of ecosystems. *BioScience* 47(8):536-543.

341

Andrews, E. D. 1980. Effective and bankfull discharges of streams in the Yampa River Basin, Colorado and Wyoming. *J. Hydrol.* 46:311-330.

Angermeir P. L., and J. R. Karr. 1984. Relationships between woody debris and fish habitat in a small warm water stream. *Trans. Am. Fish. Soc.* 113:716-726.

Annable, W. K. 1995. Morphological relationships of rural water courses in southwestern Ontario for use in natural channel designs. Master's thesis, The University of Guelph. Guelph, Ontario. 377 p.

Anonymous. 1910. Splashing logs through breaks of Sandy. *Amer. Lumberman.* March 19, 1990.

Anthony, R. G., E. D. Forsman, G. A. Green, G. Witmer, and S. K. Nelson. 1987. Small mammal populations in riparian zones of different aged coniferous forests. *Murrelet.* 68:94-102.

APHA (American Public Health Association). 1989. *Standard methods for the examination of water and wastewater.* (17th ed.). APHA, Washington, DC.

Arcement, G. J., Jr., and V. R. Schneider. 1989. Guide for selecting Manning's roughness coefficients for natural channels and flood plains. Water-Resour. Pap. 2339. U.S. Geological Survey. 38 p.

Archambault, L., B. V. Barnes, and J. A. Witter. 1990. Landscape ecosystems of disturbed oak forests of southeastern Michigan, USA. *Can. J. For. Res.* 20:1570-1582.

Askins, R. A. 1994. Open corridors in a heavily forested landscape: impact on shrubland and forest-interior birds. *Wildl. Soc. Bull.* 22:339-347.

Aubertin, G. M., and J. H. Patric. 1974. Water quality after clearcutting a small watershed in West Virginia. *J. Environ. Qual.* 3:243-249.

Aust, W. M. 1994. Timber harvesting considerations for site protection in southeastern forested wetlands. In *Proceedings, water management in forested wetlands.* Tech. Publ. R8-TP-20. U.S. Department of Agriculture, Forest Service, Southern Region, Atlanta, GA:5-12.

Aust, M. W., R. Lea, and J. D. Gregory. 1991. Removal of floodwater sediments by a clearcut tupelo-cypress wetland. *Water Resour. Bull.* 27(1):111-116.

Aust, M. W., M. D. Tippett, J. A. Burger, and W. H. McKee, Jr. 1995. Compaction and rutting during harvesting affect better drained soils more than poorly drained soils on wet pine flats. *South. J. Appl. For.* 18(2):72-77.

Avers, P. E., D. T. Cleland, and W. H. McNab. 1994. National hierarchical framework of ecological units. In Silviculture: From the cradle of forestry to ecosystem management, comp. L. H. Foley, Gen. Tech. Rep. SE-88. U.S. Department of Agriculture, Forest Service. Southeastern Forest Experiment Station, Asheville, NC. 258 p.

B

Babbie, E. 1998. *The practice of social research.* 8th ed. Belmont, CA: Wadsworth Publishing Co. 465 p. + appendices.

Bahls, L. L. 1993. *Periphyton bioassessment methods for Montana streams.* Helena, MT: Water Quality Bureau, Department of Health and Environmental Sciences.

Bailey, R. G. 1995. *Description of the ecoregions of the United States. 2nd ed.* Misc. Publ. No. 1391. U.S. Department of Agriculture, Forest Service, 108 p + map.

Bailey, R. G. 1996. *Ecosystem geography.* New York, NY:Springer-Verlag.

Bailey, R. G., R. D. Pfister, and J. A. Henderson. 1978. Nature of land and resource classification - a review. *J. For.* 76(10):650-655.

Bailey, S. W., and J. W. Hornbeck. 1992. Lithologic composition and rock weathering potential of forested, glacial-till soils. Res. Pap. NE-662. U.S. Department of Agriculture, Forest Service, Broomall, PA. 7p.

Baker, C. O. 1979. The impacts of logjam removal on fish populations and stream habitat in western Oregon. Master's thesis, Oregon State University, Corvallis.

Baker, C. O., and F. E. Votapka. 1990. Fish passage through culverts. Rep. FHWA-FL-90-006. U.S. Government Printing Office, Washington, DC. 67 p.

Baker, M. E. 1995. The diversity of landscape ecosystems in rivers valleys of the Huron-Manistee National Forests, northern Lower Michigan. Master's thesis, University of Michigan, Ann Arbor. 101 p.

Baker, M. E. and B. V. Barnes. 1998. Landscape ecosystem diversity of river floodplains in northwestern lower Michigan, U.S.A. *Can. J. For. Res.*:1405-1418.

Baker, R. H. 1983. *Michigan mammals.* East Lansing, MI: Michigan State University Press. 642 p.

Banasiak, C. F. 1961. Deer in Maine. Game Div. Bull. No. 6. Department of Inland Fisheries and Game, Augusta, ME. 159 p.

Barbour, C. D., and J. H. Brown. 1974. Fish species diversity in lakes. *Am. Nat.* 108:473-489.

Barbour, R. W., J. W. Hardin, J. P. Schafer, and M. J. Harvey. 1969. Home range, movements, and activity of the dusky salamander, *Desmognathus fuscus. Copeia* 1969:293-297.

Barbour, M. T., J. B. Stribling, J. Gerritsen, and J. R. Karr. 1996. Biological criteria: technical guidance for streams and small rivers. EPA 822-B-96-001. U.S. Environmental Protection Agency, Washington, DC.

Barlocher, F., R. J. Mackay, and G. B. Wiggins. 1978. Detritus processing in a temporary vernal pool in southern Ontario. *Arch. Hydrobiol.* 81:269-295.

Barnes, B. V. 1985. Forest ecosystem classification and mapping in Baden-Württemberg, West Germany. In *Forest land classification: experience, problems, perspectives.* J. G. Bockheim, ed. 49-65 NCR-102. North Central Forest Soils Committee, Society of American Foresters, USDA Forest Service, and USDA Soil Conservation Service, Madison, WI.

Barnes, B. V. 1996. Silviculture, landscape ecosystems, and the iron law of the site. *Forstarchiv.* 167:226-235.

Barnes, B. V., K. S. Pregitzer, T. A. Spies, and V. Spooner. 1982. Ecological forest site classification. *J. For.* 80:493-498.

Barnes, B. V., D. R. Zak, S. R. Denton, and S. H. Spurr. 1998. *Forest ecology.* 4th ed. New York: John Wiley and Sons, Inc. 774 p.

Barro, S. C. and N. A. Bopp. (In press). What people think about ecological restoration and related topics: A first look. In *Paper presented at the 23rd Natural Areas, 15th North American Prairie, and Indiana Dunes Ecosystems joint conferences,* St. Charles, IL.

Bartke, R. W. and S. H. Patton. 1979. Water-based recreational developments in Michigan: problems of developers. *Wayne Law Rev.* 25:1005-1063.

Barton, D. R., W. D. Taylor, and R. M. Biette. 1985. Dimensions or riparian buffer strips required to maintain trout habitat in southern Ontario streams. *North Am. J. Fish Manage.* 5:364-378.

Baumgras, J. E. 1996. Environmental concerns affecting forest operations on public lands in the central Appalachians. In *Planning and implementing forest operations to achieve sustainable forests*: Proceedings of papers presented at the joint meeting of the Council on Forest Engineering and International Union of Forest Research Organizations, eds. C. R. Blinn and M. A. Thompson. Gen. Tech. Rep. NC-186. U.S. Department of Agriculture, Forest Service, North Central Forest Experiment Station, St. Paul, MN. 282 p.

Baumgras, J. E., J. R. Sherar, and C. B. Ledoux. 1995. Environmental impacts from skyline yarding in an Appalachian hardwood stand: a case study. In *Proceedings of the Council on Forest Engineering, 18th annual meeting.* 87-96. Cashiers, NC.

Bayley, P. B., and H. W. Li. 1992. Riverine fishes. Chapter 12, The Rivers Handbook, VI, eds. P. Calow and G. E. Petts. London: Blackwell Scientific Publications. 526 p.

Bedford, B. L., and E. M. Preston. 1988. Developing the scientific basis for assessing cumulative effects of wetland loss and degradation on landscape functions: status, perspectives, and prospects. *Environ. Manage.* 12:751-771.

Bell, D. T., and F. L. Johnson. 1974. Ground-water level in the flood plain and adjacent uplands of the Sangamon River. *IL State Acad. Sci.* 67:376-383.

Bell, D. T., and S. K. Sipp. 1975. The litter stratum in the streamside forest ecosystem. *Oikos* 26:391-397.

Belovsky, G. E. 1984. Summer diet optimization by beaver. *Am. Midl. Nat.* 111:209-222.

Beneski, J. T., Jr., and D. W. Stinson. 1987. *Sorex palustris*. Mammalian species. Shippensburg, PA: American Society of Mammalogists. 296:1-6.

Bent, P. C. 1971. Influence of surface glacial deposits on streamflow characteristics of Michigan streams. U.S. Geological Survey, Lansing, MI.

Bengston, D. N. and Xu, Z. 1995. Changing National Forest values. Res. Pap. NC-323. U.S. Department of Agriculture, Forest Service, North Central Forest Experiment Station, St. Paul, MN. 29 p.

Benke, A. V., T. C. Van Arsdale, D. M. Gillespie, and F. K. Parrish. 1984. Invertebrate productivity in a subtropical blackwater river: The importance of habitat and life history. *Ecol. Monogr.* 54:25-63.

Bennett, G. W. 1971. *Management of lakes and ponds*. New York: Van Nostrad Reinhold Company.

Benson, B. J., and J. J. Magnuson. 1992. Spatial heterogeneity of littoral fish assemblages in lakes: relation to species diversity and habitat structure. *Can. J. Fish. Aquat. Sci.* 49:1493-1500.

Berg, D. R. 1995. Riparian silvicultural system design and assessment in the Pacific Northwest Cascade Mountains, USA. *Ecol. Appl.* 5: 87-96.

Berger, L., R. Speare, P. Daszak, D. E. Green, A. A. Gunningham, C. L. Goggin, R. Slocombe, M. A. Regan, A. D. Hyatt, K. R. McDonald, H. B. Hines, K. R. Lips, G. Marantelli, and H. Parkes. 1998. Chytridiomycosis causes amphibian mortality associated with population declines in the rain forests of Australia and Central America. In *Proceedings of the National Academy of Science* 95:9031-9036. Washington DC.

Berkman, H. E., and C. F. Rabeni. 1987. Effect of siltation on stream fish communities. *Environ. Biol. Fishes.* 18: 285-287.

Beschta, R.L. and R. L. Taylor. 1988. Stream temperature increases and land use in a forested Oregon watershed. *Water Resour. Bull.* 24:19-25.

Beschta, R. L., R. E. Bilby, G. W. Brown, L. B. Holtby, and T. D. Hofstra. 1987. Stream temperature and aquatic habitat: fisheries and forestry interactions. In *Streamside management: forestry and fishery interactions*, eds. E. O. Salo, and T. W. Cundy. 191-232. Seattle, WA: University of Washington. Institute of Forest Resources. Contribution 57.

Bickman, L., and D. J. Rog, ed. 1998. *Handbook of applied social research methods*. Thousand Oaks, CA: Sage Publications. 580 p.

Bidwell, D., and S. C. Barro, 1997. *Roadblocks to understanding biodiversity: an introduction to public perceptions of biodiversity and strategies to communicate biodiversity messages*. Chicago, IL: Chicago Wilderness. 11 p.

Bilby, R. E. 1981. Role of organic debris dams in regulating the export of dissolved and particulate matter from a forested watershed. *Ecology* 62:1234-1243.

Bilby, R. E., and G. E. Likens. 1980. Importance of organic debris dams in the structure and function of stream ecosystems. *Ecology* 61:1107-1113.

Bilby, R. E. and P. A. Bisson. 1987. Emigration and production of hatchery coho salmon (*Oncorhynchus kisutch*) stocked in streams draining an old-growth a clear-cut watershed. *Can. J. Fish. Aquat. Sci.* 44: 1397-1407.

Birch, T. W. 1996. Private forest owners of US 1994. Resour. Bull. NE-134. U.S. Department of Agriculture, Forest Service, Northeastern Forest Experiment Station, Radnor, PA.

Bishop, S. C. 1941. The salamanders of New York. *New York State Mus. Bull.* 324:1-365.

Bisson, P. A., J. L. Nielsen, R. A. Palmanson, and L. E. Grove. 1982. A system of naming habitat types in small streams, with examples of habitat utilization by salmonids during low streamflow. In *Aquisition and utility of aquatic inventory information*, ed. N. B. Armantrout. 62-73. Bethesda, MD: American Fisheries Society, Western Division.

Bisson, P. A., R. E. Bilby, M. D. Bryant, C. A. Dolloff, G. B. Grette, R. A. House, M. L. Murphy, K. V. Koski, and J. R. Sedell. 1987. In *Large woody debris in forested streams in the Pacific northwest: past, present, and future. Stream management; forestry and fisheries interactions*, eds. E. O. Salo, and T. W. Cundy. University of Washington, Institute of Forest Resources.

Black, P. E., and P. M. Clark. 1958. Timber, Water, and Stamp Creek. U.S. Department of Agriculture, Forest Service, Southeastern Forest Experiment Station, Asheville, NC: 12 p.

Blaustein, A. R., and Wake, D. B. 1990. Declining amphibian populations: a global phenomenon? *Trends in Ecol. and Evol.* 5:203-204.

Blaustein, A. R., P. H. Hoffman, D. G. Hokit, J. M. Kiesecker, S. C. Walls, and J. B. Hays. 1994a. UV repair and resistance to solar UV-B in amphibian eggs: a link to population declines? *Proc. Nat. Acad. Sci.*, USA. 91:1791-1795.

Blaustein, A. R., D. G. Hokit, and R. K. OHara. 1994b. Pathogenic fungus contributes to amphibian losses in the Pacific northwest. *Biol. Conserv.* 67:251-254.

Blinn, C. R., and R. A. Dahlman. 1995. Riparian harvesting with a soft footprint. In *Proceedings, At the waters edge: the science of riparian forestry*. Minnesota Extension Service BU-6637-S. University of Minnesota, St. Paul, MN:76-81.

Blinn, C. R., R. Dahlman, L. Hislop, and M. Thompson. 1998. Temporary stream and wetland crossing options for forest management. General Technical Report NC-202. U.S. Department of Agriculture, Forest Service, North Central Research Station, St. Paul, MN. 125 p.

Bloom, A. L. 1991. *Geomorphology*. New Jersey Prentice-Hall.

Boarman, W. I. 1993. When a native predator becomes a pest: a case study. In *Conservation and resource management*, eds. S. K. Majumdar, E. W. Miller, D. E. Baker, E. K. Brown, J. R. Pratt, and R. F. Schmalz. 191-206. Easton, PA: Pennsylvania Academy of Science.

Bolstad, P. V., and W. T. Swank. 1997. Cumulative impacts of land use on water quality in a southern Appalachian watershed. *J. Am. Water. Resour. Assoc.* 33(3):513-533.

Bonell, M. 1998. Selected challenges in runoff generation research in forests from the hillslope to headwater drainage basin scale. *J. Am. Water Resour. Assoc.* 34(4):765-785.

Bonin, J. M., J. Quellet, J. Rodrigue, F. DesGranges, T. F. Gagne, L. A. Sharbel, and L. A. Lowcock. 1997. Measuring the health of frogs in agricultural habitats subjected to pesticides. *Herpetol. Conserv.* 1:246-257.

Bormann, F. H., G. E. Likens, T. G. Siccama, R. S. Pierce, and J. S. Eaton. 1974. The export of nutrients and recovery of stable conditions following deforestation at Hubbard Brook. *Ecol. Monogr.* 44:255-277.

Bormann, F. H., and G. E. Likens. 1979. *Pattern and process in a forested ecosystem.* New York: Springer-Verlag, 253 p.

Bormann, F. H., G. E. Likens, and J. S. Eaton. 1969. Biotic regulation of particulate and solution losses from a forest ecosystem. *BioScience* 19:600-611.

Born, S. M., S. A. Smith, and D. A. Stephenson. 1974. *The hydroponic regime of glacial-terrain lakes, with management and planning applications.* Upper Great Lakes Reg. Comm. 73 p.

Bozek, G., M. Jennings, and G. Hatzenbeler. 1997. Changes in fish habitat and fish communities caused by development in riparian areas of lakes. *In Abstract for the 59th Midwest Fish and Wildlife Conference*, Milwaukee, WI.

Bradshaw, A. D. 1983. The reconstruction of ecosystems. *J. Appl. Ecol.* 10:1-17.

Brazier, J. R., and G. W. Brown. 1973. Buffer strips for stream temperature control. Res. Pap. 15. Forest Research Laboratory, School of Forestry, Oregon State University, Corvallis, OR. 8 p.

Briggs, R. D., J. Cormier, and A. Kimball. 1998. Compliance with forestry best management practices in Maine. *North. J. Appl. For.* 15: 57-68.

Brinson, M. M. 1990. Riverine forests. In *Forested wetlands*, ed. D. W. Goodall. 87-141. New York: Elsevier. Ecosystems of the World, Volume 15.

Brinson, M. M. 1993. A hydrogeomorphic classification of wetlands. Tech. Rep. WRP-DE-4. US Army Corps of Engineers, Waterways Experiment Station, Vicksburg, MS. 79 p.

Brinson, M. M., B. L. Swift, R. C. Plantico, and J. S. Barclay. 1981. Riparian ecosystems: their ecology and status. FWS/OBS-81/17. U.S. Fish and Wildlife Service, Biological Services Program. Kearneysville, WV. 154 p.

Brosofske, K. D. 1996. Effects of harvesting on microclimate from small stream to uplands in western Washington. Master's report, Michigan Technological University. 72 p.

Brosofske, K. D., J. Chen, R. J. Naiman, and J. F. Franklin. 1997. Harvesting effects on microclimatic gradients from small streams to uplands in western Washington. *Ecol. Appl.* 7:1188-1200.

Brown, A. M., and N. C. Collins. 1997. Let dead trees lie; impacts of coarse woody debris on small fish densities in the littoral zone. In *Abstract for the 59th Midwest Fish and Wildlife Conference*, Milwaukee, WI.

Brown, G., and C. C. Harris, Jr. 1991. National forest management and the tragedy of the commons: A multidisciplinary perspective. *Soc. and Nat. Resour.* 5:67-83.

Brown, G. W. 1980. *Forestry and water quality.* Corvallis, OR: Oregon State Book Stores, Corvallis. 124 p.

Brown, G. W., A. R. Gahler, and R. B. Marston. 1973. Nutrient losses after clear-cut logging and slash burning in the Oregon Coast Range. *Water Resour. Res.* 9:1450-1453.

Brush, G. S. 1986. Geology and paleoecology of Chesapeake Bay: A long-term monitoring tool for management. *J. Wash. Acad. Sci.* 76:146-160.

Brussock, P. P., A. V. Brown, and J. C. Dixon. 1985. Channel form and stream ecosystem models. *Water Resour. Bull.* 21(5):859-866.

Bryan, M. D., and D. L. Scarnecchia. 1992. Species richness, composition, and abundance of fish larvae and juveniles inhabiting natural and developed shorelines of a glacial Iowa lake. *Environ. Biol. Fishes* 35:329-341.

Buchanan. D. V., P. S. Tate, and J. R. Moring. 1976. Acute toxicities of spruce and hemlock extracts to some estuarine organisms in southeastern Alaska. *J. Fish. Res. Board Canada* 33:1188-1192.

Buchholz, K. 1981. Effects of minor drainages on woody species distributions in a successional floodplain forest. *Can. J. For. Res.* 11:671-676.

Buck, D. H. 1956. Effects of turbidity on fish and fishing. *Transactions of the 21ˢᵗ North American Wildlife Conference.* 21:249-261.

Buck, S. V. 1997. Florida's herp trade: a collector's paradise and ...a land exploited. *Reptile & Amphibian Mag.* Jan/Feb:72-81.

Buech, R. R. 1995. The wood turtle: its life history, status and relationship with forest management. In Proceedings of the 1995 NCASI Central-Lake States Regional meeting. 118-127. Spec. Rep. 95-14. National Council Paper Industry for Air and Stream Improvement.

Buech, R. R., L. G. Hanson, and M. D. Nelson. 1997. Identification of wood turtle nesting areas for protection and management. In *Proceedings Conservation, restoration, and management of tortoises and turtles -- an international conference*, eds. J. Van Abbema, and C. H. Peter. 383-391. Purchase, NY: New York Turtle and Tortoise Society.

Buhlmann, K. A., and J. W. Gibbons. 1997. Imperiled aquatic reptiles of southeastern United States: historical review and current conservation status. In *Aquatic fauna in peril: The southern perspective*, eds. G. W. Benz, and D. E. Collins. 201-231. Spec. Publ. 1. Southern Aquatic Research Institute, Lenz Designs and Communications, Decatur, GA.

Buhlmann, K. A., J. C. Mitchell, and C. A. Pague. 1994. Amphibian and small mammal abundance and diversity in saturated forested wetlands and adjacent uplands of southeastern Virginia. In *Proceedings of a workshop on Saturated forested wetlands in the Mid-Atlantic region: The state of science*, eds. S. D. Eckles, A. Jennings, A. Spingarn, and C. Wienhold. 1-8. Annapolis, MD: U.S Fish and Wildlife Service.

Bull, J. J., and R. C. Vogt. 1979. Temperature-dependent sex determination in turtles. *Science* 206:1186-1188.

Burke, V. J., and J. W. Gibbons. 1995. Terrestrial buffer zones and wetland conservation: a case study of freshwater turtles in a Carolina bay. *Conserv. Biol.* 9:1365-1369.

Burroughs, Edward R., Jr., and John G. King. 1989. Reduction of soil erosion on forest roads. Gen. Tech. Rep. INT-264. U.S. Department of Agriculture, Forest Service, Intermountain Research Station, Ogden UT. 21 p.

Burton, T. A. 1997. Effects of basin-scale timber harvest on water yield and peak streamflow. *J. Am. Water Res. Assoc.* 33(6):1187-1196.

Burton, T. M., and G. E. Likens. 1975. Salamander populations and biomass in the Hubbard Brook Experimental Forest, New Hampshire. *Copeia* 1975:541-546.

Burton, T. M., R. R. Turner, and R. C. Harriss. 1977a. Nutrient export from three north Florida watersheds in contrasting land use. In *Watersheds research in eastern North America*, ed. D. L. Correll. Edgewater, MD: Smithsonian Institution, Vol. 1.

Burton, T. M., R. R. Turner, and R. C. Harriss. 1977b. Suspended and dissolved solids exports from three North Florida watersheds in contrasting land use. *Watersheds research in eastern North America*, ed. D. L. Correll. Edgewater, MD: Smithsonian Institution. Vol. 1.

Bury, R. B. 1988. Habitat relationships and ecological importance of amphibians and reptiles. In *Streamside management: riparian wildlife and forestry interactions*: Proceedings of a symposium, ed. K. J. Raedeke, Contrib. 59:61-76. Seattle, WA: Institute of Forest Resources, University of Washington.

Bury, R. B., and P. S. Corn. 1988. Responses of aquatic and streamside amphibians to timber harvest: a review. In S *Streamside management: riparian wildlife and forestry interactions*: Proceedings of a symposium, ed. K. J. Raedeke, Contrib. 59:165-181. Seattle, WA: Institute of Forest Resources, University of Washington.

C

Canadian Forest Practice Codes. 1995. Riparian management area guidebook. Ministry of Forestry, British Columbia. www.for.gov.bc.ca/tasb/legsregs/fpc/fpcguide/riparian/.

Carlson, R. E. 1975. A trophic state index for lakes. *Limnol. Oceanogr.* 22:361-369.

Carraway, B., L. Clendenen, and D. Work. 1998. *Voluntary compliance with forestry best management practices in East Texas: Results from Round 3 of BMP compliance monitoring.* Texas Forest Service. 28 p.

Cassidy, G. J., J. B. Aron, and M .J. Tremblay. 1996. *Best management practices for Rhode Island.* Water Quality Protection and Forest Management Guidelines.

Castelle, A. J., A. W. Johnson, and C. Conolly. 1994. Wetland and stream buffer size requirements - review. *J. Environ. Qual.* 23:878-882.

Chamberlin, T. W., R. D. Harr, and F. H. Everest. 1991. Timber harvesting, silviculture, and watershed processes. In *Influences of forest and rangeland management on salmonid fishes and their habitats*, ed. W. R. Meehan, Chapter 6 Spec. Publ. 19. Bethesda, MD: American Fisheries Society.

Chapman, J. A., and G. A. Feldhamer, eds. 1982. *Wild mammals of North America: biology, management, and economics.* Baltimore, MD: Johns Hopkins University Press. xiii + 1146 p.

Chase, V. P., L. S. Deming, and F. Latawiec. 1997. *Buffers for wetlands and surface waters: a guidebook for New Hampshire municipalities, revised.* Concord, NH: Audubon Society of New Hampshire. 80 p.

Chasko, G. G., and G. E. Gates. 1982. Avian habitat suitability along a transmission-line corridor in an oak-hickory forest region. *Wildl. Monogr.* 82. 41 p.

Cherry, J., and R. L. Beschta. 1989. Coarse woody debris and channel morphology: a flume study. *Water Resour. Bull.* 25:1031-1036.

Chick, J. H., and C. C. Mclvor. 1994. Patterns in the abundance and composition of fishes among beds of different macrophytes: viewing a littoral zone as a landscape. *Can. J. Fish. Aquat. Sci.* 51:2873-2882.

Chisolm, J. L., and S. C. Down. 1978. *Stress and recovery of aquatic organisms as related to highway construction along Turtle Creek, Boone County, West Virginia.* Water Resour. Pap. 2055. U.S. Geological Survey, Washington, DC.

Christensen, D. L., B. J. Herwig, D. E. Schindler, and S. R. Carpenter. 1996. Impacts of lakeshore residential development on coarse woody debris in north temperate lakes. *Ecol. Appl.* 6:1143-1149.

Christiansen, J. L., and B. J. Gallaway. 1984. Raccoon removal, nesting success, and hatchling emergence in Iowa turtles with special reference to *Kinosternon flavescens* (Kinosternidae). *Southwest. Nat.* 29:343-348.

Church, M. 1992. Channel morphology and typology. In The Rivers Handbook, eds. P. Calow, and G.E. Petts. Chapter 6. London: Blackwell Scientific Publications. 526 p.

Cline, M. G. 1949. Basic principles of soil classification. *Soil Sci.* 67:81-91.

Cole, D. W., S. P. Gessel, and J. Turner. 1978. Comparative mineral cycling in red alder and Douglas-fir. In *Utilization and management of Alder*, eds. D. G. Briggs, D. S. DeBell, and W. A. Atkinson, 327-336. Gen. Tech. Rep. PNW-70. U.S. Department of Agriculture, Forest Service, Pacific Northwest Research Station, Portland, OR.

Conant, R., and J. T. Collins. 1991. A field guide to reptiles and amphibians of eastern and central North America. Boston, MA: Houghton Mifflin Co. 450 p.

Congdon, J. D., and R. E. Gatten. 1989. Movements and energetics of nesting *Chrysemys picta. Herpetologica* 45:94-100.

Conner, R. N., R. G. Hooper, H. S. Crawford, and H. S. Mosby. 1975. Woodpecker nesting habitat in cut and uncut woodlands in Virginia. *J. Wildl. Manage.* 39:144-150.

Conners, M. E., and R. J. Naiman. 1984. Particulate allochthonous inputs: relationships with stream size in an undisturbed watershed. *Can. J. Fish. Aquat. Sci.* 41:1473-1483.

Copstead, R. 1997. The water/road interaction technology series: an introduction. U.S. Department of Agriculture, Forest Service, San Dimas Technology and Development Center, San Dimas, CA. 9777 1805-SDTDC.

Corn, P. S., and R. B. Bury. 1989. Logging in western Oregon: responses of headwater habitats and stream amphibians. *For. Ecol. Manage.* 29:39-57.

Covich, A. P., T. A. Crowl, S. L. Johnson, D. Varza, and D. L. Certain. 1991. Post-Hurricane Hugo increases in atyid shrimp abundance in a Puerto Rican montane stream. *Biotropica* 23:448-454.

Covington, W. W. 1981. Changes in forest floor organic matter and nutrient content following clearcutting in northern hardwoods. *Ecology* 62:41-48.

Cowardin, L. M., V. Carter, F. C. Golet, E. T. LaRoe. 1979. Classification of wetlands and deepwater habitats of the United States. U.S. Department of Interior, Fish and Wildlife Service. Washington, DC. 131 p.

Coy, F. E., Jr., T. Fuller, L. G. Meadows, and D. Fig. 1992. Splash dam construction in eastern Kentucky's Red River Drainage Area. *For. Conserv. Hist.* 36:179-184.

Cronan, C. S., and C. L. Schofield. 1990. Relationships between aqueous aluminum and acidic deposition in forested watersheds of North America and northern Europe. *Environ. Sci. Technol.* 24:1100-1105.

Crow, T. R. 1991. Landscape ecology: the big picture approach to resource management. In *Challenges in the Conservation of Biological Resources.* eds. D. J. Decker, M. K. Krasny, G. R. Goff, C. R. Smith, and D. W. Gross. 55-65. Boulder, CO: Westview Press.

Crow, T. R., A. Haney, and D. M. Waller. 1994. Report on the scientific roundtable on biological diversity convened by the Chequamegon and Nicolet National Forests. Gen. Tech. Rep. NC-166. U.S. Department of Agriculture, Forest Service, North Central Forest Experiment Station, St. Paul, MN. 55 p.

Crowther, B. (ed.). *in press*. Standard English and scientific names of amphibians and reptiles of North America north of Mexico. Society for the Study of Amphibians and Reptiles, Herpetolgical Circular.

Cuffney, T. F. 1988. Input, movement and exchange of organic matter within a sub-tropical coastal blackwater river-floodplain system. *Freshwater Biol.* 19:305-320.

Cullen, J. B. 1996. *Best management practices for erosion control on timber harvesting operations in New Hampshire*. Durham, NH: University of New Hampshire Cooperative Extension. 65 p.

Cummins, K. W., M. A. Wilzbach, D. M. Gates, J. B. Perry, and W. B. Taliaferro. 1989. Shredders and riparian vegetation. *BioScience* 39:24-30.

Cummins, K. W. 1973. Trophic relations of aquatic insects. *Ann. Rev. Entomol.* 18:183-206.

Cummins, K. 1977. Form headwater streams to rivers. *Am. Bio. Teacher*. May 1977:305-312.

Cummins, K. W., and M. J. Klug. 1979. Feeding ecology of stream invertebrates. *Annu. Rev. Ecol. Syst.* 10:147-172.

Cummins, K. W., J. R. Sedell, F. J. Swanson, G. W. Minshall, S. G. Fisher, C. E. Cushing, R. C. Petersen, and R. L. Vannote. 1983. Organic matter budgets for stream ecosystems: problems in their evaluation. In *Stream ecology*, eds. J. R. Barnes, and G. W. Minshall, 299-353. New York: Plenum Press.

Curtis, J. T. 1951. Hardwood woodlot cover and its conservation. In Report to the people of Wisconsin on cover destruction, habitat improvement and watershed problems of the state in 1950. ed. N. H. Hoveland. *Wisconsin Conserv. Bull.* 16(2):11-15.

Curtis, J. T. 1956. The modification of mid-latitude grasslands and forests by man. In *Man's Role in changing the face of the earth*, ed. W. L. Thomas Jr., 721-736. Chicago, IL: University of Wisconsin Press.

D

Daniels, S. E., R. L. Lawrence, and R. J. Alig. 1996. Decision-making and ecosystem-based management: applying the Vroom-Yetton model to public participation strategy. *Environ. Impact Assess. Rev.* 16:13-29.

Darlington, P. J. 1957. Zoogeography; the geographical distribution of animals. New York: John Wiley & Sons. 673 p.

Darveau, M., P. Beauchesne, L. Belanger, J. Huot, and P. Larue. 1995. Riparian forest strips as habitat for breeding birds in boreal forest. *J. of Wildl. Manage.* 59:67-78.

Davis, M. B. 1976. Erosion rates and land-use history in southern Michigan. *Environ. Conserv.* 3:139-148.

Davis, M. 1996. Getting rid of the stumps: Wisconsin's land clearing program--the experience of the northern lake country, 1900-1925. *Trans. Wisconsin Acad. Sci., Art, and Letter* 84:11-22.

Davis, W. S. 1995. Biological assessment and criteria: building on the past. *Biological assessment and criteria: tools for water resource planning and decision making*, eds. W. S. Davis, and T. P. Simon, 15-29. Boca Raton, FL: Lewis.

Dawson, T. E., and J. R. Ehleringer. 1991. Streamside trees that do not use stream water. *Nature* 350: 335-337.

Dean, J. F., W. T. Momot, and S. A. Stephenson. 1991. *GEOFISHERIES a new concept for determining fish habitat suitability as applied to brook trout.* Part I. Ontario Ministry of the Environment. Natural Resources, Fisheries and Tourism Research Unit, Lakehead University. 77 p.

DeBano, L. F., and L. J. Schmidt. 1989. Interrelationships between watershed condition and health of riparian areas in the southwestern United States. In *Proceedings of practical approaches to riparian management workshop*, eds, R. E Gresswell, B. A. Barton, and J. L. Kershner, 45-52. Billings, MT: U.S. Bureau of Land Management.

Décamps, H., J. Joachim, and J. Lauga. 1987. The importance for birds of the riparian woodlands within the alluvial corridor of the River Garonne, s.w. France. *Regul. Rivers: Res. and Manage.* 1: 301-316.

DeFerrari, C. M., and R. J. Naiman. 1994. A multi-scale assessment of the occurrence of exotic plants on the Olympic Peninsula, Washington. *J. Veg. Sci.* 5: 247-258.

DeGraaf, R. M., M. Yamasaki, W. B. Leak, and J. W. Lanier. 1992. New England wildlife: management of forested habitats. Gen. Tech. Rep. NE-144. U.S. Department of Agriculture, Forest Service, Northeastern Forest Experiment Station. Radnor, PA. 271 p.

DeGraaf, R. M., D. P. Snyder, and B. J. Hill. 1991. Small mammal habitat associations in poletimber and sawtimber stands of four forest cover types. *For. Ecol. Manage.* 46:227-242.

DeGraaf, R. M. 1995. Nest predation rates in managed and reserved extensive northern hardwood forests. *For. Ecol. Manage.* 79:227-234.

DeGraaf, R. M., and A. L. Shigo. 1985. Managing cavity trees for wildlife in the northeast. Gen. Tech. Rep. NE-101. U.S. Department of Agriculture, Forest Service, Northeastern Forest Experiment Station, Broomall, PA. 21 p.

DeGraaf, R. M., and D. D. Rudis. 1986. New England wildlife: habitat, natural history, and distribution. Gen. Tech. Rep. NE-108. U.S. Department of Agriculture, Forest Service, Northeastern Forest Experiment Station, Broomall, PA. 491 p.

DeLorme. 1992. Wisconsin atlas and gazetteer. 3rd ed. Freeport, MA: DeLorme Mapping. p 27.

deMaynadier, P. G., and M. L. Hunter, Jr. 1995. The relationship between forest management and amphibian ecology: a review of the North American literature. *Environ. Rev.* 3:230-261.

DeWalle, D. R., P. J. Edwards, B. R. Swistock, R. Araven, and R. J. Drimmie. 1997. Seasonal isotope hydrology of three Appalachian forest catchments. *Hydrol. Process.* 11:1895-1906.

Dick, R. C. 1989. Cartographic modeling of riparian buffers in New Brunswick. Thesis, University of New Brunswick, Fredericton, New Brunswick, Canada. 75 p.

Dickson, J. G., ed. 1992. *The wild turkey: biology and management.* Harrisburg, PA: Stackpole Books. 463 p.

Diefenbach, D. R., S. J. Lovett, R. B. Owens, Jr. 1988. Beavers and wetlands. In *A forester's guide to managing wildlife habitats in Main*, ed. C. A. Elliott, 38-40. Orono, ME: University of Maine Cooperative Extension and Maine Chapter of The Wildlife Society.

Dillon, P. J., and F. H. Rigler. 1975. A simple method for predicting the capacity of a lake for development based on lake trophic status. *J. Fish. Res. Board Can.* 32:1519-1531.

Dillon, P. J., and L. A. Molot. 1997. Dissolved organic and inorganic carbon mass balances in central Ontario lakes. *Biogeochemistry* 36:29-42.

Dillon, P. J., W. A. Scheider, R. A. Reid, and D. S. Jeffries. 1994. Lakeshore capacity study: part 1 - test of effects of shoreline development on the trophic status of lakes. *Lake Reserv. Manage.* 8:121-129.

Dissmeyer, G. E. 1994. Evaluating the effectiveness of forestry best management practices in meeting water quality goals or standards. Southern Region, U.S. Forest Service, Internal Report, Atlanta, GA. 179 p.

Dissmeyer, G. E. and B. Foster. 1987. Some economic benefits of protecting water quality. In *Managing southern forests for wildlife and fish: a proceedings.* Gen. Tech. Rep. SO-65, U.S. Department of Agriculture, Southern Forest Experiment Station, New Orleans, LA: 6-11.

Dodd, C. K. 1997. Imperiled amphibians: a historical perspective. In *Aquatic fauna in peril: the southern perspective*, eds. G. W. Benz, and D. E. Collins. Spec. Publ. 1. Decatur, GA: Southern Aquatic Research Institute, Lenz Designs and Communications, 165-200.

Dodd, C. K., Jr., and L. V. LeClaire. 1995. Biogeography and status of the striped newt (*Notophthalmus perstriatus*). *Herpetol. Nat. Hist.* 3:37-46.

Dodd, C. K., Jr. 1996. Use of terrestrial habitats by amphibians in the sandhill uplands of north-central Florida. *Alytes* 14:42-52.

Dodd, C. K., Jr, and B. S. Cade. 1998. Movement patterns and the conservation of amphibians breeding in small, temporary wetlands. *Conserv. Biol.* 12:331-339.

Dodson, S. I. 1970. Complementary feeding niches sustained by size-selective predation. *Limnol. Oceanogr.* 15:131-137.

Dodson, S. I., and V. E. Dodson. 1971. The diet of *Ambystoma tigrinum* larvae from western Colorado. *Copeia* 1971:614-624.

Dolloff, C. A., P. A. Flebbe, and M. D. Owen. 1994. Fish habitat and fish populations in a southern Appalachian watershed before and after Hurricane Hugo. *Trans. Am. Fish. Soc.* 123: 668-678.

Dolloff, C. A. 1986. Effects of stream cleaning on juvenile coho salmon and Dolly Varden in southeast Alaska. *Trans. Am. Fish. Soc.* 115:743-755.

Dolloff, C. A. 1994. Large woody debris: the common denominator for integrated environmental management of forest streams. In Implementing integrated environmental management, eds. J. Cairns, T. V. Crawford, and H. Salwasser, 93-108. Blacksburg, VA: Virginia Tech Press.

Douglas, M. E., and B. L. Monroe, Jr. 1981. A comparative study of topographical orientation in *Ambystoma* (Amphibia: Caudata). *Copeia* 1981:460-463.

Downs, F. L. 1989. *Ambystoma jeffersonianum* (Green). In *Salamanders of Ohio*, eds. R.A. Pfingston, and Floyd L. Downs, 88-101. Bulletin of the Ohio Biological Society 7. Columbus: Ohio State University.

Doyle, A. T. 1990. Use of riparian and upland habitats by small mammals. *J. Mammal.* 71:14-23.

Dresen, M. D., and R. M. Kozak. 1995. *Law of the land: A citizen's guide: Influencing local land use decisions that affect water quality.* University of Wisconsin-Extension and WI Department of Natural Resources Lake Management Program.

Dunaway, D., S. R. Swanson, J. Wendel, and W. Clary. 1994. The effect of herbaceous plant communities and soil textures on particle erosion of alluvial streambanks. *Geomorphology* 9: 47-56.

Dunn, C. P. and F. Stearns. 1987. Comparison of vegetation and soils in floodplain and basin forested wetlands of southeastern Wisconsin. *Am. Midl. Nat.* 118:375-385.

Dunne, T. 1983. Relation of field studies and modeling in the prediction of runoff. *J. Hydrol.* 65:25-48.

Dunne, T., and L. B. Leopold. 1978. *Water in Environmental Planning.* San Francisco: W. H. Freeman Co. 818 p.

Dunne, T., and R. D. Black. 1970. Partial area contributions to storm runoff in a small New England watershed. *Water Resour. Res.* 6(5):1296-1297.

Dunson, W. A., and R. R. Martin. 1973. Survival of brook trout in a bog-derived acidity gradient. *Ecology* 54(6):1370-1376.

Dunson, W. A., R. L. Wyman, and E. S. Corbett. 1992. A symposium on amphibian declines and habitat acidification. *J. Herpetol.* 26:349-352.

Dury, G. H. 1969. Hydraulic geometry. In Water, earth and man, ed. R. J. Chorley. London, Methuen: 319-330.

Dwyer, J. F., H. W. Schroeder, J. L. Louviere, and D. H. Anderson. 1989. Urbanites willingness to pay for trees and forests in recreation areas. *J. Arboric.* 15:247-252.

E

Eagle, T. C., and J. S. Whitman. 1987. Mink. In *Wild furbearer management and conservation in North America.* eds. M. Novak, J. A. Baker, M. E. Obbard, B. Malloch, 614-624. Toronto, Ontario: Ontario Ministry of Natural Resources and Ontario Trappers Association.

Edde, J. 1985. Managing the riparian area of two stands for fish and old-growth dependent wildlife. U.S. Delpartment of Agriculture, Forest Service *Unpubl. Proj. Rep.*

Edwards, P. J., and F. Woods. 1994. Centroid lag time changes resulting from harvesting, herbiciding, and stand conversion. Proceedings American Water Resources Association Annual Summer Symposium. Jackson Hole, WY. June 26-29, 1994. p 727-734.

Ehrman, T. P., and G. A. Lamberti. 1992. Hydraulic and particulate matter retention in a 3rd-order Indiana stream. *J. North. Am. Bentholog. Soc.* 11:341-349.

Ellefson, P. V., and P. D. Miles. 1985. Protecting water quality in the mid-west: impact on timber harvesting costs. *North. J. Appl. For.* 2:57-61.

Elliott, C. A., ed. 1988. *A forester's guide to managing wildlife habitats in Maine.* Orono, ME: University of Maine Cooperative Extension and Maine Chapter of The Wildlife Society.

Elliott, S. T. 1986. Reduction of a Dolly Varden population and macrobenthos after removal of logging debris. *Trans. Am. Fish. Soc.* 115: 392-400.

Elowe, K. D. 1984. Home range, movements, and habitat preferences of black bears (*Ursus americanus*) in western Massachusetts. Master's thesis. University of Massachusetts Amherst, MA. 112 p.

Elwood, J. W., J. D. Newbold, A. F. Trimble, and R. W. Stark. 1981. The limiting role of phosphorus in a woodland stream ecosystem: Effects of P enrichment on leaf decomposition and primary producers. *Ecology* 62:146-158.

Emerson, P. M. 1996. Cultural values in riparian areas. In *At the water's edge: the science of riparian forestry conference proceedings,* ed. S. B. Laursen. Minnesota Extension Service BU-6637-S:36-39. 160 p.

Engstrom, D. R. 1987. Influence of vegetation and hydrology on the humus budgets of Labrador lakes. *Can. J. Fish. Aquat. Sci.* 44:1306-1314.

Erdmann, G. E., T. R. Crow, R. M. Peterson, and C. D. Wilson. 1987. Managing black ash in the Lake States. Gen. Tech. Rep. NC-115. U.S. Department of Agriculture, Forest Service, North Central Forest Experiment Station, St. Paul, MN. 10 p.

Ericsson, K. A., and D. J. Schimpf. 1986. Woody riparian vegetation of a Lake Superior tributary. *Can. J. Bot.* 64:769-773.

Erman, D. C., J. D. Newbold, and K. R. Roby. 1977. Evaluation of streamside buffer strips for protecting aquatic organisms. Contrib. No. 165. California Water Resources Center, University.

Ernst, C. H., and R. W. Barbour. 1989. *Snakes of eastern North America*. Fairfax, VA: George Mason University Press. 282 p.

Ernst, C. H., J. W. Lovich, and R. W. Barbour. 1994. *Turtles of the United States and Canada.* Washington, DC: Smithsonian Institution Press. 578 p.

Errington, P. L. 1963. *Muskrat populations.* Ames, IA: Iowa State University Press. 655 p.

Erwin, R. M. 1996. The status of forested wetlands and waterbird conservation in North and Central America. In *Conservation of faunal diversity in forested landscapes.* eds. R. M. DeGraaf; R. I. Miller, London: Chapman and Hall. 633 p.

Evans, K. E., and R. N. Conner. 1979. Snag management. In *Management of northcentral and northeastern forests for nongame birds*, comp. R. M. DeGraaf, and K. E. Evans, 214-245. Gen. Tech. Rep. NC-51. U.S. Department of Agriculture, Forest Service, North Central Forest Experiment Station, St. Paul, MN. 268 p.

Ewert, M. A. 1969. Seasonal movements of the toads *Bufo americanus* and *B. cognatus* in northwestern Minnesota. Ph.D. Dissertation, University of Minnesota. 193 p.

F

Fausch, K. D. and T. G. Northcote. 1992. Large woody debris and salmonid habitat in a small coastal British Columbia stream. *Can. J. Fish. Aquat. Sci.* 49:682-693.

Federer, C. A., L. D. Flynn, C. W. Martin, J. W. Hornbeck, and R. S. Pierce. 1990. Thirty years of hydrometeorological data at the Hubbard Brook Experimental Forest, New Hampshire. Gen. Tech. Rep. NE-GTR-141. U.S. Department of Agricultre, Forest Service, Northeastern Forest Experiment Station, Radnor, PA. 44 p.

Federer, C. A. 1995. BROOK90: a simulation model for evaporation, soil water, and streamflow, Version 3.1. Computer freeware and documentation. U.S. Department of Agriculture, Forest Service, Northeastern Forest Experiment Station, Durham, NH.

Federer, C. A., J. W. Hornbeck, L. M. Tritton, C. W. Martin, R. S. Pierce, and C. T. Smith. 1989. Long-term depletion of calcium and other nutrients in eastern U.S. forests. *Environ. Manage.* 13:593-601.

Fee, E. J., R. E. Hecky, S .E. M. Kaisan, and D. R. Cruikshank. 1996. Effects of lake size, water clarity, and climactic variability on mixing depths in Canadian Shield lakes. *Limnol. Oceanogr.* 41:912-920.

Fellers, G. M., and K. L. Freel. 1995. A standardized protocol for surveying aquatic amphibians. Tech. Rep. NPS/WRUC/NRTR-95-01. National Park Service, Cooperative Park Studies Unit, Davis, CA. 117 p.

FEMAT. 1993. Aquatic ecosystem assessment: riparian ecosystem components. *In Forest ecosystem management: an ecological, economic, and social assessment.* V-25--V-29 Report of the Forest Ecosystem Management Assessment Team. Washington, DC: U.S. Department of Agriculture, Forest Service.

FEMAT. 1993. *Forest ecosystem management: an ecological, economic, and social assessment.* Chapters I-IX. Washington, DC: U.S. Department of Agriculture, U.S. Department of Commerce, U.S. Department of the Interior, Environmental Protection Agency.

Fenn, M. E., M. A. Poth, J. D. Aber, J. S. Baron, B. T. Bormann, D. W. Johnson, A. D. Lemly, S. G. McNulty, D. F. Ryan, and R. Stottlemyer. 1998. Nitrogen excess in north American ecosystems: predisposing factors, ecosystem responses, and management strategies. *Ecol. Appl.* 8(3):706-733.

Ferguson, Rob. 1987. Hydraulic and sedimentary controls of channel pattern. In *River Channels: Environment and Process*, ed. K. S. Richards. 129-158. New York: Basil Blackwell. 129-158.

Fetherston, K. L., R. J. Naiman, and R. E. Bilby. 1995. Large woody debris, physical process, and riparian forest development in montane river networks of the Pacific Northwest. *Geomorphology* 13: 133-144.

Fiebig, D. M., M. A. Lock, and C. Neal. 1990. Soil water in the riparian zone as a source of carbon for a headwater stream. *J. Hydrol.* 116: 217-237.

Fischer, R. A., and N. R. Holler. 1991. Habitat use and relative abundance of gray squirrels in southern Alabama. *J. Wildl. Manage.* 55: 52-59.

Fisher, S. G., and G. E. Likens. 1973. Energy flow in Bear Brook, New Hampshire: an integrative approach to stream ecosystem metabolism. *Ecol. Monogr.* 43: 421-439.

Fisher, R. A. 1935. *The design of experiments.* London: Oliver & Boyd. 248 p.

Fitch, H. S. 1958. *Home ranges, territories, and seasonal movements of vertebrates of the Natural History Reservation.* University of Kansas Publications, Museum of Natural History. 11:63-326.

Fitch, H. S. 1965. An ecological study of the garter snake, *Thamnophis sirtalis.* University of Kansas Publications, Museum of Natural History. 15:493-564.

Flebbe, P. A., and C. A. Dolloff. 1995. Trout use of woody debris and habitat in Appalachian wilderness streams of North Carolina. *North Am. Fish. Manage.* 15:579-591.

Florida. 1993. Silviculture Best Management Practices. Florida Department of Agriculture and Consumer Services. 98 p.

Foley, D. H., III. 1994. Short-term response of herpetofauna to timber harvesting in conjunction with streamside-management zones in seasonally-flooded bottomland-hardwood forests of southeast Texas. Master's Thesis, Texas A&M University, College Station, TX. xii+93 p.

Fore, L. S., J. R. Karr, and L. L. Conquest. 1994. Statistical properties of an index ob biotic integrity used to evaluate water resources. *Can. J. Fish. Aquat. Sci.* 51:1077-1087.

Forested Wetlands Task Force. 1993. Best management practices for silvicultural activities in Pennsylvania's forest wetlands. 54 p.

France, R. L. 1997a. Land-water linkages: influences of riparian deforestation on lake thermocline depth and possible consequences for cold stenotherms. *Can. J. Fish. Aquat. Sci.* 54:1299-1305.

France, R. L. 1997b. Macroinvertebrate colonization of woody debris in Canadian Shield lakes following riparian clearcutting. *Conserv. Biol.* 11:513-521.

Franklin, J. F. 1989. Towards a new forestry. *Am. For.* Nov./Dec. 37-44.

Franzmann, A. W., and C. C. Schwartz, comp. 1997. *Ecology and management of the North American moose: A Wildlife Management Institute book.* Washington, DC: Smithsonian Institution Press. 733 p.

Frayer, W. E., L. S. Davis, and P. G. Risser. 1978. Uses of land classification. *J. For.* 76:647-649.

Freda, J., W. J. Sadinski, and W. A. Dunson. 1991. Long term monitoring of amphibian populations with respect to the effects of acidic deposition. *Water, Air, and Soil Pollut.* 55:445-462.

Frelich, L. E. 1995. Old forest in the lake states today and before European settlement. *Nat. Areas J.* 15:157-167.

Frey, D. G. 1990. What is a lake? *Verh. Internat. Verein. Limnol.* 24:1-5.

Frissel, C. A., and R. K. Nawa. 1992. Incidence and causes of physical failure of artificial structures in streams of western Oregon and Washington. *North Am. J. Fish. Manage.* 12:182-197.

Frissell, C. A., W. L. Liss, C. E. Warren, and M. C. Hurley. 1986. A hierarchical framework for stream habitat classification, viewing streams in a watershed context. *Environ. Manage.* 10:199-214.

Frye, R. J., II, and J. A. Quinn. 1979. Forest development in relation to topography and soils on a floodplain of the Raritan River, New Jersey. *Bull. Torrey Bot. Club* 106:334-345.

G

Gammon, J. R. 1970. The effects of inorganic sediment on stream biota. U.S. Environmental Protection Agency. Water Pollution Control Research Series. December 1970. 141 p.

Garcia, M. W. 1989. Forest Service experience with interdisciplinary teams developing integrated resource management plans. *Environ. Manage.* 13:583-592.

Gardner, J. E., J. D. Garner, and J. E. Hofmann. 1991. Summer roost selection and roosting behavior of *Myotis sodalis* (Indiana bat) in Illinois. Champaign, IL: Illinois Natural History Survey. Unpublished report.

Garn, H. S., and H. A. Parrott. 1977. Recommended methods for classifying lake condition, determining lake sensitivity and predicting lake impacts. Hydro. Pap. 2. U. S. Department of Agriculture, Forest Service, Eastern Region. 49 p.

Gates, J. E., and L. W. Gysel. 1978. Avian nest dispersion and fledgling success in field forest ecotones. *Ecology* 58:871-883.

Gathman, J., M. J. Phillips, N. Troelstrop, Jr., and J. Perry. 1992. Nonindustrial private forest owners and best management practices in Minnesota. A report to the Minnesota Department of Natural Resources, Division of Forestry, St. Paul, MN. 20 p.

Geier, A. R., and L. B. Best. 1980. Habitat selection by small mammals of riparian communities: evaluating effects of habitat alterations. *J. Wildl. Manage.* 44:16-24.

Georgia Forestry Commission. 1995. Recommended forestry best management practices for forestry in Georgia. 24 p.

Gergel, S. E. 1996. Scale-dependent landscape effects on north temperate lakes and rivers. Master's thesis, The University of Wisconsin-Madison. 46 p.

Gibbons, J. W. 1988. The management of amphibians, reptiles, and small mammals in North America: the need for an environmental attitude adjustment. In *Management of amphibians, reptiles, and small mammals in North America*, eds. R. C. Szaro, K. E. Severson, and D. R. Patton, 4-10. Gen. Tech. Rep. RM-166. 4-10. U.S. Department of Agriculture, Forest Service, Rocky Mountain Forest and Range Experiment Station, Denver, CO.

Gilmore, J. S. L. 1951. The development of taxonomic theory since 1851. *Nature.* 168:400-402.

References 357

Gingras, J. F. 1994. A comparison of full-tree versus cut-to-length systems in the Manitoba model forest. Spec. Rep. SR-92. Forest Engineering Research Institute of Canada (FERIC), Pointe-Claire, Quebec. 16 p.

Gobster, P. H., and L. M. Westphal. eds. 1998a. *People and the river.* Milwaukee, WI: National Park Service Rivers, Trails, and Conservation Assistance Program. 192 p.

Gobster, P. H. and L. M. Westphal. 1998b. People and the river: summary and conclusions. In *People and the river, eds.* P. H. Gobster, and L. M. Westphal. 183-189. Milwaukee, WI: National Park Service Rivers, Trails, and Conservation Assistance Program.

Goebel, P. C., B. J. Palik, L. K. Kirkman, and L. West. 1996. Geomorphic influences on riparian forest composition and structure in a karst landscape of southwestern Georgia. In *Proceedings of the Southern forested wetlands ecology and management conference,* ed. K. M. Flynn. 110-114. Clemson, SC: Consortium for Research on Southern Forested Wetlands, Clemson University.

Golden, M. S., C. L. Tuttle, J. S. Kush, and J. M. Bradley. 1984. Forestry activities and water quality in Alabama: effects, recommended practices, and an erosion-classified system. Bull. 555. Auburn University Agricultural Experiment Station, Auburn, AL.

Golladay, S. W., J. R. Webster, and E. F. Benfield. 1989. Changes in stream benthic organic matter following watershed disturbances. *Holarctic Ecol.* 12:96-105.

Golladay, S. W., J. R. Webster, E. F. Benfield, and W. T. Swank. 1992. Changes in stream stability following forest clearing as indicated by storm nutrient budgets. *Arch. Hydrobiol.* 1:1-33.

Gomez, D. M., and R. G. Anthony. 1996. Amphibian and reptile abundance in riparian and upslope areas of five forest types in western Oregon. *Northwest Sci.* 70:109-119.

Goodman, L. A. 1961. Snowball sampling. *Ann. Math. Stat.* 32:148-170.

Gorham, E., W. E. Dean, and J. E. Sanger. 1983. The chemical composition of lakes in the north-central United States. *Limnol. Oceanogr.* 28:287-301.

Greeley, W. B. 1925. The relation of topography to timber supply. *Econ. Geogr.* 1(1):1-14.

Green, N. B., and T. K. Pauley. 1987. *Amphibians and reptiles in West Virginia.* Pittsburgh, PA: University of Pittsburgh Press. 241 p.

Green, D. M. 1997. Amphibians in decline, Canadian studies of a global problem. Saint Louis, MO: Society for the Study of Amphibians and Reptiles. *Herpetol. Conserv.* 1:338 p.

Green, J. C. 1995. *Birds and forests: a management and conservation guide.* St. Paul, MN: Minnesota Department of Natural Resources. 182 p.

Gregory, K. J. 1992. Vegetation and river channel process interactions. In *River conservation and management,* eds. P. J. Boon, P. Calow, and G. E. Petts, 255-269. Chichester. John Wiley & Sons.

Gregory, S. V., F. J. Swanson, W. A. McKee, and K. W. Cummins. 1991. An ecosystem perspective of riparian zones. *Bioscience* 41:540-551.

Gregory, S. V., G. A. Lamberti, D. C. Erman, K. V. Koski, M. L. Murphy, and J. R. Sedell. 1987. Influence of forest practices on aquatic production. In *Streamside management: forestry and fishery interactions,* eds. E. O. Salo, and T. W. Cundy. 233-255. Seattle, WA: University of Washington, Institute of Forest Resources.

Gregory, S. V. 1997. Riparian management in the 21st century. In *Creating a forestry for the 21st century: the science of ecosystem management*, eds. K. A. Kohm, and J. F. Franklin, 69-86. Washington, DC: Island Press.

Grimm, N. B., and S. G. Fisher. 1986. Nitrogen limitation in a Sonoran desert stream. *J. North Am. Bentholog. Soc.* 5:2-15.

Gosselink, J. G., M. M. Brinson, L. C. Lee, and G. T. Auble. 1990. Human activities and ecological processes in bottomland hardwood ecosystems: the report of the ecosystem workgroup. In *Ecological processes and cumulative impacts: illustrated by bottomland hardwood wetland ecosystems,* eds. J. G. Gosselink, L. C. Lee, and T. A. Muir, 549-598. Chelsea, MI Lewis Publishers.

Grost, R. T., W. A. Hubert, and T. A. Wesche. 1991. Field comparison of three devices used to sample substrate in small streams. *North Am. J. Fish. Manage.* 11:347-351.

Grubaugh, J. W., and R. V. Anderson. 1989. Upper Mississippi River: seasonal and floodplain forest influences on organic matter transport. *Hydrobiologia* 174:235-244.

Gurnell, A., and G. Petts, eds. 1995. *Changing river channels.* Chichester: John Wiley & Sons. 442 p.

Gurtz, M. E., and J. B. Wallace. 1986. Substratum-production relationships in net-spinning caddisflies (Trichoptera) in disturbed and undisturbed hardwood catchments. *J. North Am. Bentholog. Soc.* 5:230-236.

Guyette, R. P., and W. G. Cole. 1997. White pine and hemlock coarse woody debris in the littoral zones of several Canadian Shield lakes. In *Abstract for the 59th Midwest Fish and Wildlife Conference*, Milwaukee, WI.

Gysel, L. W. 1961. An ecological study of tree cavities and ground burrows in forest stands. *J. Wildl. Manage.* 25:12-20.

H

Hack, John T., and John C. Goodlett. 1960. Geomorphology and forest ecology of a mountain region in the central Appalachians. Geological Survey Prof. Pap. 347. U.S. Gov. Printing Office, Washington, DC. 66 p.

Harris, R. R. 1985. Relationships between fluvial geomorphology and vegetation on Cottonwood Creek, Tehama and Shasta Counties, California. Ph.D. dissertation. Department of Forestry, University of California, Berkeley. 329 p.

Hagan, J. M., P. S. McKinley, A. L. Meehan, S. L. Grove. 1997. Diversity and abundance of landbirds in a northeastern industrial forest. *J. Wildl. Manage.* 61:718-735.

Haines-Young, R., and J. R. Petch. 1986. *Physical geography: Its nature and method.* London: Harper & Row. 230 p.

Haines-Young, R., and J. R. Petch. 1980. The challenge of critical rationalism for methodology in physical geography. *Prog. in Phys. Geog.* 4:63-7.

Hairston, N. G., Sr. 1987. *Community ecology and salamander guilds.* Cambridge: Cambridge University Press. 230 p.

Hall, E. R. 1951. *American weasels.* University of Kansas Publication of Museum of Natural History. 4:1-466.

Hall, R. O., G. E. Likens, S. B. France, and G. R. Hendrey. 1980. Experimental acidification of a stream in the Hubbard Brook Experimental Forest, New Hampshire. *Ecology* 61:976-989.

Hamel, P. B. 1992a. Cerulean warbler *Dendroica cerulea*. In *Migratory nongame birds of management concern in the Northeast*, eds. K. J. Schneider, and D. M. Pence, 385-400. Newton Corner, MA: U. S. Fish and Wildlife Service. 400 p.

Hamel, P. B. 1992b. *Land managers guide to the birds of the south*. Chapel Hill, NC. The Nature Conservancy. 367 p. and Appendices.

Hanes, W. T., and R. A. LaFayette. 1987. Analyzing watershed condition: symptoms vs. causes. In *Proceedings: conference XX, International Erosion Control Association*, Vancouver, British Columbia, Canada. 11 p.

Hankin, D. G., and G. H. Reeves. 1988. Estimating total fish abundance and total habitat area in small streams based on visual estimation methods. *Can J. Fish. Aquat. Sci.* 45:834-844.

Harding, J. H. 1997. *Amphibians and reptiles of the Great Lakes Region*. Ann Arbor, MI: University of Michigan Press. 378 p.

Harmon, M. E., J. F. Franklin, F. J. Swanson, P. Sollins, S. V. Gregory, J. D. Lattin, N. H. Anderson, S. P. Cline, N. G. Aumen, J. R. Sedell, G. W. Lienkaemper, K. Cromack, Jr., and K.W. Cummins. 1986. Ecology of coarse woody debris in temperate ecosystems. In *Advances in ecological research*, eds. A. Macfadyen, and E. D. Ford. New York: Academic Press, Harcourt Brace Jovanovich. *Adv. in Ecol. Res.* 15:133-302.

Harr, R. D., and F. M. McCorison. 1979. Initial effects of clearcut logging on size and timing of peak flows in a small watershed in western Oregon. *Water Resour. Res.* 15(1):90-94.

Harr, R. D., W. C. Harper, and J. T. Krygier. 1975. Changes in storm hydrographs after road building and clear-cutting in the Oregon Coast Range. *Water Resour. Res.* 15(1):436-444.

Harrelson, C. C., C. L. Rawlins, and J. P. Potyondy. 1994. Stream channel reference sites: an illustrated guide to field technique. Gen. Tech. Rep. RM-245. U.S. Department of Agriculture, Forest Service, Rocky Mountain Forest and Range Experiment Station, Fort Collins, CO. 61 p.

Harris, D. D. 1973. Hydrologic changes after clear-cut logging in a small Oregon coastal wateshed. *J. of Res. USGS* 1(4):487-491.

Harris, R. R. 1987. Occurrence of vegetation on geomorphic surfaces in the active floodplain of a California alluvial stream. *Am. Midl. Nat.* 118: 393-405.

Harris, R. R. 1988. Associations between stream valley geomorphology and riparian vegetation as a basis for landscape analysis in the eastern Sierra Nevada, California, USA. *Environ. Manage.* 12:219-228.

Harrison, D. J., J. A. Bissonette, and J. A. Sherburne. 1989. Spatial relationships between coyotes and red foxes in eastern Maine. *J. Wildl. Manage.* 53:181-185.

Hart, D. D., and C. T. Robinson. 1990. Resource limitation in a stream community: phosphorus enrichment effects on periphyton and grazers. *Ecology* 71(4):1494-1502.

Harvey, A. M. 1975. Some aspects of the relation between channel characteristics and riffle spacing in meandering streams. *Am. J. Sci.* 275:470-478.

Harvey, A. M. 1969. Channel capacity and the adjustment of streams to hydrologic regime. *J. Hydrol.* 8:82-98.

Hawkins, C. P., J. L. Kerchner, P. A. Bisson, M. D. Bryant, L. M. Decker, S. V. Gregory, D. A. McCullugh, C. K. Overton, G. H. Reeves, R. J. Steedman, and M. K. Young. 1993. A hierarchical approach to classifying stream habitat features. *Fisheries* 18(6):3-11.

Hawkinson, C. F., and E. S. Verry. 1975. Specific conductance identifies perched and groundwater lakes. Res. Pap. NC-120. U.S. Department of Agriculture, Forest Service, North Central Forest Experiment Station, St. Paul, MN. 5 p.

Healy, W. M., and R. T. Brooks. 1988. Small mammal abundance in northern hardwood stands in West Virginia. *J. Wildl. Manage.* 52:491-496.

Healy, W. R. 1975. Terrestrial activity and home range in efts of *Notophthalmus viridescens* (Rafensque). *Am. Midl. Nat.* 93:131-138.

Hedgpeth, J. W. 1993. Foreign invaders. *Science* 261:34-35.

Hedin, L. O., M. S. Mayer, and G. E. Likens. 1988. The effect of deforestation on organic debris dams. *Verhandlung der Internationalen Vereinigung für Theoretische und Angewandte Limnologie* 23:1135-1141.

Heede, B. H. 1980. Stream dynamics: An overview for land managers. Gen. Tech. Rep. RM-72. U.S. Department of Agriculture, Forest Service, Rocky Mountain Forest and Range Experiment Station, Fort Collins, CO.

Heede, B. H. 1972. Influences of a forest on the hydraulic geometry of two mountain streams. *Water Resour. Bull.* 8:523-530.

Heiskary, S. A. 1989. Lake assessment program: a cooperative lake study program. *Lake and Reserv. Manage.* 5:85-94.

Heiskary, S. A., J. Lindbloom, and C. B. Wilson. 1994. Detecting water quality trends with citizen volunteer data. *Lake and Reserv. Manage.* 9:4-9.

Helvey, J. D., and Kochenderfer, J. N. 1988. Culvert sizes needed for small drainage areas in the central Appalachians. *North. J. Appl. For.* 5:123-127.

Henson, M. 1996. *Best management practices implementation and effectiveness survey on timber operations in North Carolina - 1996.* Raleigh, NC: North Carolina Division of Forest Resources, Department of Environment, Health and Natural Resources, 25 p.

Henson, M. 1995. *Forest practice guidelines and best management practices implementation and effectiveness survey on timber operations in North Carolina.* Raleigh, NC: North Carolina Division of Forest Resources.

Herricks, E. E., and J. Cairns, Jr. 1974. *The recovery of stream macrobenthos from low pH stress.* Blacksburg, VA: Virginia Polytechnic Institute and State University, Department of Biology, 14 p.

Hewlett, J. D., and J. D. Helvey. 1970. Effects of forest clearfalling on the storm hydrograph. *Water Resour. Res.* 6(3):768-782.

Hewlett, J. D. 1961. Soil moisture as a source of base flow from steep mountain watersheds. Paper 132. U.S. Department of Agriculture, Forest Service, Southeastern Forest Experiment Station, Asheville, NC. 11 p.

Hewlett, J. D., and J. E. Douglass. 1968. Blending forest uses. Res. Pap. SE-37. U.S. Department of Agriculture, Forest Service, Southeastern Forest Experiment Station, Asheville, NC. 15 p.

Hewlett, J. D., and R. A. Hibbert. 1967. Factors affecting the response of small watersheds to precipitation in humid areas. In Proceedings, *Forest hydrology*, eds. W. Sopper, and H. W. Lull, 275-290. Oxford: Pergamon Press.

Heyer, W. R., M. A. Donnelly, R. W. McDiarmid, L. C. Hayek, and M. S. Foster. 1994. *Measuring and monitoring biological diversity, standard methods for amphibians.* Washington, DC: Smithsonian Institution Press. 364 p.

Hicks, B. J., J. D. Hall, P. A. Bisson, and J. R. Sedell. 1991. Response pf salmonids to habitat changes. In *Influences of forest and rangeland management on salmonid fishes and their habitats*, ed. W. R. Meehan. Spec. Publ. 19. Bethesda MD: American Fisheries Society. Chapter 14.

Higgins, D. A., and S. Reinecke. 1995. Water quality, fisheries and mussel interpretations for tentative valley types on national forests in Wisconsin. Unpublished tables. U.S. Department of Agriculture, Forest Service, Chequamegon and Nicolet National Forests, Park Falls, WI. 4 p.

Higgins, J., M. Lammert, M. Bryer, M. DePhilip, and D. Grossman. 1998. *Freshwater conservation in the Great Lakes basin: development and application of an aquatic community classification framework*. Chicago, IL: The Nature Conservancy. 276 p.

Hilderbrand, R. H., A. D. Lemly, C. A. Dolloff, and K. L. Harpster. 1997. Effects of large woody debris placement on stream channels and benthic invertebrates. *Can. J. Fish. Aquat. Sci.* 54: 931–939.

Hill, E. P. 1982. Beaver (*Castor canadensis*). In *Wild mammals of North America -- biology, management, and economics*, eds. J. A. Chapman, and G. A. Feldhamer, 256-281. Baltimore, MD: Johns Hopkins University Press.

Hillbricht-Ilkowska, A. 1995. Managing ecotones for nutrients and water. *Ecol. Internat.* 22:73-93.

Hocutt, C. H. and E. O. Wiley, eds. 1986. *The zoogeography of North American freshwater fishes*. New York: John Wiley and Sons. 1015 p.

Hodgdon, K. W., and J. H. Hunt. 1953. Beaver management in Maine. Augusta, ME; Maine Department of Inland Fisheries and Game. *Game Div. Bull.* 3:1-102.

Hodges, J. D. 1998. Minor alluvial floodplains. In *Southern forested wetlands*, eds. M. G. Messina, and W. H. Conner, 325-341. Boca Raton, FL: Lewis Publishers.

Holaday, S., J. Filbert, and N. Merryfield. 1995. *Wisconsin's forestry best management practices for water quality, the 1995 BMP monitoring report: baseline data*. Milwaukee, WI: Wisconsin Department of Natural Resources, Bureau of Forestry.

Holomuzki, J. P. 1997. Habitat-specific life-histories and foraging by stream-dwelling American toads. *Herpetologica* 53:445-453.

Hook, D. D., W. McKee, T. Williams, B. Baker, L. Lundquist, R. Martin, and J. Mills. 1991. *A survey of voluntary compliance of forestry Best Management Practices*. Columbia, SC: South Carolina Forestry Commission. 27 p.

Hooper, S. T. 1991. Distribution of songbirds in riparian forests of central Maine. Master's thesis, University of Maine, Orono, ME. 90 p.

Hopey, M. E., and J. W. Petranka. 1994. Restriction of wood frogs to fish-free habitats: how important is adult choice. *Copeia* 1994:1023-1025.

Hornbeck, J. W. 1973. Storm flow from hardwood-forested and cleared watersheds in New Hampshire. *Water Resour. Res.* 9:346-354.

Hornbeck, J. W., M. B. Adams, E. S. Corbett, E. S. Verry, and J. A. Lynch. 1993. Long-term impacts of forest treatments on water yield: a summary for northeastern USA. *J. Hydrol.* 150:323-344.

Hornbeck, J. W., E. S. Verry, and C. A. Dolloff. 1994. Managing riparian areas in northeastern forests: a quest for desired future conditions. In *Proceedings 1994 Society of American Foresters Convention*, 299-304. Bethesda, MD: Society of American Foresters.

Hornbeck, J. W., C. W. Martin, R. S. Pierce, F. H. Borman, G. E. Likens, and J. S. Eaton. 1987. The northern hardwood forest ecosystem: ten years of recovery from clearcutting. NE-RP-596. U. S. Department of Agriculture, Forest Service, Northeastern Forest Experiment Station, Broomall, PA. 29 p.

Hornbeck, J. W., and W. T. Swank. 1992. Watershed ecosystem analysis as a basis for multiple-use management of eastern forests. *Ecol. Appl.* 2:238-247.

Hornbeck, J. W., and K. G. Reinhart. 1964. Water quality and soil erosion as affected by logging in steep terrain. *J. Soil Water Conserv.* 19(1):23-27.

Hornbeck, J. W., C. W. Martin, and C. Eager. 1997. A summary of water yield experiments at Hubbard Brook Experimental Forest, New Hampshire. *Can. J. For. Res.* 27:2043-2052.

Horton, R. E. 1933. The role of infiltration in the hydrologic cycle. *Trans. Am. Geophys. Union.* 14:446-460.

Host, G. E., and K. S. Pregitzer. 1992. Geomorphic influences on ground flora and overstory composition in upland forests of northwestern Lower Michigan. *Can. J. For. Res.* 22:1547-1555.

Host, G. E., K. S. Pregitzer, C. W. Ramm, D. P. Lusch, and D. T. Cleland. 1988. Variation in overstory biomass among glacial landforms and ecological land units in northwestern Lower Michigan. *Can. J. For. Res.* 18:659-668.

Houle, D., R. Carignan, M. Lachance, and J. Dupont. 1995. Dissolved organic carbon and sulfur in southwestern Quebec lakes: Relationships with catchment and lake properties. *Limnol. Oceanogr.* 40:710-717.

Howard, R. J., and J. S. Larson. 1985. A stream habitat classification system for beaver. *J. Wildl. Manage.* 49:19-25.

Hunt, J. 1997. Assessing habitat availability in the littoral zone: techniques and applications. In *Abstract for the 59th Midwest Fish and Wildlife Conference*, Milwaukee, WI.

Hunter, W. C., Pashley, D. N., Hamel, P. B., and J. G. Dickson. In prep. Bird Communities and conservation issues in southern forests. In *Wildlife of Southern Forests: Habitat and Management*, eds. J. G. Dickson.

Hunter, C. J. 1991. *Better trout habitat, a guide to stream restoration and management.* Montana Land Reliance. Washington, DC: Island Press. 320 p.

Hunter, M. L., Jr. 1990. Shores. In *Wildlife, forests and forestry: principles of managing forests for biological diversity.* Chapter 9, 139-153. Englewood, NJ: Prentice Hall. 370 p.

Hunter, M. L., Jr. 1990. *Wildlife, forests, and forestry: principles of managing forests for biological diversity.* Englewood Cliffs, NJ: Prentice Hall. 370 p.

Hupp, C. R. 1982. Stream-grade variation and riparian forest ecology along Passage Creek, Virginia. *Bull. Torrey Bot. Club* 109: 488-499.

Hupp, C. R. 1988. Plant ecological aspects of flood geomorphology and paleoflood history. *Flood Geomorphology*, eds. Baker *et al.* Reston, Virginia: Wiley.

Hupp, C. R., and W. R. Osterkamp. 1985. Bottomland vegetation distribution along Passage Creek, Virginia, in relation to fluvial landforms. *Ecology* 66:670-681.

Hupp, C. R., and W. R. Osterkamp. 1996. Riparian vegetation and fluvial geomorphic processes. *Geomorphology* 14: 227-295.

Hutchinson, G. E. 1957. *A treatise on limnology: V1, Geography, physics, and chemistry.* New York: John Wiley and Sons. 1015 p.

Hutchinson, N. J., B. P. Neary, and P. J. Dillon. 1991. Validation and use of Ontario's trophic status model for establishing lake development guidelines. *Lake Reserv. Manage.* 7:13-23.

Hynes, H. B .N. 1970. *The ecology of running waters*. Toronto: University of Toronto Press. 555 p.
Hynes, H. B. N. 1975. The stream and its valley. Edgardo Baldi Memorial Lecture. *Verh. Internat. Verein. Limnol.* 19:1-15.

I

Ikeda S., and N. Izumi. 1990. Width and depth of self-formed straight gravel rivers with bank vegetation. *Water Resour. Res.* 26(10):2353-2364.
Imhof, J. G., J. Fitzgibbon, and W. K. Annable. 1996. A hierarchical evaluation system for characterizing watershed ecosystems for fish habitat. *Can. J. Fish. Aquat. Sci.* 53 (Suppl. 1):312-326.
Interagency Task Force. 1995. The ecosystem approach: healthy ecosystems and sustainable economies. Vol. I - Overview report of the interagency ecosystem management task force, June 1995.
Irwin, E. R., R. L. Noble, and J. R. Jackson. 1997. Distribution of age-0 largemouth bass in relation to shoreline landscape features. *North Am. J. Fish. Manage.* 17:882-893.

J

Jackson, D. R., and R. N. Walker. 1997. Reproduction in the Suwannee cooter, *Pseudemys concinna suwanniensis*. *Bull. Florida Mus. Nat. Hist.* 41:69-167.
Jakes, P. J. 1998. Why study values? In *Proceedings of the 1997 Northeastern recreation research symposium*, ed. Hans G. Vogelson, 147-150. Gen. Tech. Rep. NE-241. U.S. Department of Agriculture, Forest Service, Northeastern Forest Experiment Station, Radnor, PA. 147-150.
Jakes, P. J., T. Fish, D. Carr, and D. Blahna. 1998a. Functional communities: a tool for national forest planning. *J. For.* 96(3):33-36.
Jakes, P. J., T. Fish, D. Carr, and D. Blahna. 1998b. Practical social assessments for National Forest planning. Gen. Tech. Rep. NC-198. U.S. Department of Agriculture, Forest Service, North Central Forest Experiment Station, St. Paul, MN. 44 p.
Jakes, P. J., and J. Harms. 1995. Report on the socioeconomic roundtable convened by the Chequamegon and Nicolet National Forests. Gen. Tech. Rep. NC-177. U.S. Department of Agriculture, Forest Service, North Central Forest Experiment Station. St. Paul, MN. 62 p.
Jennings, M., M. Bozek, and G. Hatzenbeler. 1997. Riparian habitat and fish assemblage structure in lentic systems: limiting factors and effects of scale. In *Abstract for the 59th Midwest Fish and Wildlife Conference*, Milwaukee, WI.
Jennings, M., M. Bozek, G. Hatzenbeler, and D. Fago. 1996. *Fish and habitat*. Shoreline Protection Study, Wisconsin Department of Natural Resources PUBL-RS-921-96.
Johnson, B. L., Richardson W. B., and T. J. Naimo. 1995. Past, present, and future concepts in large river ecology. *BioScience* 45:134-141.
Johnson, F. L., and D. T. Bell. 1976. Plant biomass and net primary production along a flood-frequency gradient in the streamside forest. *Castanea* 41:156-165.
Johnson, W. N., and P. W. Brown. 1990. Avian use of a lakeshore buffer strip in an undisturbed lakeshore in Maine. *North. J. Appl. For.* 7:114-117.
Johnson, L. B., and S. H. Gage. 1997. Landscape approaches to the analysis of aquatic ecosystems. *Freshwater Biol.* 37: 113-132.

Johnson, L. B., C. Richards, G. E. Host, and J. Arthur. In Press. Influence of landscape factors on water chemistry in agricultural catchments. *Freshwater Biol.*

Johnson, M. G., J. H. Leach, C. K. Mins, and C. K. Oliver. 1977. Limnological characteristics of Ontario Lakes in relation to associations of walleye, northern pike, lake trout, and small mouth bass. *J. Fish. Board Canada.* 34:1592-1601.

Johnson, L. J., C. R. Richards, F. Kutka, and C. A. Hagley. 1994. Role of wetlands in mitigating upland management effects on stream water quality. Final Report to the U.S. Forest Service, Tech. Rep. No. NRRI-TR-94-32.

Johnson, R. R., L. T. Haight, and J. M. Simpson. 1977. Endangered species vs. endangered habitats: a concept. In Importance, management and preservation of riparian habitats: a Symposium. Gen. Tech. Rep. RM-43. U.S. Department of Agriculture, Forest Service, Rocky Mountain Forest and Range Experiment Station, Fort Collins, CO: 68-79.

Jones, K. B. 1988. Comparison of herpetofaunas of a natural and altered riparian ecosystems. In *Management of amphibians, reptiles, and small mammals in North America*, eds. R. C. Szaro, K. E. Severson, and D. R. Patton, 222-227. Gen. Tech. Rep. RM-166. U.S. Department of Agriculture, Forest Service, Rocky Mountain Forest and Range Experiment Station, Fort Collins, CO.

Junk, W. J., P. B. Bayley, and R. E. Sparks. 1989. The flood pulse concept in river-flood plain systems. In *Proceedings International symposium on large rivers*, ed. D. P. Dodge. *Can. Spec. Publ. of Fish. and Aquat. Sci.* 106:110-127.

K

Kalliola, R., and M. Puhakka. 1988. River dynamics and vegetation mosaicism: a case study of the River Kamajohka, northernmost Finland. *J. Biogeogr.* 15:703-719.

Kaplan, R. 1977. Down by the riverside: Informational factors in waterscape preferences. In *Proceedings: River recreation management and research symposium.* Gen. Tech. Rep. NC-28. U.S. Department of Agriculture, Forest Service, North Central Forest Experiment Station, St. Paul, MN: 285-289.

Karr, J. R. 1981. Assessment of biotic integrity using fish communities. *Fisheries* 6(6): 21-27.

Karr, J. R. 1991. Biological integrity: a long-neglected aspect of water resource management. *Ecol. Appl.* 1: 66-84.

Karr J. R., and E. W. Chu. 1999. *Restoring life in running waters.* Washington, DC: Island Press. 206 p.

Karr, J. R., and O. T. Gorman. 1975. Effects of land treatment on the aquatic environment. In *Non-point source pollution seminar.* EPA-905/9-75-007. U.S. Environmental Protection Agency, Chicago, IL: 120-150.

Karr, J. R., and I. J. Schlosser. 1978. Water resources and the land-water interface. *Science* 201:209-234.

Karr, J. R., P. L. Angermeier, and I. J Schlosser. 1983. Habitat structure and fish communities of warmwater streams. EPA-600/D-83-094. U.S. Environmental Protection Agency. Corvallis, OR. 6 p.

Karr, J. R., K. D. Fausch, P. L. Angermeier, P. R. Yant, and I. J. Schlosser. 1986. Assessment of biological integrity in running waters: a method and its rationale. *Illinois Nat. Hist. Surv. Spec. Publ.* 5.

Kauffman, J. B., and W. C. Krueger. 1984. Livestock impacts on riparian ecosystems and streamside management implications...a review. *J. Range Manage.* 37:430-438.

Kaufmann, J. H. 1992. Habitat use by wood turtles in central Pennsylvania. *J. Herpetology* 26:315-321.

Keller, E. A., and F. J. Swanson. 1979. Effects of large organic material on channel form and fluvial processes. *Earth Surf. Process.* 4:361-380.

Kellerhals, R., C. R. Neill, and D. I. Bray. 1972. *Hydraulic geomorphic characteristics of rivers in Alberta*. Research Council of Alberta. River Engineering and Surface Hydrology Report 72-1. 52 p.

Kellison, R. C., M. J. Young, R. R. Braham, and E. J. Jones. 1998. Major alluvial floodplains. In *Southern forested wetlands*, eds. M. G. Messina, and W. H. Conner, 291-323. Boca Raton, FL: Lewis Publisher.

Kentucky Division of Forestry. 1992. *Kentucky forest practice guidelines for water quality management*. Kentucky Division of Forestry, Frankfort. 55 p.

Keys, James E., Jr., Constance A. Carpenter, S. Hooks, F. Koenig, W. H. McNab, W. W. Russell, and M. L. Smith. 1995. Ecological units of the Eastern United States: first approximation. U.S. Department of Agriculture, Forest Service. 83 p. and map.

Kirkland, G. L., Jr., and F. J. Jannett, Jr. 1982. *Microtus chrotorrhinus*. Mammalian Species. Shippensburg, PA: American Society of Mammalogists. 18:1-8.

Kittredge, D. B., and M. Parker. 1995. Massachusetts forestry best management practices manual. Massachusetts Department of Environmental Protection, Office of Watershed Management and U.S. Environmental Protection Agency.

Kleeberger, S. R., and J. K. Werner. 1983. Post-breeding migration and summer movement of *Ambystoma maculatum*. *J. Herpetol.* 17:176-177.

Klemens, M. W. 1993. Amphibians and reptiles of Connecticut and adjacent regions. Bull. No. 112. State Geological and Natural History Survey of Connecticut, Hartford, CT. 318 p.

Klemm, D. J., P. A. Lewis, F. Fulk, and J. M. Lazorchak. 1993. Fish field and laboratory methods for evaluating the biological integrity of surface waters. EPA-600-R-92-111. U.S. Environmental Protection Agency, Environmental Monitoring and Support Laboratory, Cincinnati, OH.

Klessig, L., B. Sharp, and R. Wedepohl. 1989. Using a special purpose unit of government to manage lakes. *Lake Reserv. Manage.* 5:37-44.

Kovalchik, B. L., and Chitwood, L.A. 1990. Use of geomorphology in the classification riparian plant associations in mountainous landscapes of central Oregon, U.S.A. *For. Ecol. Manage.* 33/34:405-418.

Knox, J. C. 1977. Human impacts on Wisconsin stream channels. *Ann. Assoc. Am. Geograph.* 67:323-342.

Knox, J. C. 1987. Historical valley floor sedimentation in the Upper Mississippi Valley. *Ann. Assoc. Am. Geograph.* 77:224-244.

Kochenderfer, J. N., P. J. Edwards, F. Wood. 1997. Hydrologic impacts of logging an Appalachian watershed using West Virginia's best management practices. *North. J. Appl. For.* 14:207-218.

Kochenderfer, J. N., J. D. Helvey, and G. W. Wendel. 1987. Sediment yield as a function of land use in central Appalachian forests. In *Proceedings: Sixth Central Hardwood Forest Conference*. Gen. Tech. Rep. NC-132. U.S. Department of Agriculture, Forest Service, North Central Forest Experiment Station, St. Paul, MN: 497-502.

Kochenderfer, J. N., and J. D. Helvey. 1987. Using gravel to reduce soil losses from minimum-standard forest roads. *J. Soil Water Conserv.* 42:46-50.

Kochenderfer, J. N., and G. W. Wendel. 1983. Plant succession and hydrologic recovery on a deforested and herbicided watershed. *For. Sci.* 29:545-558.

Kochenderfer, J. N. 1970. Erosion control on logging roads in the Appalachians. Res. Pap. NE-158. U.S. Department of Agriculure, Forest Service, Northeastern Forest Experiment Station, Upper Darby, PA. 28 p.

Koger, J. L., C. Ashmore, and B. J. Stokes. 1984. Ground skidding wetlands with dual-tired skidders: a South Carolina case study. ASAE Pap. 84-1618. American Society of Agricultural Engineers, St. Joseph, MI. 15 p.

Kortelainen, P. 1993. Content of total organic carbon in Finnish lakes and its relationship to catchment characteristics. *Can. J. Fish. Aquat. Sci.* 50:1477-1483.

Korth, R. M., and L. L. Klessig. 1990. Overcoming the tragedy of the commons: alternative lake management institutions at the community level. *Lake Reserv. Manage.* 6:219-225.

Kovalchik, B. L., and L. A. Chitwood. 1990. Use of geomorphology in the classification of riparian plant associations in mountainous landscapes of central Oregon, U.S.A. *For. Ecol. Manage.* 33/34:405-418.

Kovalchik, B. L., 1987. Riparian zone associations of the Deschutes, Ochoco, Fremont and Winema National Forsts. ECOL TP-279-87. U.S. Department of Agriculture, Forest Service, Pacific Northwest Region 6, Portland, OR. 171 p.

Kramer, D. C. 1974. Home range of the western chorus frog *Pseudacris triseriata triseriata*. *J. Herpetol.* 8:245-246.

Kratz, T. K., K. E. Webster, C. J. Bowser, J. J. Magnuson, and B. J. Benson. 1997. The influence of landscape position on lakes in northern Wisconsin. *Freshwater Biol.* 37:209-217.

Krusic, R. A. 1995. Habitat use and identification of bats in the White Mountain National Forest. Master's thesis. University of New Hampshire, Durham, NH. 86 p.

Krusic, R. A., M. Yamasaki, C. D. Neefus, and P. J. Pekins. 1996. Bat habitat use in White Mountain National Forest. *J. Wildl. Manage.* 60:625-631.

Krusic, R. A., and C. D. Neefus. 1996. Habitat associations of bat species in the White Mountain National Forest. In *Bats and forests symposium*, eds. R. M. R. Barclay, and R. M. Brigham, 185-198. Work. Pap. 23/1996. Victoria, BC: Research Branch, British Columbia Ministry of Forestry. 292 p.

Küchler, A. W. 1964. Potential natural vegetation of the conterminous United States. *Am. Geogr. Soc. Spec. Publ.* 36.

Kurta, A. 1995. *Mammals of the Great Lakes region.* Ann Arbor, MI: University of Michigan Press. 246 p.

Kurta, A., D. King, J. A. Teramino, J. M. Stribley, and K. J. Williams. 1993. Summer roosts of the endangered Indiana bat (*Myotis sodalis*) on the northern edge of its range. *Am. Midl. Nat.* 129:132-138.

Kutka, F., and C. Richards. 1996. Relating diatom assemblage structure to stream habitat quality. *J. North Am. Bentholog. Soc.* 15:469-480.

L

LaFayette, R. A., and L. F. DeBano. 1990. Watershed condition and riparian health: linkages. In *Proceedings–1990 ASCE Irrigation and Drainage Division Meeting.* 473–484. Durango, CO. 473-484.

Lamberti, G. A., and M. B. Berg. 1995. Invertebrates and other benthic features as indicators of environmental change in Juday Creek, Indiana. *Nat. Areas J.* 15:249-258.

Lamberti, G. A., and S. V. Gregory. 1996. Transport and retention of CPOM. In *Methods in stream ecology.* eds. F. R. Hauer, and G. A. Lamberti. San Diego, CA: Academic Press.

Lamberti, G. A., and V. H. Resh. 1983. Stream periphyton and insect herbivores: an experimental study of grazing by a caddisfly population. *Ecology* 64:1124-1135.

Lamberti, G. A., S. V. Gregory, L. R. Ashkenas, A. D. Steinman, and C. D. McIntire. 1989. Productive capacity of periphyton as a determinant of plant-herbivore interactions in streams. *Ecology* 70:1840-1856.

Lamberti, G. A., S. V. Gregory, L. R. Ashkenas, R. C. Wildman, and K. M. S. Moore. 1991. Stream ecosystem recovery following a catastrophic debris flow. *Can. J. Fish. Aquat. Sci.* 48:196-208.

Lannoo, M. J., K. Lang, T. Waltz, and G. S. Phillips. 1994. An altered amphibian assemblage: Dickinson County, Iowa, 70 years after Frank Blanchard's survey. *Am. Midl. Nat.* 131:311-319.

Lapin, M., and B. V. Barnes. 1995. Using the landscape ecosystem approach to assess species and ecosystem diversity. *Conserv. Biol.* 9:1148-1158.

Large, A. R. G., and G. E. Petts. 1994. Rehabilitation of river margins. In *The rivers handbook*, Volume Two, eds. P. Calow, and G. E. Petts, 401-437. Blackwell Scientific Publishers.

Larimore, R. W., W. F. Childers, and C. Heckrotte. 1959. Destruction and re-establishment of stream fish and invertebrates affected by drought. *Trans. Am. Fish. Soc.* 88:261-285.

Larue, P., L. Belanger, and J. Huot. 1995. Riparian edge effects on boreal balsam fir bird communities. *Can. J. For. Res.* 25:555-566.

Laursen, S. B. 1996. A description of the riparian forest resource in the state of Minnesota. In At the water's edge: the science of riparian forestry, ed. S. B. Laursen, 15-20. BU-6637-S: Minnesota Extension Service, University of Minnesota.

Leak, W. B. 1982. Habitat mapping and interpretation in New England. Res. Pap. NE-496. U.S. Department of Agriculture, Forest Service, Northeastern Forest Experiment Station, Broomall, PA. 28 p.

Leland, H. V. 1995. Distribution of phytobenthos in the Yakima River basin, Washington, in relation to geology, land use, and other environmental factors. *Can. J. Fish. Aquat. Sci.* 52:1108-1129.

Lenat, D. R. 1988. Water quality assessment of streams using a qualitative collection method for benthic macroinvertebrates. *J. North Am. Bentholog. Soc.* 7:222-233.

Lenat, D. R. 1984. Agriculture and stream water quality, a biological evaluation of erosion control practices. *Environ. Manage.* 8:333-344.

Leopold, A. 1949. *A Sand County Almanac.* Oxford University Press.

Leopold, L. B. 1994. *A view of the river.* Cambridge, MA: Harvard University Press. 298 p.

Leopold, L. B. 1997. *Water, rivers, and creeks.* Sausalito, CA: University Science Books. 185 p.

Leopold, L. B., and T. Maddock. 1953. The hydraulic geometry of stream channels and some physiographic implications. Prof. Pap. 252. U.S. Geological Survey, Washington, DC.

Leopold, L. B., M. G. Wolman, and J. P. Miller. 1964. Fluvial processes in geomorphology. San Francisco, CA: Freeman Press.

Leopold, L. B., and M. G. Wolman. 1957. River channel patterns; braided meandering and straight. Prof. Pap. 282-B. U.S. Geologic Survey, Washington, DC.

Leptich, D. J., Gilbert, J. R. 1989. Summer home range and habitat use by moose in northern Maine. J. Wildl. Manage. 53:880-885.

Levell, J. P. 1997. A field guide to reptiles and the law. Lanesboro, MN: Serpent's Tale Natural History Books Distributors. 270 p.

Lickwar, P., C. Hickman, and F. W. Cubbage. 1991. Costs of protecting water quality during harvesting on private forestlands in the southeast. South. J. Appl. For. 16:13-20.

Likens, G. E., and R. E. Bilby. 1982. Development maintenance and role of organic debris dams in New England streams. In Sediment budgets and routing in forested drainage basins, eds. F. J. Swanson, R. J. Janda, T. Dunne, and D. N. Swanston, 122-128. Res. Pap. PNW-141. U.S. Department of Agriculuture, Forest Service, Pacific Northwest Research Station, Portland, OR.

Lindsey, A. A., R. O. Petty, D. K. Sterlin, and W. Van Asdall. 1961. Vegetation and environment along the Wabash and Tippecanoe Rivers. Ecol. Monogr. 31:105-156.

Lisle, T. E. 1986. Stabilization of a gravel channel by large streamside obstructions and bedrock bens, Jacoby Creek, northwestern California. Geol. Soc. Am. Bull. 97:999-1011.

Liu, Chiun-Ming. 1987. Source of increased stormwaters after forest operations. In Proceedings, Forest hydrology and watershed management, eds. R. H. Swanson, P. Y. Bernier, and P. D. Woodward, 539-544. Vancouver, BC: IAHS-AISH Publ. No. 167.

Longfellow, H. W. 1855. The song of Hiawatha. Boston: ticknor and Fields. 316 p.

Lövel, G. L. 1997. Global change through invasion. Nature 388:627-628.

Lowrance, R., R. Todd, J. Fail, Jr., O. Hendrickson, Jr. R. Leonard, and L. Asmussen. 1984. Riparian forests as nutrient filters in agricultural watersheds. BioScience 34:374-377.

Lu, Shiagn-Yue. 1994. Forest harvesting effects on streamflow and flood frequency in the northern Lake States. PhD Thesis. University of Minnesota, St. Paul, MN. 112.

Lundbeck, J. 1934. "Uber den primar oligotrophen seetypus und den Wollingster See als dessen mitteleuropaischen Vetreter." Arch. Hydrobiol. 27:331-250.

Lynch, J. A., E. S. Corbett, and R. J. Hutnik. 1975. Chapter 5, Water Resources. In Clearcutting in Pennsylvania. University Park, PA: School of Forest Resources, Pennsylvania State University: 51-63.

Lynch, J. A., W. E. Sopper, and D. B. Partridge. 1972. Changes in streamflow following partial clearcutting on a forested watershed. In Proceedings, Watershed in transition, Am.Water Resour. Assoc. Proc. Series No. 14:313-320.

Lynch, J. A., and E. S. Corbett. 1990. Evaluation of best management practices for controlling nonpoint pollution from silvicultural operations. Water Resour. Bull. 26(1):41-52.

Lyons, J., L. Wang, and T. D. Simonson. 1996. Development and validation of an index of biotic integrity for coldwater streams in Wisconsin. North. Am. J. Fish Manage. 16:241-256.

Lyons, J. 1992. Using the index of biotic integrity (IBI) to measure environmental quality in warm water streams in Wisconsin. Gen. Tech. Rep. NC-49. U.S. Department of Agriculture, Forest Service, North Central Forest Experiment Station, St. Paul, MN. 51 p.

Lyons, J., S. Navarro-Perez, P. A. Cochran, E. C. Santana, and M. Guzman-Arroyo. 1995. Index of biotic integrity based on fish assemblages for the conservation of streams and rivers in west-central Mexico. *Conserv. Biol.* 9:569-584.

M

Maas, R. P., D. J. Kucken, and P. F. Gregutt. 1991. Developing a rigorous water quality database through a volunteer monitoring network. *Lake Reserv. Manage.* 7:123-126.

MacCulloch, R. D., and J. R. Bider. 1975. Phenology, migrations, circadian rhythm and the effect of precipitation of the activity of *Eurycea b. bislineata* in Quebec. *Herpetologica* 31:433-439.

MacDonald, L. H., A. W. Smart, and R. C. Wissmar. 1991. Monitoring guidelines to evaluate effects of forestry activities on streams in the Pacific Northwest and Alaska. EPA 910/9-91-001, Environmental Protection Agency, Region 10, Seattle, WA. 166 p.

MacKenzie, S. H. 1997. Toward integrated resource management: lessons about ecosystem approach from the Laurentian Great Lakes. *Environ. Manage.* 21: 173-183.

Madison, D. M. 1997. The emigration of radio-implanted spotted salamanders, *Ambystoma maculatum. J. Herpetol.* 31:542-551.

Magnuson, J. J., and C. J. Bowser. 1990. A network of long-term ecological research in the United States. *Freshwater Biol.* 23:137-147.

Magnuson, J. J., W. M. Tonn, A. Banerjee, J. Toivonen, O. Sanchez, and M. Park. 1998. Isolation vs. extinction in the assembly of fishes in small northern lakes. *Ecology* 2911-2956.

Maguire, D. J. 1998. ARC/INFO Version 8: object-component GIS. *ARC News* 20(4):1-5.

Malanson, G. P. 1993. Riparian landscapes. New York: Cambridge University Press. 296 p.

Malterer, T. J., E. S. Verry, and J. Erjavec. 1992. Fiber content and degree of decomposition in peats. *Soil Sci. Soc. Am. J.* 56(6):1200-1211.

Marcum, C., Jr. 1994. Ecology and natural history of four plethodontid species in the Fernow Experimental Forest, Tucker County, West Virginia. Master's thesis, Marshall University, Huntington, WV. 254 p.

Marshall, T. R., and P. A. Ryan. 1987. Abundance and community attributes of fishes relative to environmental gradients. *Can. J. Fish. Aquat. Sci.* 44:196-215.

Marshall, P. 1949. *Mr. Jones: meet the master.* Richmond, VA: John Knox Press.

Marsh, G. P. 1864. *Man and nature or physical geography as modified by human action*, ed. D. Lowenthal. Cambridge, MA: Harvard University Press.

Martin, C. W., and J. W. Hornbeck. 1994. Logging in New England need not cause sedimentation in streams. *North. J. Appl. For.* 11(1):17-23.

Martin, C. W., R. S. Pierce, G. E. Likens, and F. H. Bormann. 1986. Clearcutting affects stream chemistry in the White Mountains of New Hampshire. Res. Pap. NE-579. U.S. Department of Agriculture, Forest Service, Broomall, PA. 12 p.

Martin, C. W., and J. W. Hornbeck. 1989. Revegetation after strip cutting and block clearcutting in northern hardwoods: a 10-year history. Res. Pap. NE-625. U.S. Department of Agriculture, Forest Service, Broomall, PA. 17 p.

Maryland Department of Natural Resources, Resource Conservation Service, Forestry Division. 1992. *Maryland's guide to forest harvest operations and best management practices.* Delaware, MD: Maryland Department of Natural Resources.

Maser, C., and J. Sedell. 1994. From the forest to the sea: the ecology of wood in streams, rivers, estuaries, and oceans. Delray Beach, FL: St. Lucie Press. 195 p.

Mason, D. P. 1982. Physical and hydrochemical effects on stream insect communities in the White Mountain National Forest of New Hampshire. Masters thesis, University of New Hampshire, Durham, NH. 106 p.

Maxwell, J. R., C. J. Edwards, M. E. Jensen, S. J. Paustian, H. Parrott, and D. M. Hill. 1995. A hierarchical framework of aquatic ecological units in North America (nearctic zone). Gen. Tech. Rep. NC-176. U.S. Department of Agriculture, Forest Service, North Central Forest Experiment Station, St. Paul, MN. 72 p.

Mayr, E. 1976. *Evolution and the diversity of life.* Cambridge, MA: Belknap Press of Harvard University Press. 721 p.

McComb, W. C., K. McGarigal, and R. G. Anthony. 1993. Small mammal and amphibian abundance in streamside and upslope habitats of mature Douglas-fir stands, western Oregon. *Northwest Sci.* 67:7-15.

McCormick, J. F. 1979. A summary of the national riparian symposium. In *Proceedings, Strategies for protection and management of floodplain wetlands and other riparian ecosystems.* Gen. Tech. Rep. WO-12. U.S. Department of Agriculuture, Forest Service, Washington, DC: 362-363.

McCoy, C. J. 1985. Amphibians and reptiles. In *Species of special concern in Pennsylvania,* eds. H. H. Genoways, and F. J. Brenner, 259-295. Pittsburgh, PA: Special Publication of Carnegie Museum of Natural History, Number 11.

McGurrin, J., and H. Forsgren. 1997. What works, what doesn't, and why. Pages 459 – 471. In Watershed restoration: principles and practices, 459-471. Bethesda, MD: American Fisheries Society.

McKee, A., S. Gregory, and L. Ashkenas. 1996. Draft riparian management reference. National Forests in Minnesota. U.S. Department of Agriculture, Chippewa National Forest, Cass Lake, MN. 110 p.

McKevlin, M. R., D. D. Hook, and A. A. Rozelle. 1997. Adaptations of plants to flooding and soil waterlogging. In *Southern forested wetlands: ecology and management.* eds. M. G. Messina, and W. H. Conner, 173-204. New York: Lewis Publishers.

McNab, W. H., and P. E. Avers. comp. 1994. Ecological subregions of the United States: section descriptions. Admin. Publ. WO-WSA-5. U.S. Department of Agriculture, Forest Service, Washington, DC. 267 p.

Meek, P. 1994. Thinning in streamside protection strips with the Valmet 901. Forest Engineering Research Institute of Canada (FERIC), Pointe-Claire, Quebec. Field Note No. Felling-22. 2 p.

Megahan, W. F. 1972. Logging, erosion, sedimentation — are they dirty words? *J. For.* 70:403-407.

Melquist, W. E., and A. E. Dronkert. 1987. *River otter.* Wild furbearer management and conservation in North America, eds. M. Novak, J. A. Baker, M. E. Obbard, and B. Malloch. Toronto, Ontario: Ontario Ministry of Natural Resources and Ontario Trappers Association.

Merritt, R. W., and D. L. Lawson. 1992. The role of leaf litter macroinvertebrates in stream-floodplain dynamics. *Hydrobiologia* 248:65-77.

Minnesota Forest Resources Council. 1999. Sustaining Minnesota forest resources: voluntary site-level forest management guidelines for landowners, loggers and resource managers. Minnesota Forest Resources Council, St. Paul, MN. 492p.

MIDNR (Michigan Department of Natural Resources). 1994. Water quality management practices on forest land. Lansing, MI: Michigan Department of Natural Resources. 77 p.

MNDNR (Minnesota Department of Natural Resources). 1995. Protecting water quality and wetlands in forest management, best management practices in Minnesota. St. Paul, MN. 140 p.

Miller, K. L., and 13 others. 1988. Regional applications of an index of biotic integrity for use in water resource management. *Fisheries* 13(5):12-20.

Miller, D. H., Getz, L. L. 1977. Factors influencing local distribution and species diversity of forest small mammals in New England. *Can. J. Zool.* 55:806-814.

Minshall, G. W., K. W. Cummins, R. C. Peterson, C. E. Cushing, D. A. Bruns, J. R. Sedell, and R. L. Vannote. 1985. Developments in stream ecosystem theory. *Can. J. Fish. Aquat. Sci.* 42:1045-1055.

Mitchell, J. C., S. Y. Erdle, J. F. Pagels, and C. S. Hobson. 1994. The effect of short-term sampling on descriptions of terrestrial vertebrate communities in saturated forested wetlands, and comment on estimating wetland value. In Proceedings of a workshop on Saturated forested wetlands in the Mid-Atlantic region : the state of science. eds. S. D. Eckles, A. Jennings, A. Spingarn, and C. Wienhold, 24-28. Annapolis, MD: U.S Fish and Wildlife Service.

Mitchell, J. C. 1994. *The Reptiles of Virginia.* Washington, DC: Smithsonian Institution Press. 352 p.

Mitchell, J. C. 1997b. Life in a pothole II. *Virginia Wildl.* 58(June):23-28.

Mitchell, J. C. 1974. Statistics of *Chrysemys rubriventris* hatchlings from Middlesex County, Virginia. *Herpetol. Rev.* 5:71.

Mitchell, J. C. 1991. Amphibians and reptiles. In *Virginia's endangered species*, coord. K. Terwilliger, 411-423. Blacksburg, VA: McDonald & Woodward Publishing Co.

Mitchell, J. C. 1997a. Life in a pothole. *Virginia Wildl.* 58(April):5-9.

Mitchell, J. C., and M. W. Klemens. In press. Primary and secondary effects of habitat alteration on turtles. In *Conservation biology of turtles: decline of an ancient lineage.* ed., M. W. Klemens. Washington DC: Smithsonian Institution Press.

Mitsch, W. J., and J. G. Gosselink. 1993. *Wetlands.* Second edition. New York: Van Nostrand Reinhold. 722 p.

Molles, M. C., Jr. 1982. Trichopteran communities of streams associated with aspen and conifer forests: long term structural change. *Ecology* 63:1-6.

Molot, L. A., and P. J. Dillon. 1997. Colour-mass balances and colour-dissolved organic carbon relationships in lakes and streams in central Ontario. *Can. J. Fish. Aquat. Sci.* 54:2789-2795.

Montgomery, D. R., G. E. Grant, and K. Sullivan. 1995. Watershed analysis as a framework for implementing ecosystem management. *Water Resour. Bull.* (*J. Am. Water Resour. Assoc.*) 31(3):369-386.

Montgomery D. R., and J. M. Buffington. 1993. Channel classification, prediction of channel response, and assessment of channel condition. TFW-SH10-93-022. University of Washington, Department of Geological Sciences, Seattle, WA. 84p.

Moore, C. M. 1987. Group techniques for idea building. Applied Social Research Methods Series Vol. 9. Sage Publications, Inc. 143 p.

Moriarty, J. J., and A. M. Bauer. 1999. State and provincial amphibian and reptile publications for the United States and Canada. *Soc. Study Amphibians and Reptiles, Herpetol. Circ. 28.* (in press).

Mosley, M. P. 1981. The influence of organic debris on channel morphology and bedload transport in a New Zealand forest stream. *Earth Surf. Process. Landforms* 6:571-579.

Mou, P. U., T. J. Fahey, and J. W. Hughes. 1993. Effects of soil disturbance on vegetation recovery and nutrient accumulation following whole-tree harvest of a northern hardwood ecosystem. *J. Appl. Ecol.* 30:661-675.

Murosky, D. L., and A. E. Hassan. 1991. Impact of tracked and rubber-tired skidders traffic on a wetland site in Mississippi. *Trans. Am. Soc. Agricul. Engin.* 34(1):322-327.

Murphy, M. L., and K. V. Koski. 1989. Input and depletion of woody debris in Alaska streams and implications for streamside management. *North Am. J. Fish. Manage.* 9:427-436.

Murphy, B. R., and D. W. Willis. 1996. *Fisheries techniques.* Second edition. Bethesda, MD: American Fisheries Society.

Murray, N. L., and Stauffer, D. F. 1995. Nongame bird use of habitat in central Appalachian riparian forests. *J. Wildl. Manage.* 59:78-88.

N

Naiman, R. J., and H. DeCamps. 1997. The ecology of interfaces: riparian zones. *Annu. Rev. Ecol. Syst.* 28:621-658.

Naiman, R. J., H. Decamps, and M. Pollock. 1993. The role of riparian area corridors in maintaining regional diversity. *Ecol. Appl.* 3:209-212.

Naiman, R. J., H. Decamps, J. Pastor, and C. A. Johnston. 1988. The potential importance of boundaries to fluvial ecosystems. *J. North Am. Bentholog. Soc.* 7: 289-306.

Naiman, R. J., D. G. Lonzarch, T. J. Beechie, and S. C. Ralph. 1992a. General principles of the classification and assessment of the conservation potential of rivers. In *River conservation and management.* eds., P. J. Boon, C. Calow, and G. E. Petts, 93-122. New York: John Wiley & Sons Ltd.

Naiman, R. J., T. J. Beechie, L. E. Benda, D. R. Berg, P. A. Bisson, L. H. MacDonald, M. D. O'Connor, P. L. Olson, and E. A. Steel. 1992b. Fundamental elements of ecological healthy watersheds in the Pacific Northwest coastal region. In *Watershed management: balancing sustainability and environmental change.* ed., R. J. Naiman, 122-188. New York: Springer.

Naiman, R. J., K. L. Fetherston, S. McKay, and J. Chen. 1997. Riparian forests. In *River ecology and management: lessons from the pacific coastal ecoregion*, eds. R. J. Naiman, and R. E. Bilby, 289-323. New York: Springer.

Nanson, G. C., and H. F. Beach. 1977. Forest succession and sedimentation on a meandering-river floodplain, northeastern British Columbia, Canada. *J. Biogeogr.* 4: 229-251.

Nanson, G. C., and Croke, J. C. 1992. A genetic classification of floodplains. *Geomorphology* 4:459-486.

Natural Resources Conservation Service. 1996. *Interim conservation practice standard for CRP. riparian forest buffers.* Code 391, MN. 6 p.

Nature Conservancy. 1997. A classification framework for freshwater communities. The Nature Conservancy aquatic community classification workshop proceedings. New Haven, MO. 16 p.

Naumann, E. 1919. "Nagra synpvnkter angaende planktons okoloti Medd. Sarskild hantsyn fytoplankton." *Svensk. bot. Titskr.* 13:129-158.

Newbold, J. D., D. C. Erman, and K. B. Roby. 1980. Effects of logging on macroinvertebrates in streams with and without bufferstrips. *Can. J. Fish. Aquat. Sci.* 37:1076-1085.

NCASI (National Council of the Paper Industry for Air and Stream Improvement). 1994. Southern regional review of state nonpoint source control programs and best management practices for forest management operations. Tech. Bull. 686. National Council of the Paper Industry for Air and Stream Improvement, New York. 175 p.

NH Div. Forest and Lands, DRED and SPNHF (compilers). 1997. Good forestry in the Granite State: recommended voluntary forest management practices for New Hampshire. Tilton, NH: Sant Bani Press. 65 p.

Nilsson, C. 1992. Conservation management of riparian communities. In *Ecological principles of nature conservation,* ed. L. Hansson, 352-372. Amsterdam: Elsevier Science.

Nilsson, C., G. Grelsson, M. Dynesius, M. E. Johansson, and U. Sperens. 1991. Small rivers behave like large rivers: effects of postglacial history on plant species richness along riverbanks. *J. Biogeogr.* 18:533-541.

Noble, S. M. 1993. Evaluating predator distributions in Maine forest riparian zones using a geographic information system. Master's thesis. University of Maine, Orono, ME. 54 p.

Noel, D. S., C. W. Martin, and C. A. Federer. 1986. Effects of forest clearcutting in New England on stream macroinvertebrates and periphyton. *Environ. Manage.* 10(5) 661-670.

North Carolina DNER (Department of Environmental and Natural Resources). 1998. Forestry Best Management Practices for water quality protection. North Carolina Division of Forest Resources, Department of Environment and Natural Resources, Raleigh, NC.

Noss, R. F., and A. Y. Cooperrider. 1994. *Saving nature's legacy.* Washington, DC: Island Press. 416 p.

Nyland, R. D. 1996. *Silviculture: concepts and applications.* New York: McGraw-Hill.

O

O'Connor, W. P., R. M. Roth, and L. L. Klessig. 1995. *A guide to Wisconsin's lake management law.* 9th edition. Wisconsin Department of Natural Resources.

O'Laughlin, J., and G. H. Belt. 1995. Functional approaches to riparian buffer strip design. *J. For.* 93:29-32.

O'Neil, H. E. 1910. Preservation of Adirondak Forest in relation to water supply. *Am. Lumberman.* March 12:36-37.

Odum, E. P. 1979. Ecological importance of the riparian zone. In *Strategies for protection and management of floodplain wetlands and other riparian ecosystems.* Gen. Tech. Rep.WO-12. U.S. Department of Agriculture, Forest Service, Washington, DC: 2-4.

Ohio EPA (Environmental Protection Agency). 1988. *Biological criteria for the protection of aquatic life,* volumes 1-3. Ecological Assessment Section, Division of Water Quality Monitoring and Assessment, Ohio Environmental Protection Agency, Columbus, OH.

Ohio Department of Natural Resources, Division of Forestry. 1992. *BMPs for erosion control on logging jobs in Ohio.* Ohio Department of Natural Resources, Division of Forestry. 53 p.

Ohmart, R. D., and B. W. Anderson. 1986. Riparian habitat. In Inventory and monitoring of wildlife habitat. eds. A.Y. Cooperrider, R. J. Boyd, and H. R. Stuart. BLM/YA/PT-87/001+6600. Bureau of Land Management Service Center, Denver, CO: 169-199.

Oldfield, B., and J. J. Moriarty. 1994. Amphibians and reptiles native to Minnesota. Minneapolis, MN: University of Minnesota Press. 237 p.

Oliver, C. D., and T. M. Hinckley. 1987. Species, stand structures, and silvicultural manipulation patterns for the streamside zone. In Streamside management: forestry and fishery interactions. eds. E. O. Salo, and T. W. Cundy, 259-276. University of Washington Institute of Forest Resources.

Olson, D. H., W. P. Leonard, and R. B. Bury. 1997. Sampling amphibians in lentic habitats. *Northwest Fauna* 4:1-134.

Omernick, J. M. 1987. Map supplement: ecoregions of the conterminous United States. *Ann. Assoc. Am. Geogr.* 77(1):118-125.

Omernik, J. M., A. R. Abernathy, and L. M. Male. 1981. Stream nutrient levels and proximity of agricultural and forest land to streams: Some relationships. *J. Soil Water Conserv.* 36:227-231.

Omernik, J. M., and R. G. Bailey. 1997. Distinguishing between watersheds and ecoregions. *J. Am. Water Resour. Assoc.* 33(5):935-949.

OMNR. 1978. Designation of assessment units. Ontario Ministry of Natural Resources, final report of SPOF working Group One, Ontario, Canada.

Ordiway, L. O. 1994. Factors influencing the spatial distribution and natural history of *Desmognathus ochrophaeus* in West Virginia. Master's thesis, Marshall University, Huntington, WV. 147 p.

OR/WA Wildlife Committee (Oregon/Washington Interagency Wildlife Committee). 1979. Managing riparian ecosystems (zones) for fish and wildlife in eastern Oregon and western Washington. Unpubl. Rep. 77 p.

Organ, J. A. 1961. Life history of the pygmy salamander, *Desmognathus wrighti*, in Virginia. *Am. Midl. Nat.* 66:384-390.

Osborne, L. L., S. L. Kohler, P. B. Bayley, D. M. Day, W. A. Bertrand, M. J. Wiley, and R. Sauer. 1992. Influence of stream position in a drainage network on the index of biological integrity. *Trans. Am. Fish. Soc.* 121:635-643.

Osborne, L. L., and D. A. Kovacic. 1993. Riparian vegetated buffer strips in water-quality restoration and stream management. *Freshwater Biol.* 29:243-258.

Osborne, L. L., and M. J. Wiley. 1988. Empirical relationships between land use/cover patterns and stream water quality in an agricultural watershed. *J. Environ. Manage.* 26:9-27.

Osterkamp, W. R., and C. R Hupp. 1984. Geomorphic and vegetative characteristics along three northern Virginia streams. *Geol. Soc. Am. Bull.* 95:1093-1101.

Ovaska, K. 1997. Vulnerability of amphibians in Canada to global warming and increased solar ultraviolet radiation. *Herpetol. Conserv.* 1:206-225.

P

Palone, R. S., and A. H. Todd, eds. 1997. Chesapeake Bay riparian handbook: a guide for establishing and maintaining riparian forest buffers. NA-TP-02-97. U.S. Department of Agriculture, Forest Service, Northeastern Area State and Private Forestry, Radnor, PA.

Packer, P. E. 1967. Forest treatment effects on water quality. In *International symposium on forest hydrology*. Oxford: Pergamon Press: 677-687.

Pais, R. C., S. A. Bonney, and W. C. McComb. 1988. Herpetofaunal species richness and habitat associations in an eastern Kentucky forest. *Proc. Ann. Conf. Southeast. Assoc. Fish and Wildl. Agencies* 42:448-455.

Palik, B. J., S. W. Golladay, P. C. Goebel, and B. W. Taylor. 1998. Geomorphic variation in riparian tree mortality and stream coarse woody debris recruitment from record flooding in a coastal plain stream. *EcoScience* 5:551-560.

Palmer, W. M., and A. L. Braswell. 1995. *Reptiles of North Carolina*. Chapel Hill, NC: University of North Carolina. 412 p.

Pardo, R. 1980. What's forestry's contribution to nonpoint source pollution? In *U.S. forestry and water quality: what course in the 80s?* Proceedings of the Water Pollution Control Federation Seminar, 31-41. Richmond, VA.

Parker G. 1978a. Self-formed rivers with equilibrium banks and mobile bed, Part I, the sand-silt river. *J. Fluid Mech.* 89:109-125.

Parker, G. 1978b. Self-formed straight rivers with equilibrium banks and mobile bed, Part II, the gravel river. *J. Fluid Mech.* 89:127-148.

Parola, A. C. 1987. HY8 culvert analysis: Microcomputer Program Applications Guide. Rep. FHWA-ED-87-101. USDOT, Washington, DC. 84 p.

Parrott, H. A., D. A. Marions, and R. D. Perkinson. 1989. A four-level hierarchy for organizing stream resources information. In Proceedings, *Headwater hydrology symposium*. American Water Resources Association, Missoula, MT: 41-54.

Patric, J. H., J. O. Evans, and J. D. Helvey. 1984. Summary of sediment yield data from forested land in the United States. *J. For.* 82:101-104.

Patric, J. H. 1976. Soil erosion in the eastern forest. *J. For.* 74(10):671-677.

Patric, J. H. 1978. Harvesting effects on soil and water in the eastern hardwood forest. *South. J. Appl. For.* 2:66-73.

Paustian, S., J. Holcomb, D. Marion, H. Parrott, and D. Perkinson. 1997. Ecosystem characterization and analysis: Aquatic biophysical environment. U.S. Department of Agriculture, Forest Service. unpublished report. 45 p. plus glossary.

Pautou, G., and H. Décamps. 1985. Ecological interactions between the alluvial forests and hydrology of the Upper Rhone. *Arch. Hydrobiol.* 104:13-37.

Pautou, G., J. Girel, and J. L. Borel. 1992. Initial repercussions and hydroelectric development in the French Upper Rhone Valley: a lesson for predictive scenarios propositions. *Environ. Manage.* 16:231-242.

PCSD. 1996. Presidential Commission on Sustainable Development, Chap 5. Natural Resources Stewardship. 27 p.

Pearce, A. J., M. K. Stewart, and M. G. Sklash. 1986. Storm runoff generation in humid headwater catchments 1. Where does the water come from? *Water Resour. Res.* 22(8):1263-1272.

Pearsall, D. R. 1995. Landscape ecosystems of the University of Michigan Biological Station: ecosystem diversity and ground-cover diversity. Ph.D. thesis, University of Michigan, Ann Arbor.

Pearson, P. G. 1955. Population ecology of the spadefoot toad, *Scaphiopus h. holbrooki* Harlan. *Ecol. Monogr.* 25:233-267.

Pechmann, J. H., and H. M. Wilbur. 1994. Putting declining amphibians populations in perspective: natural fluctuations and human impacts. *Herpetologica* 50:65-84.

Pechmann, J. H., D. E. Scott, R. D. Semlitsch, J. P. Caldwell, L. J. Vitt, and J. W. Gibbons. 1991. Declining amphibian populations: the problem of separating human impacts from natural fluctuations. *Science* 253:892-895.

Pechmann, J. H. K., D. E. Scott, J. W. Gibbons, and R. D. Semlitsch. 1989. Influence of wetland hydroperiod on diversity and abundance of metamorphosing juvenile amphibians. *Wetlands Ecol. Manage.* 1:3-11.

Pennsylvania Forested Wetlands Task Force. 1993. Best management practices for silvicultural activities in Pennsylvania's forest wetlands. A pocket guide for foresters, loggers, and other forest land managers.

Perillo, K. M. 1997. Seasonal movements and habitat preferences of spotted turtles (*Clemmys guttata*) in north central Connecticut. Linnaeus Fund Research Report. *Chelonian Conserv. Biol.* 2:445-447.

Perkey, A. W., B. L. Wilkins, and H. C. Heath. 1993. Crop tree management in eastern hardwoods. NA-TP-19-93. U.S. Department of Agriculture, Forest Service, Northeast Area State and Private Forestry, Broomall, PA.

Perkins, D. L., and C. C. Krueger. 1993. Heritage brook trout in northeastern USA: genetic variability within and among populations. *Trans. Am. Fish. Soc.* 122(4):515-532.

Peterjohn, W. T., and D. L. Correll. 1984. Nutrient dynamics in an agricultural catchment: Observations on the role of a riparian forest. *Ecology* 65:1466-1475.

Petersen, K. E., and T. L. Yates. 1980. *Condylura cristata.* Mammalian Species. Shippensburg, PA: *Am. Soc. Mammal.* 129:1-4.

Peterson, D. L., and G. L. Rolfe. 1982. Nutrient dynamics and decomposition of litterfall in floodplain and upland forests of central Illinois. *For. Sci.* 28:667-681.

Petranka, J. W. (In press). *Salamanders of the United States and Canada.* Washington, DC: Smithsonian Institution Press.

Pfankuch, D. J. 1975. Stream reach inventory and channel stability evaluation. U.S. Department of Agriculture, Forest Service, R1-75-002. Government Printing Office #696-260/200, Washington, DC. 22 p.

Phillips, M. J., R. Rossman, and R. Dahlman. 1994. Best management practices for water quality, evaluating BMP compliance of forest lands in Minnesota: a three-year study. Minnesota Department of Natural Resources, Division of Forestry, St. Paul, MN. 50 p.

Phillips, M. J. 1995. Regional application of Best Management Practices for the northeastern states. In *Proceedings of the 1995 Society of American Foresters convention*, October 28-November 1, 1995. Portland, ME: 154-162.

Pierce, G. J. 1980. The influence of flood frequency on wetlands of the Allegheny River floodplain in Cattaraugus Co., New York. *Wetlands* 2:87-104.

Pierce, R. S., J. W. Hornbeck, C. W. Martin, L. M. Tritton, C. T. Smith, C. A. Federer, and H. W. Yawney. 1993. Whole-tree clearcutting in New England: manager's guide to impacts on soils, streams, and regeneration. Gen. Tech. Rep. NE-172. U.S. Department of Agriculture, Forest Service, Northeastern Forest Experiment Station, Broomall, PA. 23 p.

Pitty, A. F. 1979. Conclusions. In *geographical approaches to fluvial processes*, ed. A. F. Pitty. Norwich, UK: Geo Books: 261-280.

Plafkin, J. L., M. T. Barbour, K. D. Porter, S. K. Gross, and R. M. Hughes. 1989. Rapid bioassessment protocols for use in streams and rivers: macroinvertebrates and fish. EPA/444/4-89-001. U.S. Environmental Protection Agency, Washington, DC. 174 p.

Platts, W. S., C. Amour, G. D. Booth, M. Bryant, J. L. Bufford, P. Cuplin, S. Jensen, et al. 1987. Methods for evaluating riparian habitats with applications to management. Gen. Tech. Rep. INT-221. U.S. Department of Agriculture, Forest Service, Intermountain Forest and Range Experiment Station, Ogden, UT. 177 p.

Poff, N. L., and J. V. Ward. 1989. Implications of streamflow variability and predictability for lotic community structure: A regional analysis of streamflow patterns. *Can. J. Fish. Aquat. Sci.* 46:1805-1816.

Poff, N. L., and J. V. Ward. 1990. Physical habitat template of lotic systems: Recovery in context of historical pattern of spatiotemperal heterogeneity. *Environ. Manage.* 14(5):629-645.

Poff, N. L., and J. D. Allan. 1995. Functional organization of stream fish assemblages in relation to hydrological variability. *Ecology* 76(2):606-627.

Poff, N. L. et al. 1997. The natural flow regime, a paradigm for river conservation and restoration. *BioScience* 47: 769-784.

Ponce, S. L. 1980. Water quality monitoring programs. WSDG-TP-00002. Watershed Systems Development Group. U.S. Department of Agriculture, Forest Service. Fort Collins, CO. 66 p.

Ponce, S. L., K. D. Sundeen, and W. D. Striffler. 1979. Effects of selected geology-soil complexes on water quality of the Little Black Fork Creek. Department of Earth Resources, Colorado State University. Fort Collins, CO. 96 p.

Poole, G. C., C. A. Frissell, and S. C. Ralph. 1997. In-stream habitat unit classification: inadequacies for monitoring and some consequences for management. *J. Am. Water Resour. Assoc.* 33(4):879-896.

Popper, K. R. 1959. *The Logic of Scientific Discovery.* London: Hutchinson. 480 p.

Porter, J. D. 1887. *Forestry in Europe: Reports from the Consuls of the United States.* U.S. Government Printing Office, Washington, DC. 315 p.

Possardt, E. E., and Dodge, W. E. 1978. Stream channelization impacts on songbirds and small mammals in Vermont. *Wildl. Soc. Bull.* 6:18-24.

Post, W. M., and J. Pastor. 1996. LINKAGES - an individual-based forest ecosystem model. *Clim. Change* 34:253-361.

Powell, D. S., J. L. Faulkner, D. R. Darr, Z. Zhu, and D. W. MacCleery. 1993. Forest resources of the United States, 1992. Gen. Tech. Rep. RM-234. U.S. Department of Agriculture, Forest Service, Rocky Mountain Forest and Range Experiment Station, Fort Collins, CO. 132 p.

Pregitzer, K. S., and B. V. Barnes. 1984. Classification and comparison of the upland hardwood and conifer ecosystems of the Cyrus H. McCormick Experimental Forest, Upper Peninsula, Michigan. *Can. J. For. Res.* 14: 362-375.

Pringle, C. M. 1987. Effects of water and substratum nutrient supplies on lotic periphyton growth: and integrated bioassay. *Can. J. Fish. Aquat. Sci.* 44:619-629.

Pyburn, W. F. 1958. Size and movements of a local population of cricket frogs (*Acris crepitans*). *Texas J. Sci.* 10:325-342.

Q

Quellet, J., J. Bonin, J. Rodrigue, J-L. DesGranges, and S. Lair. 1997. Hindlimb deformities (ectromelia, ectrodactyly) in free-living anurans from agricultural habitats. *J. Wildl. Dis.* 33:95-104.

R

Rabon, M. W. 1994. Management of the riparian zone to maximize accumulation of large woody debris in streams. Master's thesis, University of New Hampshire, Durham, NH. 97 p.

Rahel, F. J. 1984. Factors structuring fish assemblages along a bog lake successional gradient. *Ecology* 65:1276-1289.

Raikow, D. F., S. A. Grubbs, and K. W. Cummins. 1995. Debris dam dynamics and coarse particulate organic matter retention in an Appalachian Mountain stream. *J. North Am. Bentholog. Soc.* 14:535-546.

Raney, E. C. 1940. Summer movements of the bullfrog, *Rana catesbeiana* Shaw, as determined by the jaw-tag method. *Am. Midl. Nat.* 23:733-745.

Rask, M., L. Arvola, and K. Salonen. 1993. Effects of catchment deforestation and burning on the limnology of a small forest lake in Finland. *Verh. Internat. Verein. Limnol.* 25:525-528.

Rasmussen, J. B., L. Godbout, and M. Schallenberg. 1989. The humic content of lake water and its relationship to watershed and lake morphometry. *Limnol. Oceanogr.* 34:1336-1343.

Reay, R. S., D. W. Blodgett, B. S. Burns, S. J. Weber, and T. Frey. 1990. Management guide for deer in Vermont. Vermont Department of Forests, Parks and Recreation, and Fish and Wildlife, Montpelier, VT. 35 p.

Rector, W. G. 1951. The development of log transportation in the Lake States lumber industry 1840-1918. Ph.D. thesis, University of Minnesota, St. Paul, MN. 379 p.

Regan, R. J., and G. Anderson. 1995. A landowner's guide to wildlife habitat management for Vermont's woodlands.

Reiners, W. A. 1981. Nitrogen cycling in relation to ecosystem succession. *Ecol. Bull.* 33:507-528.

Reinhardt, K. F., A. R. Escher, and G. R. Trimble. 1963. Res. Pap. NE-1. Effect of streamflow of four forest practices in the mountains of West Virginia. U.S. Department of Agriculture, Forest Service, Northeast Forest Experiment Station, Broomall, PA. 79 p.

Reisinger, T. W., and W. M. Aust. 1990. Specialized equipment and techniques for logging wetlands. ASAE Paper 90-7570. American Society of Agricultural Engineers, St. Joseph, MI. 12 p.

Richards, C., and G. W. Minshall. 1988. The influence of periphyton abundance on *Baetis bicaudatus* distribution and colonization in a small stream. *J. North. Am. Bentholog. Soc.* 7:77-86.

Richards, C., R. J. Haro, L. B. Johnson, and G. E. Host. 1997. Catchment and reach-scale properties as indicators of macroinvertebrate species traits. *Freshwater Biol.* 37: 219-230.

Richards, K. S. 1987. Rivers: Envrionment, Process and Form. Chapter 1. In (ed.). *River channels, environment and process*, ed. K. Richards, 1-13. Oxford, UK: Institute of British Geographers, Basil and Blackwell.

Richards, R. P. 1990. Measure of flow variability and a new flow-based classification of Great Lakes tributaries. *J. Great Lakes Res.* 16:53-70.

Richards, C., and G. Host. 1994. Examining land use influences on stream habitats and macroinvertebrates: a GIS approach. *Water Resour. Bull.* 30(4):729-738.

Richards, C., G. E. Host, and J. W. Arthur. 1993. Identification of predominant environmental factors structuring stream macroinvertebrate communities within a large agricultural catchment. *Freshwater Biol.* 29:285-294.

Richards, C., L. B. Johnson, and G. E. Host. 1996. Landscape-scale influences on stream habitats and biota. *Can. J. Fish. Aquat. Sci.* 53(Suppl):295-311.

Richardson, C. J., and J. W. Gibbons. 1993. Pocosins, Carolina bays, and mountain bogs. In *Biodiversity of the southeastern United States, lowland terrestrial communities*, eds. W. H. Martin, S. G. Boyce, and A. C. Echternacht, 257-310. New York, NY: John Wiley and Sons, Inc.

Richardson, R., and I. Makkonen. 1994. The performance of cut-to-length systems in eastern Canada. Tech. Rep. No. TR-109. Forest Engineering Research Institute of Canada (FERIC), Pointe-Claire, Quebec. 16 p.

Ringler, N. H., and J. D. Hall. 1975. Effects of logging on water temperature and dissolved oxygen in spawning beds. *Trans. Am. Fish. Soc.* 104:11-121.

Roble, S. M. 1989. Life in fleeting waters. *Massachusetts Wildl.* 34(February):22-28.

Rosemond, A. D., P. J. Mulholland, and J. W. Elwood. 1993. Up-down and bottom-up control of stream periphyton: effects of nutrients and herbivores. *Ecology* 74:1264-1280.

Rose, W. M. 1984. Biomass, net primary production and successional dynamics of a virgin white pine (*Pinus strobus*) stand in northern Michigan. Ph.D. thesis, Michigan State University Ann Arbor, MI.

Rosgen, D. 1996. *Applied river morphology.* Pagosa Springs, CO. Wildland Hydrology Press. 376 p.

Rosgen, D. 1994. A classification of natural rivers. *Catena.* 22:169-199.

Roth, N. E., J. D. Allan, and D. L. Erickson. 1996. Landscape influences on stream biotic integrity assessed at multiple spatial scales. *Landscape Ecol.* 11:141-156.

Rothacher, J. S. 1973. Does harvest in west slope douglas-fir increase peak flow in small forest streams. Res. Pap. PNW-163. U.S. Department of Agriculture, Forest Service, Pacific Northwest Forest and Range Experiment Station, Portland, Oregon. 13 p.

Rothrock, J. T. 1902. Statement of work done by the Pennsylvania Department of Forestry during 1901 and 1902. Wm. Stanely Ray, State Printer of Pennsylvania. Chapter VIII. *The Black Willow (Salix nigra, Marsh.) As a Protector of River Banks:* 136-137.

Rowe, C. L., and W. A. Dunson. 1995. Impacts of hydroperiod on growth and survival of larval amphibians in temporary ponds of central Pennsylvania, USA. *Oecologia* 102:397-403.

Rowe, C. L., W. J. Sadinski, and W. A. Dunson. 1992. Effect of acute and chronic acidification on three larval amphibians that breed in temporary ponds. *Arch. of Environ. Contam. and Toxicol.* 23:339-350.

Rowe, J. W., and E. O. Moll. 1991. A radiotelemetric study of activity and movements of the Blanding's turtle (*Emydoidea blandingii*) from Illinois. *J. Herpetology* 25:178-185.

Rowe, J. S. 1961. The level-of-integration concept and ecology. *Ecology* 42:420-427.

Rowe, J. S. 1992. The ecosystem approach to forestland management. *For. Chron.* 68:222-224.

Rudolph, D. C., and J. G. Dickson. 1990. Streamside zone width and amphibian and reptile abundance. *Southwest. Nat.* 35:472-476.

Rumery, C., and J. G. Vennie. 1988. Wisconsin's self-help monitoring program: a review of the first year - 1986. *Lake Reserv. Manage.* 4:81-86.

Rummer, B., B. J. Stokes, and A. Schilling. 1997. Wetland harvesting systems-developing alternatives for sustainable operation. In *Proceedings of the Council on Forest Engineering, 20th annual meeting*, Rapid City, SD: 7-11.

Ruttner, F. 1953. *Fundamentals of Limnology.* University of Toronto Press.

Ryder, R. A. 1965. A method for estimating the potential fish production of north-temperate lakes. *Trans. Am. Fish. Soc.* 94:214-218.

S

Salzberg, A. 1995. Report on import/export turtle trade in the United States. In *Proceedings of the International Congress of Chelonian Conservation*, ed. B. Devaux, 314-322. Gonfaron, France: SOPTOM.

Sanderson, G. C. 1987. Raccoon. In *Wild furbearer management and conservation in North America*, ed. M. Novak, J. A. Baker, M. E. Obbard, and B. Malloch, 486-499. Toronto, Ontario: Ontario Ministry of Natural Resources and Ontario Trappers Association.

Sartz, R. S. 1969. Folklore and bromides in watershed management. *J. of For.* 67:366-371.

Sasse, D. B., and R. J. Pekins. 1996. Summer roosting ecology of northern long-eared bats *(Myotis septentrionalis)* in the White Mountain National Forest. In *Bats and forests symposium*, eds. R. M. R. Barclay, and R. M. Brigham, 99-101. Work. Pap. 23/1996. Research Branch, British Columbia Ministry of Forestry, Victoria, BC.

Schaefer, J. M., and M. T. Brown. 1992. Designing and protecting river corridors for wildlife. *Rivers* 3:14-26.

Schindler, D. W., S. E. Bayley, B. R. Parker, K. G. Beaty, D. R. Cruikshank, E. J. Fee, E. U. Schindler, and M. P. Stainton. 1996. The effects of climactic warming on the properties of boreal lakes and streams at the Experimental Lakes Area, northwestern Ontario. *Limnol. Oceanogr.* 41:1004-1017.

Schlesinger, D. R., and H. A. Regier. 1982. Climate and morpho-edaphic indices of fish. *Trans. Am. Fish. Soc.* 4:214-218.

Schlosser, I. J. 1982. Fish community structure and function along two habitat gradients in a headwater stream. *Ecol. Monogr.* 52: 395-414.

Schlosser, I. J., and J. R. Karr. 1981b. Water quality in agricultural catchments: impact of riparian vegetation during base flow. *Water Resour. Bull.* 17: 233-240.

Schlosser, I. J., and J. R. Karr. 1981a. Riparian vegetation and channel morphology impact on spatial patterns of water quality in agricultural catchments. *Environ. Manage.* 5: 233-243.

Schneider, D. W., and T. M. Frost. 1996. Habitat duration and community structure in temporary ponds. *J. North Am. Bentholog. Soc.* 15:64-86.

Schneider, W. J., and G. R. Ayer. Effect of reforestation on streamflow in central New York. Water Supply Paper 1602. U. S. Geological Survey, Washington, DC 61p.

Schneller, M. V. 1955. Oxygen depletion in Salt Creek, Indiana. *Investigations of Indiana Lakes and Streams* 4:163-175.

Schooley, R. L. 1990. Habitat use, fall movements, and denning ecology of female black bears in Maine. Master's thesis. University of Maine, Orono, ME. 115 p.

Schrameyer, R. B. 1997. Wisconsin's statewide lakes organizations-a history. *Laketides* 22:2-3.

Schroeder, H. W. 1996. Voices from Michigan's Black River: Obtaining information on "special places" for natural resource planning. Gen. Tech. Rep. NC-184. U.S. Department of Agriculture, Forest Service, North Central Forest Experiment Station, St. Paul, MN. 25 p.

Schroeder, H. W., Dwyer, J .F., Louviere, J. J., and Anderson, D. H. 1990. Monetary and nonmonetary tradeoffs of urban site attributes in a logit model of recreation choice. In Forest resource value and benefit measurement: Some cross-cultural perspectives, comps. B. L. Driver and G. L. Peterson. Gen. Tech. Rep. RM-197. U.S. Department of Agriculture, Forest Service, Rocky Mountain Forest and Range Experiment Station, Fort Collins, CO: 41-51.

Schumm, S. A. 1981. Evolution and response of the fluvial system, sedimentological implications. Spec. Publ. 31. Society of the Economic Paleontologists and Mineralogists 19-29.

Schumm, S. A. 1977. *The fluvial system.* New York: John Wiley & Sons.

Schumm, S. A. 1963. Sinuosity of alluvial rivers in the Great Plains. *Geol. Soc. Am. Bull.* 74:1089-1100.

Schupp, D. H. 1992. An ecological classification of Minnesota lakes with associated fish communities. Invest. Rep. 417. Minnesota Department of Natural Resources, St. Paul, MN. 27 p.

Seale, D. B. 1980. Influence on amphibian larvae on primary production, nutrient flux and competition in a pond ecosystem. *Ecology* 61:1531-1550.

Sedell, J. R., and J. L. Froggatt. 1984. Importance of streamside forests to large rivers: The isolation of the Willamette River, Oregon, U.S.A., from its floodplain by snagging and streamside forest removal. *Verhandlung der Internationalen Vereinigung fur Theoretische und Angewandte Limnologie* 22:1828-1834.

Sedell, J. R., and K. J. Luchessa. 1982. Using the historical record as an aid to salmonid habitat enhancement. In *Proceedings of the symposium on acquisition and utilization or aquatic habitat inventory information,* ed. N. B. Armantrout. 210-223. Western Division, American Fisheries Society, Bethesda, MD.

Sedell, J. R., F. H. Everest, and F. J. Swanson. 1982. Fish habitat and streamside management: past and present. In *Proceedings of the Society American Foresters Annual Meeting.* 244-255 Bethesda, MD: Society American Foresters.

Sedell, J. R., G. H. Reeves, F. R. Hauer, J. A. Stanford, and C. P. Hawkins. 1990. Role of refugia in recovery from disturbances: modern fragmented and disconnected river systems. *Environ. Manage.* 14: 711-724.

Seelbach, P. W., M. J. Wiley, J. C. Kotanchik, and M. E. Baker. 1997. A landscape based ecological classification system for river valley segments in lower Michigan. Fish. Rep. 2036. Department of Natural Resources, Fisheries Division. 50 p.

Seidensticker, J., M. A. O'Connell, and A. J. T. Johnsingh, 1987. Virginia opossum. In Wild furbearer management and conservation in North America, eds. Novak, M., J. A. Baker, M. E. Obbard, and B. Malloch, 249-263. Toronto, Ontario: Ontario Ministry of Natural Resources and Ontario Trappers Association.

Semlitsch, R. D. 1998. Biological delineation of terrestrial beffer zones for pond-breeding salamanders. *Conserv. Biol.* 12:1113-1119.

Semlitsch, R. D. 1983. Terrestrial movements of an eastern tiger salamander, *Ambystoma tigrinum.* *Herpetol. Rev.* 14:112-113.

Semlitsch, R. D. 1981. Terrestrial activity and summer home range of the mole salamander (*Ambystoma talpoideum*). *Can. J. Zool.* 59:315-322.

Settergren, C. D., R. M. Nugent, and G. S. Henderson. 1980. Timber harvest and water yield in the Ozarks. In Symposium on watershed management. New York: American Society of Civil Engineers. 2:661-668.

Sharitz, R. R., and W. J. Mitsch. 1993. Southern floodplain forests. *In Biodiversity of the southeastern United States,* eds. W. H. Martin, S. G. Boyce, and A. C. Echternacht, 311-372. New York: John Wiley & Sons.

Shaw, S. P., and C. G. Fredine. 1956. Wetlands of the United States. Circ. 39. U.S. Fish and Wildlife Service. 67 p.

Shelford, V. E. 1911. Ecological succession I. Stream fishes and method of physiographic analysis. *Biol. Bull.* 21:9-35.

Shirvell, C. S. 1990. Role of instream rootwads as juvenile coho salmon (*Oncorhynchus kisutch*) and steelhead trout (*O. Mykiss*) cover habitat under varying streamflows. *Can. J. Fish. and Aquat. Sci.* 47:852-861.

Short, H. L., J. B. Hestbeck, and R.W. Tiner. 1996. Ecosearch: a new paradigm for evaluating the utility of wildlife habitat. In *Conservation of faunal diversity in forested landscapes, eds.* R. M. DeGraaf and R. I. Miller, 569-594. New York: Cligman & Hale.

Shortle, W. C., and K. T. Smith. 1988. Aluminum-induced calcium deficiency syndrome in declining red spruce. *Science* 240:1017-1018.

Shugart, H. H., and D. C. West. 1977. Development of an Appalachian deciduous forest succession model and its application to assessment of the impact of Chestnut blight. *J. Environ. Manage.* 5:161-179.

Simpson, T. A., P. E. Stuart, and B. V. Barnes. 1990. Landscape ecosystems and cover types of the Reserve Area and adjoining lands of the Huron Mountain Club. Occas. Pap. No. 4. Huron Mountain Wildlife Foundation, Marquette, MI. 128 p.

Sirois, D. L., and A. E. Hassan. 1985. Performance of skidder tires in swamps. ASAE Pap. 85-1616. American Society of Agricultural Engineers, St. Joseph, MI. 20 p.

Slocombe, D. S. 1993. Environmental planning, ecosystem science, and ecosystem approaches for integrating environment and development. *Environ. Manage.* 17: 289-303.

Small, M. F., and W. N. Johnson, Jr. 1985. Wildlife management in riparian habitats. In *Is good forestry good wildlife management?* ed. J. A. Bissonette, 69-80. Misc. Publ. 689. Maine Agricultural Experiment Station, Orono, ME. 369 p.

Small, M. F. 1986. Response of songbirds and small mammals to powerline and river edges of Maine oak-pine forests. Master's thesis, University of Maine, Orono, ME. 58 p.

Smith, D. G. 1976. Effect of vegetation on lateral migration of anastomosed channels of a glacier meltwater river. *Geol. Soc. Am. Bull.* 87:857-860.

Smith, D. G., B. C. Larson, M. J. Kelty, and P. M. S. Ashton. 1997. *The practice of silviculture: applied forest ecology.* New York: John Wiley & Sons.

Smith, D. C. 1972. *A History of Lumbering in Maine 1861-1960.* University of Maine Press, Orono, ME. 469 p.

Smith, R. D., R. C. Sidle, and P. E. Porter. 1993. Effects on bedload transport of experimental removal of woody debris from a forest gravel-bed stream. *Earth Surf. Process. Landforms* 18:455-468.

Smock, L. A., G. M. Metzler, and J. E. Gladden. 1989. Role of debris dams in the structure and functioning of low-gradient headwater stream. *Ecology* 70:764-775.

Sparks, R. E. 1995. Need for ecosystem management of large rivers and their floodplains. *BioScience* 45:168-182.

Sokal, R. R. 1974. Classification: purposes, principles, progress, prospects. *Science* 185:1115-1123.

Solomon, D. S., D. A. Herman, and W. B. Leak. 1995. FIBER 3.0: an ecological growth model for Northeastern forest types. Gen. Tech. Rep. NE-204. U.S. Department of Agriculture, Forest Service, Northeastern Forest Experiment Station, Radnor, PA. 24 p.

South Carolina Forestry Commission. 1994. *South Carolina's best management practices for forestry.* South Carolina Forestry Commission. 64 p.

Speaker, R., K. Moore, and S. Gregory. 1984. Analysis of the process of retention of organic matter in stream ecosystems. *Verhandlung der Internationalen Vereinigung für Theoretische und Angewandte Limnologie* 22:1835-1841.

Spies, T. A., and B. V. Barnes. 1985. A multifactor ecological classification of the northern hardwood and conifer ecosystems of Sylvania Recreation Area, Upper Peninsula, Michigan. *Can. J. For. Res.* 15: 949-960.

Spurr, S. H., and B. V. Barnes. 1980. *Forest ecology.* New York: John Wiley and Sons.

St. Louis, V. L., J. W. M. Rudd, C. A. Kelly, K. G. Beaty, N. S. Bloom, and R. J. Flett. 1994. Importance of wetlands as sources of methyl mercury to boreal forest ecosystems. *Can. J. Fish. Aquat. Sci.* 51:1065-1076.

Stanley, E. H., S. G. Fisher, and N. Grimm. 1997. Ecosystem expansion and contraction in streams. *BioScience* 47(7):428-435.

Stauffer, D. F., and L. B. Best. 1980. Habitat selection by birds of riparian communities: evaluating effects of habitat alterations. *J. Wildl. Manage.* 44:1-15.

Steinblums, I. J., H. A. Froehlich, and J. K. Lyons. 1984. Designing stable buffer strips for stream production. *J. For.* 82: 49-52.

Stevens, T. 1994. Evaluation of the Rosgen stream classification system on the Chequamegon National Forest in Wisconsin. Master's thesis, College of Graduate Studies, University of Idaho. 31 p.

Stevens-Savery, T. A., G. H. Belt, and D. A. Higgins. 1998. Evaluation of the Rosgen stream classification system on national forests in Wisconsin. (Draft). Accepted *J. Am. Water Resour. Assoc.* (1999). 27 p.

Stokes, B. J., and A. Schilling. 1997. Improved harvesting systems for wet sites. *For. Ecol. Manage.* 90(1997):155-160.

Stokes, B. J. 1988. Wetland logger survey summary and production and costs of selected wetland logging systems. APA Tech. Pap. 88-A-10. American Pulpwood Association, Washington, DC. 26 p.

Stone, D. M., and J. D. Elioff. 1998. Soil properties and aspen development five years after compaction and forest floor removal. *Can. J. Soil Sci.* 78:51-58.

Strom, K. M. 1930. Limnological observations on Norwegian lakes. *Arch. Hydrobiol.* 21:97-124.

Stuart, G. W., and D. Dunshie. 1976. Effects of timber harvest on water chemistry. Hydrol. Pap. 1. U.S. Department of Agriculture, Forest Service, Eastern Region. 34 p.

Stuart, G. W. 1996. Forestry operations and water quality building on success. In *The national association of state foresters 1996 progress report: State nonpoint source pollution control programs for silviculture.* NASF, Washington, DC. 16 p.

Stynes, D. J. 1997. Recreation activity and tourism spending in the Lake States. *In Lake States regional forest resources assessment, technical papers,* eds. J. M. Vasievich, and H. H. Webster. Gen. Tech. Rep. NC-189. U.S. Department of Agriculture Forest Service, North Central Forest Experiment Station, St. Paul, MN: 139-164.

Stynes, D. J., J. J. Zheng, and S. I. Stewart. 1997. Seasonal homes and natural resources: patterns of use and impact in Michigan. Gen. Tech. Rep. NC-194. U.S. Department of Agriculture, Forest Service, North Central Forest Experiment Station, St. Paul, MN. 39 p.

Sullivan, K. 1986. Hydraulics and fish habitat relation to channel morphology. PhD. Dissertation, John Hopkins, University, Baltimore, MD. 407 p.

Swank, W. T., J. B. Waide, D. A. Crossley, Jr., and R. L. Todd. 1981. Insect defoliation enhances nitrate export from forest ecosystem. *Oceologia* 51: 297-299.

Swank, W. T., L. W. Swift, Jr., and J. E. Douglass. 1988. Streamflow changes associated with forest cutting, species conversion, and natural disturbances. In *Forest hydrology and ecology at Coweeta*. eds. W. T. Swank, and D. A. Crossley, Jr. Ecol. Stud. New York: Springer-Verlag: 66:297-312.

Swanson, F. J., and J. F. Franklin. 1992. New forestry practices from ecosystem analysis of Pacific Northwest forests. *Ecol. Appl.* 2: 262-274.

Swanson, F. J., S. V. Gregory, J. R. Sedell, and A. G. Campbell. 1982. Land-water interactions: the riparian zone. In *Analysis of coniferous forest ecosystems in the western United States, ed.* R. L. Edmonds, 267-291. Stroudsburg, PA: Hutchinson Ross Publishers.

Swanson, F. J., T. K. Kratz, N. Caine, and R. G. Woodamnsee. 1988. Landform effects on ecosystem patterns and processes. *BioScience* 38:92-98.

Sweeney, B. W. 1993. Effects of streamside vegetation on macroinvertebrate communities of White Clay Creek in Eastern North America. *Proc. Acad. Nat. Sci. Philadelphia* 144:291-340.

Swift, C. C., C. R. Gilbert, S. A. Borntone, G. H. Burgess, and R. W. Yerger. 1986. Zoogeography of the freshwater fishes of the southeastern United States: Savannah River to Lake Ponchartrain. Chapter 7, In *The zoogeography of North American freshwater fishes*, eds. C. H. Hocutt, and E. O. Wiley. New York: John Wiley and Sons. 866 p.

Swift, L. W., Jr. 1983. Duration of stream temperature increases following forest cutting in the Southern Appalachian Mountains. In *Proceedings of the international symposium on hydrometeorology*, eds. A. I. Johnson, and R. A. Clark, 273-275. *Am. Water Resour. Assoc.*

Swift, L. W., Jr. 1984a. Gravel and grass surfacing reduces soil loss from mountain roads. *For. Sci.* 30(3):657-670.

Swift, L. W., Jr. 1984b. Soil losses from roadbeds and cut and fill slopes in the Southern Appalachian Mountains. *South. J. Appl. For.* 8(4):209-215.

Swift, L.W., Jr. 1984c. Soil losses from roadbeds and cut and fill slopes in the southern Appalachians. *South. J. Appl. For.* 8(4):209-215.

Swift, L.W., Jr. 1986. Filter strip widths for forest roads in the Southern Appalachians. *South. J. Appl. For.* 10(1):27-34.

Szaro, R. C. 1980. Factors influencing bird populations in southwestern riparian forests. In *Workshop proceedings: Management of western forests and grasslands for nongame birds*, comps. R. M. DeGraaf, and N. G. Tilghman, 403-418. Gen. Tech. Rep. INT-86. U.S. Department of Agriculture, Forest Service, Intermountain Forest and Range Experiment Station, Ogden, UT.

T

Tabacchi, E., A. M. Planty-Tabacchi, and O. Décamps. 1990. Continuity and discontinuity of the riparian vegetation along a fluvial corridor. *Landscape Ecol.* 5(1):9-20.

Tank, J. T., and J. W. Webster. In press a. Interaction of substrate and nutrient availability on wood biofilm processes in streams. Accepted *Ecology*.

Tank, J. T., J. R. Webster, E. F. Benfield, and R. L. Sinsabaugh. In press b. Effects of leaf litter exclusion on microbial enzyme activity associated with wood biofilm in streams. Accepted *J. North Am. Bentholog. Soc.*

Tate, C. M. 1990. Patterns and controls of nitrogen in tallgrass prairie streams. *Ecology* 71(5):2007-2018.

Teeter, L., B. G. Lockaby, K. Flynn, M. Mackenzie, and J. Feminella. 1997. The Alabama Demonstration Watershed Project: Assessing the cumulative effects of land-uses in the South. *Examining the scientific basis for streamside management zones.* SAF national convention, Memphis, TN, October 6, 1997.

Tennessee Division of Forestry. 1993. *Guide to forestry best management practices.* Nashville, TN: Tennessee Department of Agriculture, Division of Forestry. 41 p.

Tennessee Department of Agriculture. 1996. Best management practices (BMPs) for timber harvesting in Tennessee. Division of Forestry. 20 p.

Thienemann, A. 1918. "Untersuchungen uber die Bezeihungen zwischen dem Sauerstoffgehalt des Wassers und der Zusammensetzung der Fauna in Norddeutschen Seen". *Arch. Hydrobiol.* 12:1-65.

Thomas, J. W., C. Maser, and J. E. Rodick. 1979. Riparian zones. In *Wildlife habitats in managed forests: The Blue Mountains of Oregon and Washington,* ed. J. W. Thomas, 40-47. Agric. Handb. 553. Washington, DC.

Thomasma, L. E. 1996. Winter habitat selection and interspecific interactions of American martens (*Martes americana*) and fishers (*Martes pennanti*) in the McCormick Wilderness and surrounding area. Ph.D. dissertation. Michigan Technological University, Houghton, MI. 116 p.

Thompson, F. R., III, W. D. Dijak, T. G. Kulowiec, and D. A. Hamilton. 1992. Breeding bird populations in Missouri Ozark forests with and without clearcutting. *J. Wildl. Manage.* 56:23-30.

Tinkle, D. W. 1959. Observations of reptiles and amphibians in a Louisiana swamp. *Am. Midl. Nat.* 62:189-205.

Tonn, W. M., and J. J. Magnuson. 1982. Patterns in the species composition and richness of fish assemblages in northern Wisconsin lakes. *Ecology* 63:1149-1166.

Tonn, W. M., J. J. Magnuson, and A. M. Forbes. 1983. Community analysis of fishery management: An application with northern Wisconsin Lakes. *Trans. Am. Fish. Soc.* 112:368-377.

Trimble, S. W. 1983. A sediment budget for coon creek basin in the driftless area, Wisconsin, 1853-1977. *Am. J. Sci.* 283:454-474.

Triska, F. J. 1984. Role of wood debris in modifying channel geomorphology and riparian areas of a large lowland river under pristine conditions: a historical case study. *Verhandlung der Internationalen Vereinigung fòr Theoretische und Angewandte Limnologie* 22:1876-1892.

Trotter, E. H. 1990. Woody debris, forest-stream succession, and catchment geomorphology. *J. North Am. Bentholog. Soc.* 9:141-156.

Tschaplinski, R. J., and G. F. Hartman. 1983. Winter distribution of juvenile coho salmon (*Onchorhynchus kisutch*) before and after logging in Carnation Creek, British Columbia, and some implications for overwinter survival. *Can. J. Fish. Aquat. Sci.* 40: 452-461.

Tubbs, A. A. 1980. Riparian bird communities of the Great Plains. In *Workshop proceedings: Management of western forests and grasslands for nongame birds,* comps. R. M. DeGraaf, and N. G. Tilghman, 419-434. Gen. Tech. Rep. INT-86. U.S. Department of Agriculture, Forest Service, Intermountain Forest and Range Experiment Station, Ogden, UT.

Tubbs, C. H., R. M. DeGraaf, M. Yamasaki, and W. M. Healy. 1987. Guide to wildlife tree management in New England northern hardwoods. Gen. Tech. Rep. NE-118. U.S. Department of Agriculture, Forest Service, Northeastern Forest Experiment Station, Broomall, PA. 30 p.

Tuchman, N. C., and R. H. King. 1993. Changes in mechanisms of summer detritus processing between wooded and agricultural sites in a Michigan headwater stream. *Hydrobiologia* 268:115-127.

U

U.S. GAO. 1991. Water pollution: greater EPA leadership needed to reduce nonpoint source pollution. GAO/RCED-91-10.

USDA Forest Service. 1958. Land: the yearbook of agriculture, 1958. U.S. Department of Agriculture, U.S. Government Printing Office. 605 p.

USDA Forest Service. 1971. Criteria for Management: [part of] Guide for managing the National Forests in the Appalachians. FSH 2123. U.S. Department of Agriculture, Forest Service, Eastern and Southern Regions.

USDA Forest Service. 1972. Water quality monitoring, Swift River. White Mountain National Forest Internal Report. 11 p.

USDA Forest Service. 1986. Land and resource management plan. Eastern Region, Forest Service, White Mountain National Forest, Laconia, NH.

USDA Forest Service. 1991. Ecosystem classification. Forest Service Manual 2060. Chapter 1. Washington, DC. 5 p.

USDA Forest Service. 1993a. National hierarchial framework of ecological units. Washington. DC: U.S. Department of Agriculture, Forest Service, Ecomap. 28 p.

USDA Forest Service. 1993b. Desired future condition. Southwestern Region, Apache-Sitgreaves National Forest. 17 p.

USDA Forest Service. 1994. Watershed Protection and Management. Forest Service Manual Chapter 2520. WO Amendment 2500 94-3, 26 p.

USDA Forest Service. 1997. Portable timber bridges: an eco-friendly solution for stream crossings. NA-TP-01-97. Timber Bridge Information Resource Center, Northeastern Area State and Private Forestry, Morgantown, WV. 8 p.

USDA Forest Service. 1998. Forest Inventory and Assessment. Eastern Forest Data Base. U.S. Department of Agriculture, Forest Service, North Central Forest Research Station, St. Paul, MN. Web Site.

USDI Bureau of Land Management. 1993. Riparian area management: process for assessing proper functioning condition. Tech. Rep. 1739-9. USDI-BLM Service Center, Denver, CO. 51 p.

USDI Fish and Wildlife Service. 1997. Endangered and threatened wildlife and plants; final rule to list the northern population of the bog turtle as threatened and the southern population as threatened due to similarity of appearance. Federal Register 62:59605-59623.

USEPA (Environmental Protection Agency). 1973. Processes, procedures, and methods to control pollution resulting from silvicultural activities. U.S. Environmental Protection Agency, Washington, DC. 91 p.

USEPA (Environmental Protection Agency). 1994. The quality of our Nation's water: 1992. EPA 841-S-94-002. U.S. Environmental Protection Agency, Washington, DC. 43 p.

USEPA (Environmental Protection Agency). 1997. Top 10 watershed lessons learned. EPA840-F-97-001. Office of Water. Washington, DC. 59 p.

USEPA (Environmental Protection Agency). 1998. National water quality inventory: 1996 Report to Congress. EPA841-F-97-003. Office of Water (4503F). Washington, DC.

References

References 387

USGPO (Government Printing Office). 1998. Stream corridor restoration: principles, processes, and practices. The Federal Interagency Stream Restoration Working Group. U.S. Government Printing Office, Washingtion, DC. 9 chpts.

V

Van Cleve, K., C. T. Dyrness, G. M. Marion, and R. Erickson. 1993. Control of soil development on the Tanana River floodplain, interior Alaska. *Can. J. For. Res.* 23:941-955.

Vander Haegen, W. M., and R. M. DeGraaf. 1996. Predation on artificial nests in forested riparian buffer strips. *J. Wildl. Manage.* 60:542-550.

Vannote, R., J. W. Minshall, K. W. Cummins, J. R. Sedell, and C. E. Cushing. 1980. The river continuum concept. *Can. J. Fish. Aquat. Sci.* 37:130-137.

Verme, L. J. 1965. Swamp conifer deeryards in northern Michigan. *J. For.* 63:523-529.

Vermont Department of Forest, Parks and Recreation. 1987. Acceptable management practices for maintaining water quality on logging Jobs in Vermont. Vermont Department of Forest, Parks and Recreation.

Verry, E. S. 1997. Hydrological processes of natural, northern forested wetlands. Chapter 13. Northern forested wetlands, ecology and management, eds. C. C. Trettin, et al. New York: Lewis Publishers: 163-188.

Verry, E. S. 1992. Riparian systems and management. In *Forest practice and water quality workshop: a Lake States Forestry Alliance initiative*, 1992 May 27-29, Green Bay, WI. The Lakes States Forestry Initiative, Hancock, MI B1-B24.

Verry, E. S. 1988. The hydrology of wetlands and man's influence on it. In *Symposium on the hydrology of wetlands in temperate and cold regions*, Vol. 2, 1988 June 6-8, Joensuu, Finland. Publication of the Academy of Finland, Helsinki, Finland: 5/1988:41-61.

Verry, E. S. 1986. Forest harvesting and water: the Lake States experience. *Water Res. Bull.* 22(6):1039-1047.

Verry, E. S. 1972. Effect of an aspen clearcutting on water yield and quality in northern Minnesota. In *National Symposium on Watershed in Transition*. American Water Resources Association Bethesda, MD. 276-284.

Verry, E. S., J. R. Lewis, and K. N. Brooks. 1983. Aspen clearcutting increases snowmelt and storm flow peaks in north central Minnesota. Water Resour. Bull. 19(1):59-67.

Viereck, L. A., C. T. Dyrness, and M. J. Foote. 1993. An overview of the vegetation and soils of the floodplain ecosystems of the Tanana River, interior Alaska. *Can. J. For. Res.* 23: 889-898.

Virginia Department of Forestry. 1997. Forestry best management practices for water quality in Virginia: technical guide. Virginia Department of Forestry. 45p.

Vitt, L. J., and R. D. Ohmart. 1978. Herpetofauna of the lower Colorado River: Davis Dam to the Mexican border. *Proc. Western Found. Verete. Zool.* 2:35-72.

Vogt, R. C., and J. J. Bull. 1984. Ecology of hatchling sex ratio in map turtles. *Ecology* 65:582-587.

Vogt, R. C. 1981. *Natural history of amphibians and reptiles of Wisconsin.* Milwaukee, WI: Milwaukee Public Museum. 205 p.

Vollenweider, R. A. 1975. Input-output models with special reference to phosphoris loading concept in limnology. *Schwiez. Zeitschrift fur Hydrol.* 37(1):53-84.

W

Wagner, R. G., and J. C. Zasada. 1991. Integrating plant autecology and silvicultural activities to prevent forest vegetation management problems. *For. Chron.* 67:506-513.

Wallace, J. B., J. R. Webster, and J. L. Meyer. 1995. Influence of log additions on physical and biotic characteristics of a mountain stream. *Can. J. Fish. Aquat. Sci.* 52:2120-2137.

Wallace, J. B., S. L. Eggert, J. L. Meyer, and J. R. Webster. 1997. Multiple trophic levels of a forest stream linked to terrestrial litter inputs. *Science* 277:102-104.

Walters, C. J., and C. S. Holling. 1990. Large-scale management experiments and learning by doing. *Ecology* 71:2060-2068.

Ward, G. M., and N. G. Aumen. 1986. Woody debris as a source of fine particulate organic matter in coniferous forest stream ecosystems. *Can. J. Fish. Aquat. Sci.* 43:1635-1642.

Warner, R. F. 1987. Spatial adjustments to temporal variations in flood regime in some Australian rivers. Chapter 2. In *River channels, environment and process*, ed. K. Richards, 14-40. Oxford, UK: Institute of British Geographers, Basil and Blackwell.

Waters, T. F. 1995. Sediment in streams: sources, biological effects, and control. *Am. Fish. Soc. Monogr.* 7. p. 251.

Watras, C. J., K. A. Morrison, J. S. Host, and N. S. Bloom. 1996. Concentrations of mercury species in relationships to other site-specific factors in surface waters of northern Wisconsin lakes. *Limnol. Oceanogr.* 40:556-565.

Webb, J. R., B. J. Cosby, F.A. Deviney, K. N. Eshleman, and J. N. Galloway. 1995. Change in the acid-base status of an Appalachian Mountain catchment following forest defoliation by the gypsy moth. *Water Air Soil Pollut.* 85: 535-540.

Webb, W. L., D. F. Behrend, and B. Saisarn. 1977. Effects of logging on songbird populations in a northern hardwood forest. *Wildl. Monogr.* 55. 35 p.

Weber, C. A. 1907. "Aufblau und vegetation der Moore Norddeutschlands." *Bot. Jahrb.* 40 Beibl. 90:19-34.

Weber, S. J., W. W. Mautz, J. W. Lanier, and J. E. Wiley, III. 1983. Predictive equations for deeryards in northern New Hampshire. *Wildl. Soc. Bull.* 11:331-338.

Webster, J. R., and J. B. Waide. 1982. Effects of forest clearcutting on leaf breakdown in a southern Appalachian stream. *Freshwater Biol.* 12:331-344.

Webster, J. R., and E. F. Benfield. 1986. Vascular plant breakdown in freshwater ecosystems. *Annu. Rev. Ecol. Syst.* 17: 567-594.

Webster, J. R., J. B. Wallace, and E. F. Benfield. 1995. Organic processes in streams of the eastern United States. *Ecosystems of the World 22: river and stream ecosystems*, eds., C. E. Cushing, G. W. Minshall, and K. W. Cummins, 117-187. Amsterdam: Elsevier.

Webster, J. R., A. P. Covich, J. L. Tank, and T. V. Crockett. 1994a. Retention of coarse organic particles in streams in the southern Appalachian Mountains. *J. North. Am. Bentholog. Soc.* 13:125-139.

Webster, J. R., A. P. Covich, J. L. Tank, and T. V. Crockett. 1994b. Retention of coarse organic particles in streams in the southern Appalachian Mountains. *J. North Am. Bentholog. Soc.* 13:140-150.

Webster, J. R., S. W. Golloday, E. F. Benfield, D. J. D'Angelo, and G. T. Peters. 1990. Effects of forest disturbance on particulate organic matter budgets of small streams. *J. North Am. Bentholog. Soc.* 9:120-140.

Webster, J. R., E. F. Benfield, S. W. Golladay, R. F. Kazmierczak, W. B. Perry, and G. T. Peters. 1988. Effects of watershed disturbance on stream seston characteristics. In *Forest ecology and hydrology at Coweeta*, eds. W. T. Swank, and D. A. Crossley, Jr. Ecol. Stud. New York: Springer-Verlag 66:279-294.

Webster, J. R., S. W. Golladay, E. F. Benfield, J. L. Meyer, W. T. Swank, and J. B. Wallace. 1992. Catchment disturbance and stream response: an overview of stream research at Coweeta Hydrologic Laboratory. In *River conservation and management*, eds. P. J. Boon, P. Calow, and G. E. Petts, 231-253. Chichester: Wiley.

Webster, J. R., M. E. Gurtz, J. J. Hains, J. L. Meyer, W. T. Swank, J. B. Waide, and J. B. Wallace. 1983. Stability of stream ecosystems. In *Stream ecology*, eds. J. R. Barnes, and G. W. Minshall, 355-395. New York: Plenum Press.

Well, D. E., T. E. Jordan, and D. L. Correll. 1998. Heuristic models for material discharge from landscapes with riparian buffers. *Ecolog. Applic. 8:1156-1169.*

Welsch, D. J. 1991. Riparian forest buffers: Function and design for protection and enhancement of water resources. Inf. Bull. NA-PR-07-91. U.S. Department of Agriculture, Forest Service, Northeastern Area State and Private Forestry, Radnor, PA. 20 p.

Welsch, D. J., D. L. Smart, J. N. Boyer, P. Minkin, H. C. Smith, and T. L. McCandless. 1995. Forested wetlands, functions, benefits and the use of best management practices. NA-PR-01-95. U.S. Department of Agriculture, Forest Service, Northeastern Area, Radnor, PA. 63 p.

Westphal, L. M. 1998. Use patterns and user preferences of on-site river recreationists. In. editors. In *People and the river* eds. P. H. Gobster, and L. M. Westphal, 49-78. Milwaukee, WI: National Park Service Rivers, Trails, and Conservation Assistance Program.

Wetzel, R. G. 1983. Limnology. 2nd ed. Saunders College Publishing.

Wetzel, R. G. 1990. Land-water interfaces: metabolic and limnological regulators. *Verh. Internat. Verein. Limnol.* 24:6-24.

Wetzel, R. G. 1992. Gradient dominated ecosystems: sources and regulatory functions of dissolved organic matter in freshwater ecosystems. *Hydrobiologia* 229:181-198.

Wharton, C. H., V. W. Lambour, J. Newson, P. V. Winger, L. L. Gaddy, and R. Mancke. 1981. The fauna of bottomland hardwoods in southeastern United States. In *Wetlands of bottomland hardwood forests.* Proceedings of a workshop on bottomland hardwood forest wetlands of the southeastern United States, eds. J. R. Clark and J. Benforado, 87-160. New York: Elsevier Scientific Co.

Whipkey, R. Z. 1965. Subsurface stormflow from forested slopes. *Int. Assoc. Sci. Hydrol. Bull.* 10(2):74-85.

Whitney, G. G. 1994. From coastal wilderness to fruited plain: a history of environmental change in temperate North America 1500 to the present. Cambridge University Press, Cambridge, MA.

Wigley, T. B., and M. A. Melchiors. 1994. Wildlife habitat and communities in streamside management zones: a literature review for the eastern United States. In: Riparian ecosystems in the humid U.S., functions, values, and management, 100-121. Washington, DC: National Association of Conservation Districts.

Wiley, M. J., Kohler, S. L., and Seelbach, P. W. 1997. Reconciling landscape and local views of aquatic communities: lessons from Michigan's trout streams. *Freshwater Biol.* 37:133-148.

Wiley, M. J., L. L. Osborne, and R. W. Larimore. 1990. Longitudinal structure of an agricultural prairie river system and its relationship to current stream ecosystem theory. *Can. J. Fish. Aquat. Sci.* 47:373-384.

Williams, M. 1983. Pioneer farm life and forest use. In *Encyclopedia of American Forest and Conservation History*, ed. R. C. Davis, 2:529-534. New York: Macmillan Publ. Co.

Williams, M. 1989. *Americans and their forests: a historical geography*. Cambridge, MA: Cambridge Univ. Press.

Williams, S. 1794. The natural and civil history of Vermont. Walpole, H. H., I. Thomas, and D. Carlisle.

Williamson, C. E., S. L. Metzgar, P. A. Lovera, and R. E. Moeller. 1997. Solar ultraviolet radiation and the spawning habitat of yellow perch (*Perca flavescens*). *Ecol. Appl.* 7:1017-1023.

Williamson, C. E., R. S. Stemberger, D. P. Morris, T. M. Frost, and S. G. Paulson. 1996. Ultraviolet radiation in North American lakes: attenuation estimates from DOC measurements and implications for plankton communities. *Limnol. Oceanogr.* 41:1024-1034.

Winchell, N. H. and Warren Upham. 1884. *The geological and natural history survey of Minnesota* 1872-1882. Vol. I of the final report. p5.

Winter, T. C. 1992. A physiographic and climatic framework for hydrogeologic studies of wetlands. In *Proceedings of the aquatic ecosystems in semi-arid regions: Implications for resources management*. Symp. Series 7. Saskatoon: Env. Canada, 127-148.

Winter, T. C. and M. K. Woo. 1990. Hydrology of lake and wetlands. In *Surface water hydrology*, eds. M. G. Wolman, H. C. Riggs, 159-187. The Geology of North America v 0-1. Boulder, CO: The Geological Society of America.

WI DNR (Department of Natural Resources, Bureau of Forestry). 1995. *Wisconsin's forestry best management practices for water quality*. FR 093. Bureau of Forestry, Wisconsin Department of Natural Resources, Madison, WI. 76 p.

WI DNR (Wisconsin Department of Natural Resources). 1996. Northern Wisconsin's lakes and shorelands: a report examining a resource under pressure.

WI DNR (Wisconsin Department of Natural Resources). 1982. Public or private I: navigability. Wisconsin Department of Natural Resources, Publication number 5-3500.

WI DNR (Wisconsin Department of Natural Resources). 1971. A basic guide to water rights in Wisconsin. Wisconsin Department of Natural Resources, Publication number 1302-71.

Wolman, M. G., and J. P. Miller. 1960. Magnitude and frequency of forces in geomorphic processes. *J. Geol.* 68:54-74.

Wood, C. A., J. E. Williams, and M. P. Dombeck. 1997. Learning to live within the limits of the land: lessons from watershed restoration case studies. In *Watershed Restoration: Principles and Practices*, 445-458. Bethesda, MD: American Fisheries Society.

Wood, J. T. 1944. Fall aggregation of the queen snake. *Copeia* 1944:253.

Woodall, W. R., and J. B. Wallace. 1972. The benthic fauna in four small southern Appalachian streams. *Am. Midl. Nat.* 88:393-407.

Woodard, S. E., and C. A. Rock. 1995. Control of residential storm water by natural buffer strips. *Lake Reserv. Manage.* 11:37-45.

Y

Yaffee, S. L. 1997. Why environmental policy nightmares recur. *Conserv. Biol.* 11:328-337.

Yamasaki, M., T. M. McLellan, R. M. DeGraaf, and C. A. Costello. 1999. In press. Effects of land-use and management practices on the presence of brown-headed cowbirds in the White Mountains of New Hampshire and Maine. In *The ecology and management of cowbirds: studies in the conservation of North American passerine birds*, eds, R. L. Cook, S. Robinson, S. Rothstein, J. Smith, and S. Sealy. University of Texas Press.

Yarnal, B., D. L. Johnson, B. J. Frakes, G. I. Bowels, and P. Pascale. 1997. The flood of '96 and its socioeconomic impacts in the Susquehana River basin. *J. Am. Water Resour. Assoc.* 33(6):1299-1312.

Yeakley, J. A., J. L. Meyer, and W. T. Swank. 1994. Hill-slope nutrient flux during near-stream vegetation removal I. A multi-scaled modeling design. *Water Air Soil Poll.* 77:229-246.

Yoder, C. O., and E. T. Rankin. 1995. Biological response signatures and the area of degradation value: new tools for interpreting multimetric data. In *Biological Assessment and Criteria: Tools for Water Resource Planning and Decision Making*, eds. W. S. Davis and T. P. Simon, 263-286. Boca Raton, FL. Lewis Publishing.

Z

Zasada, J. C. 1995. Silviculture in the western boreal forest-some considerations. In *Yukon forests: a sustainable resource* Part 2, 210-225. Whitehorse, Yukon: Yukon College.

Zentner, J. 1997. Soil considerations in riparian restoration. *Restor. Manage. Notes* 15:56-59.

Ziemer, R. R. 1997. Temporal and spatial scales. *In Watershed restoration: principles and practices.* Bethesda, MD: American Fisheries Society: 80-95.

Zogg, G. P., and B. V. Barnes. 1995. Ecological classification and analysis of wetland ecosystems, northern lower Michigan, U.S.A. *Can. J. For. Res.* 25:1865-1875.

Rushing waters in a Virginia stream.

Fine sediment in streams is the oldest insult to fish habitat.

End dumping of rock or rip rap rarely works without first considering what channel shape and size will be stable over time.

Where possible, manage what the people want. A rock ford whose surface does not extend above the channel bottom would work well here.

Function and value are a rich mixture in riparian areas.

Index

A

Allochthonous inputs
    definition of, 127
    disturbances to, 134
    functions of, 131–133
    hurricane effects on, 130
    loads of, 127–131
    mechanisms, 127
    tornado effects on, 130
    for trophic processes, 131
    wood inputs, 129–130
Alluvial valleys, 79
Amphibians, *see also specific amphibian*
    for assessing environmental change, 169–170
    endangered status of, 170
    overview of, 169–170
    riparian area conservation and management for,
        183, 186, 331–332
    scientific names for, 186–190
    terrestrial movement of, 180–182
    threats to
        habitat loss, 182–183
        pet trade, 183
Aquatic ecosystem, *see also* Ecosystems; Fish
    classification of, 68–69, 87–88
    description of, 24–25
    lakes, *see* Lakes
    of river basins, 74
    stream, *see* Stream
    terrestrial ecosystem and, 44, 163–166
    vegetation and, interrelationship between, 44, 64–65
    water temperature effects, 93
    wood pieces in, 131–132
Autochthonous input, 127
Avian communities, *see* Birds

B

Bank, of stream
    condition assessments, 116
    elevation of, *see* Bankfull elevation
    erosion of, 91, 115
    riparian forest effects, 240
    stability of, studies regarding, 33–34
Bankfull elevation
    calculation of, 106
    description of, 105–106
    in highly entrenched channel, 113
    illustration of, 107
    in moderately entrenched channel, 113
    occurrence of, 106–108
    in poorly entrenched channel, 113–114
    vegetation above or below, 114

Basins
    classification of, 73
    cutting on, effect on unstable streams, 118
Beaver ponds, amphibians and reptiles that use, 171–175
Beavers, 150–151
Belt width ratio, 32
Best management practices
    compliance with, 274, 284–285
    definition of, 23
    description of, 95, 273–274
    development of, 284–285
    effect on sedimentation, 92
    for erosion, 275
    function of, 273–274, 281
    guidelines for, 95, 282–283, 285
    identification of, 280–281
    implementation costs, 280
    for lake riparian areas, 212
    nonpoint source pollution, 274–275
    planning of, 280–281
    pollution control, 281–282
    programs, 273–274
    for riparian forests, 237–238
    site-specific information for, 95
    stakeholder groups' involvement in, 285
    summary overview of, 285–286, 336–337
    for temporary crossings, 282
    for timber harvesting, 256, 270
    for water bodies, 276
Biodiversity
    definition of, 43
    description of, 74
    preservation of, 43
Birds
    abundance of, 144
    in Eastern United States, 144–146
    effect of timber harvesting in riparian areas
        cavity trees, 149–150
        debris, 150
        nesting/perch sites, 149
        softwood composition, 148–149
        waterside marsh vegetation, 150–151
    nest sites for, 149
    perch sites for, 149
    riparian management considerations for, 330
    scientific names of, 154–156
    vegetation types and, 143–144
Black bears, 142–143
Bobcats, 142
Bogs
    amphibians and reptiles that use, 171–175
    description of, 178
Bottomland hardwood forests, 47–48
Bridges, for temporary use, 269
Buffer zones, for riparian ecosystems
    for amphibians and reptiles, 185